A Novel Green Treatment for Textiles

Sustainability: Contributions through Science and Technology

Series Editor: Michael C. Cann, Ph.D.
Professor of Chemistry and Co-Director of Environmental Science
University of Scranton, Pennsylvania

Preface to the Series

Sustainability is rapidly moving from the wings to center stage. Overconsumption of non-renewable and renewable resources, as well as the concomitant production of waste has brought the world to a crossroads. Green chemistry, along with other green sciences technologies, must play a leading role in bringing about a sustainable society. The **Sustainability: Contributions through Science and Technology** series focuses on the role science can play in developing technologies that lessen our environmental impact. This highly interdisciplinary series discusses significant and timely topics ranging from energy research to the implementation of sustainable technologies. Our intention is for scientists from a variety of disciplines to provide contributions that recognize how the development of green technologies affects the triple bottom line (society, economic, and environment). The series will be of interest to academics, researchers, professionals, business leaders, policy makers, and students, as well as individuals who want to know the basics of the science and technology of sustainability.

Michael C. Cann

Published Titles

Green Chemistry for Environmental Sustainability
Edited by Sanjay Kumar Sharma, Ackmez Mudhoo, 2010

Microwave Heating as a Tool for Sustainable Chemistry
Edited by Nicholas E. Leadbeater, 2010

Green Organic Chemistry in Lecture and Laboratory
Edited by Andrew P. Dicks, 2011

A Novel Green Treatment for Textiles:
Plasma Treatment as a Sustainable Technology
C. W. Kan, 2014

Sustainability: Contributions through Science and Technology

Series Editor: Michael C. Cann

A Novel Green Treatment for Textiles

Plasma Treatment as a Sustainable Technology

Chi-wai Kan

CRC Press
Taylor & Francis Group
Boca Raton London New York

CRC Press is an imprint of the
Taylor & Francis Group, an **informa** business

CRC Press
Taylor & Francis Group
6000 Broken Sound Parkway NW, Suite 300
Boca Raton, FL 33487-2742

First issued in paperback 2019

ISBN-13: 978-1-4398-3944-7 (hbk)
ISBN-13: 978-0-367-26249-5 (pbk)

Library of Congress Cataloging-in-Publication Data

Kan, Chi-wai, author.
A Novel green treatment for textiles : plasma treatment as a sustainable technology / Chi-wai Kan.
pages cm -- (Sustainability contributions through science and technology)
Summary: "Focusing on green chemistry and sustainability, this book discusses how plasma treatment has been used to modify textile properties. The book highlights the benefits of generating plasma and the reaction mechanisms between the surface of a textile and plasma species. The text addresses factors such as the nature of plasma gas, gas flow rate, system pressure, and discharge power that affect the final results of plasma treatments. An opening chapter presents current "brown" methods of treating textiles, exploring the environmental, economic and social costs of these methods. Throughout the book, the author presents the twelve principles of green chemistry and how they can be applied to the textile industry. "-- Provided by publisher.
Includes bibliographical references and index.
ISBN 978-1-4398-3944-7 (hardback)
1. Textile fibers--Etching. 2. Bleaching. 3. Dyes and dyeing. 4. Textile chemistry. 5. Green chemistry. 6. Plasma chemistry--Industrial applications. I. Title.

TP893.K28 2014
660'.044--dc23 2014016993

Visit the Taylor & Francis Web site at
http://www.taylorandfrancis.com

and the CRC Press Web site at
http://www.crcpress.com

To God and my family

Agnes and Hiu-lam

Contents

List of Figures

List of Tables

Series Preface

Sustainability is rapidly moving from the wings to center stage. Overconsumption of nonrenewable and renewable resources, as well as the concomitant production of waste has brought the world to a crossroads. Green chemistry, along with other green sciences technologies, must play a leading role in bringing about a sustainable society. The Sustainability: Contributions through Science and Technology series focuses on the role science can play in developing technologies that lessen our environmental impact. This highly interdisciplinary series discusses significant and timely topics ranging from energy research to the implementation of sustainable technologies. Our intention is for scientists from a variety of disciplines to provide contributions that recognize how the development of green technologies affects the triple bottom line (society, economic, and environment). The series will be of interest to academics, researchers, professionals, business leaders, policy makers, and students, as well as individuals who want to know the basics of the science and technology of sustainability.

Michael C. Cann
Series Editor

Foreword

Industries across the globe are being pushed and pulled by environmental regulations, economics, competitors, customers, and ultimately consumers, to develop sustainable products and processes, and the textile industry is no exception. The preparation, dyeing and finishing of textile fibres requires large amounts of water and other chemicals which may be toxic or otherwise hazardous, produces copious amounts of waste, and is energy intensive. This processing of textile fibres generally requires harsh conditions which often injure/reduce fibres while subjecting workers to a hazardous environment. In contrast to wet processing of textiles, plasma treatment drastically reduces the water used and other chemicals, while also reducing energy consumption and the production of waste. It also allows for the reduction of hazardous conditions and the concomitant improvement of working conditions. Not only does plasma treatment result in lower environmental and social footprints, but it improves the third leg of the triple bottom line by lowering costs.

This book provides a thorough review of the various natural and synthetic textile fibres, processes for treating these fibres (including preparation, dyeing and finishing), a thorough introduction to plasma treatment of textile fibres and the economic, environmental and social benefits of this technique. This book is a must for those in the textile industry, those who want to understand textile fibres and their treatments, and those who want to understand and/or apply green textile technologies.

Michael C. Cann
Distinguished Professor of Chemistry
Co-Director Environmental Science
University of Scranton
Scranton, Pennsylvania

Preface

The concept of plasma was introduced by Langmuir in the 1920s. Plasma is defined as a state where a significant number of atoms and/or molecules are electrically, thermally, or magnetically charged or ionised. Plasma generally refers to a gaseous state consisting of atoms, molecules, ions, metastable compounds and their excited states, and electrons such that the concentration of positively and negatively charged species is roughly the same. The ionised gas system displays significantly different physical and chemical properties compared to its neutral condition. Theoretically, plasma is referred to as a "fourth state of matter" and is characterised in terms of the average electron temperature and the charge density within the system.

Plasma treatment was initially used in the microelectronics industries as a kind of surface treatment and then migrated to the treatment of metals and polymers. In the 1980s, research was conducted into the application of plasma to textiles to enhance or improve surface properties such as adhesion, wettabililty, and chemical affinity. Those studies were promising in that the plasma treatment could induce beneficial changes in textile surface properties without much effect on their bulk properties. However, the plasma equipment/machines were mainly operated under vacuum, which limited their application in textile wet-processing operations, which are typically carried out as continuous processes at atmospheric pressure. Since that time, advances in technology have made plasma equipment/machines available for operation under atmospheric conditions, and plasma systems can now be integrated with conventional continuous wet processing of textiles. The plasma treatment requires only gas as a medium for interacting with the textile materials, thereby eliminating the need for water in modifying important surface properties. Thus this "dry" plasma treatment can be considered as an environmentally friendly treatment for textiles.

The concept of *sustainability* has become a core marketplace value in recent years. The trend in textile wet processing is toward a zero discharge of harmful chemicals to the environment, and plasma treatment of textile fabrics shows great promise in this regard. The main advantages of plasma treatments are as follows:

Electrons in low-temperature plasmas are able to cleave covalent chemical bonds, thereby producing beneficial physical and chemical modifications of the surface of the treated substrate without changing the fibre properties.

There is a minimal consumption of chemicals, and no drying process is required.

Plasma processes are environmentally friendly and are compatible with the goal of sustainable production processes.

Plasma treatment can be applied to almost all kinds of fibres.

The discharge of harmful chemicals can be eliminated with the use of plasma treatment, which can thus be considered as a sustainable treatment for textiles.

I met Prof. Michael Cann in August 2008 at the 236th American Chemical Society National Meeting in Philadelphia, where he introduced the Sustainability Series published by CRC Press (an imprint of Taylor & Francis Group, LLC). We initially discussed the possibility of my contributing a book chapter in the series, but with the encouragement of Prof. Cann, I ultimately agreed to write an entire book instead of a book chapter. However, the writing was harder than I expected, and due to a number of unexpected circumstances, it has taken me almost four years to finish this book. A number of publications were selected as basic resources, and the information from these representative publications is included in this book. I consider my role as the author to be that of a messenger conveying to the reader the latest information related to the application of plasma treatment in textiles. I have tried my best to include and cite useful information throughout the book. If I have overlooked any relevant developments, or if I have misattributed any citations and references, I extend my humble apologies. My hope is that this book has adequately covered the most important topics related to the application of plasma treatment in different textile processes, especially those applications related to the reduced use of harmful chemicals.

This book consists of nine chapters: Chapter 1 is an introduction that covers the general idea of using plasma treatment in textile wet processing; Chapter 2 reviews the general properties of textiles; Chapter 3 provides a general description of the wet processes that are typically involved in the treatment of textile fibres; Chapter 4 introduces the general concept of plasma and its application in treating textile materials; Chapter 5 summarises the application of plasma as a pretreatment for textiles; Chapter 6 summarises the application of plasma treatment in the dyeing of textiles; Chapter 7 summarises the application of plasma treatment in the printing of textiles; and Chapter 8 summarises the application of plasma treatment in the finishing of textiles. Finally, Chapter 9 explores the concept of sustainability and its role in the development of plasma treatments in textile wet processing.

I would like to take this opportunity to express my sincere appreciation to all of the researchers working worldwide in the field of plasma treatment. Without their contributions, this book would not be possible. It is my hope that this book will promote the understanding of plasma treatment as a sustainable technology for textiles.

Acknowledgements

First of all, I thank CRC Press and Taylor & Francis Group for giving me the opportunity to write this book in the Sustainability Series. Special thanks to the Series Editor, Professor Michael Cann, who allowed me the time to finish this book. Also thank you to the colleagues of CRC Press and Taylor & Francis Group, Hilary Rowe, Iris Fahrer, Marsha Pronin and Cheryl Wolf, who provided me with editorial advice as well as administrative support. Without their patience, this book would not have been possible.

I would like to express my sincere appreciation to my family, Agnes and Hiu-lam, for their love, encouragement and patience during my career in the academic world and permitting me to take time away from them during all those weekends and evenings while I was preparing this book.

Finally, I wish to acknowledge with gratitude the facilities and financial support of the Institute of Textiles and Clothing, The Hong Kong Polytechnic University for my plasma-related research work in textiles. The preparation of this book was financially supported by the Research Grants Council of The Hong Kong Special Administrative Region, China (Project No. PolyU 5173/11E) and The Hong Kong Polytechnic University.

About the Author

C.W. Kan, PhD, graduated from The Hong Kong Polytechnic University with a BSc and PhD in textile chemistry. Dr. Kan is now an associate professor in the Institute of Textiles and Clothing, The Hong Kong Polytechnic University. Dr. Kan's research interests are in the area of textile colouration and finishing. His research experience focusses on the surface modification of textile material with the use of novel technology such as plasma and laser. Dr. Kan has published more than 300 refereed journal papers and conference proceedings.

1 Introduction

'Sustainable', 'greener' and 'cleaner' production have recently become important issues in textile manufacturing processes. The supply chain of textiles includes fibre production, yarn spinning, fabric manufacturing, textile wet processing, final products distribution (retailing, marketing and merchandising) and disposal. Among the different steps in the supply chain, the textile wet processing involves the use of large amounts of energy, chemicals and water, etc. Thus, the industry is now seeking solutions to achieve 'sustainable', 'greener' and 'cleaner' production methods in its daily operations.

1.1 TWELVE PRINCIPLES OF GREEN CHEMISTRY FOR TEXTILE WET PROCESSING

What is textile wet processing (Leung, Lo, and Yeung 1996)? After textile materials have been made, by being spun into yarn or woven into fabric, they still contain impurities which make them undesirable for immediate use. Such textiles are usually referred to as 'grey textiles' or 'grey goods'—they are unattractive to consumers because of their appearance, handle (feel), and lack of serviceability and durability. Textile wet processing is the collective term for the processes that are used to improve the textiles in terms of these properties. The most common way to examine textile wet processing is to split it into the following three stages:

1. Pretreatment or preparation
2. Colouration (dyeing and printing)
3. Finishing

The pretreatment consists of a series of chemical and other treatments which are applied to textiles at the grey stage. At this stage, the textiles cannot be dyed or printed. The pretreatment processes improve the textiles so that they are able to accept dyes and chemicals in the later stages of textile wet processing. Colouration includes dyeing which adds colour to the textiles, which would otherwise be white only, and printing which provides special design to suit consumer requirements. Without the colouration process, there would be no place for textile designers (Leung, Lo, and Yeung 1996). In addition, the consumer also expects textiles to fulfil certain end-use requirements. For example, a raincoat should at least be waterproofed, and a woollen sweater should be mothproofed. In order to meet such expectations, textiles must undergo various finishing processes. These consist of mechanical and/or chemical treatment, and take place after colouration and before the material is made up into a garment. Moreover, in the textile wet processing, various chemicals and chemical reactions are involved and have been applied to the textile materials. If the chemicals such as dyes are not all picked up by the fibres or the chemical reactions are not fully completed, a residual amount of chemicals will be discharged. Although the amount of chemicals can be estimated and eliminated, there is still a risk of having a harmful effect on the environment.

The concept of 'green chemistry' was introduced by Anastas and Warner in 1998, who listed 12 principles (Anastas and Warner 1998). The concept of green chemistry can provide a solution to achieve 'sustainable', 'greener' and 'cleaner' production in textile wet processing.

1. *Prevention*: It is better to prevent waste than to treat or clean up waste after it has been created.
2. *Atom economy*: Synthetic methods should be designed to maximise the incorporation of all materials used in the process into the final product.
3. *Less hazardous chemical syntheses*: Wherever practicable, synthetic methods should be designed to use and generate substances that possess little or no toxicity to human health and the environment.
4. *Designing safer chemicals*: Chemical products should be designed to effect their desired function while minimising their toxicity.
5. *Safer solvents and auxiliaries*: The use of auxiliary substances (e.g., solvents, separation agents, etc.) should be made unnecessary wherever possible and innocuous when used.
6. *Design for energy efficiency*: Energy requirements of chemical processes should be recognised for their environmental and economic impacts and should be iseminimised. If possible, synthetic methods should be conducted at ambient temperature and pressure.
7. *Use of renewable feedstocks*: A raw material or feedstock should be renewable rather than depleting whenever technically and economically practicable.
8. *Reduce derivatives*: Unnecessary derivatization (use of blocking groups, protection/deprotection, temporary modification of physical/chemical processes) should be minimised or avoided if possible, because such steps require additional reagents and can generate waste.
9. *Catalysis*: Catalytic reagents (as selective as possible) are superior to stoichiometric reagents.
10. *Design for degradation*: Chemical products should be designed so that at the end of their function they break down into innocuous degradation products and do not persist in the environment.
11. *Real-time analysis for pollution prevention*: Analytical methodologies need to be further developed to allow for real-time, in-process monitoring and control prior to the formation of hazardous substances.
12. *Inherently safer chemistry for accident prevention*: Substances and the form of a substance used in a chemical process should be chosen to minimise the potential for chemical accidents, including releases, explosions and fires.

1.2 IMPORTANCE OF GREEN CHEMISTRY IN TEXTILE WET PROCESSING

The 12 principles in Section 1.1 introduce the further trend of developing the chemical principles for textile wet processing. The following sections give some of the basic ideas behind the recent practices of textile wet processing. These basics can help us to modify textile wet processing in greener, cleaner and more sustainable ways.

1.2.1 Pretreatment Process

The aim of pretreatment processes in textile wet processing is to treat the goods by standard procedures so that they are brought to a state where they can be dyed, printed or finished without showing any fault or damage on the material (Leung, Lo, and Yeung 1996). The pretreated textile materials should have the following properties:

1. Uniform power of absorption for dyes and chemicals in subsequent processes
2. An even water imbibitions value
3. Removal of all types of impurities, including broken seed
4. Absence of creases and wrinkles
5. High whiteness value

The pretreatment process is a non-added-value stage of the colouration process, and for this reason, the pretreatment stage of the process is often not optimised. Frequently, the pretreatment process is excessive, and high quantities of chemicals, auxiliaries and utilities such as water, steam, electricity and time are unnecessarily consumed. This can result in a high carryover of pretreatment residues (cotton impurities and pretreatment auxiliaries) that will have a negative influence on subsequent colouration processes and will require long multi-stage intermediate wash-off procedures which will consequently harm the environment. Therefore, the pretreatment process must balance the requirements of the colouration and finishing stages and the intended end uses of the textile material with the optimised use of chemicals and chemical reaction.

1.2.2 Colouration Process

Colouration in board refers to dyeing and printing. The objective of dyeing is the uniform colouration of the textile materials, usually to match a pre-specified colour. Any significant difference in colour from that requested by the customer, and any variation in the levelness of the colour of a fabric will be immediately apparent. Dyeing of a textile material can be achieved in a number of different ways (Broadbent 2001):

1. Direct dyeing, in which the dye in the aqueous solution in contact with the material is gradually absorbed into the fibres because of the inherent substantivity
2. Dyeing with a soluble precursor of the dye, which forms an insoluble pigment deep within the fibres upon treatment after dyeing
3. Direct dyeing followed by chemical reaction of the dye with appropriate groups in the fibre
4. Adhesion of dye or pigment to the surface of the fibres using an appropriate binder

All of these methods require that the fibres, at some stage, absorb the dye, or an appropriate precursor, from an aqueous solution. This process is essentially reversible. However, the precipitation of a pigment and reaction with the fibre are irreversible chemical processes. The discharge of pollution from dyeing processes occurs in two critical ways (Broadbent 2001):

1. When dye is applied to the fabric, the colouring agent, such as dye, is not completely taken up by the textile materials.
2. There are inevitably some residual amounts of dyes that cannot be absorbed, and although efforts are currently made to recycle them, there are large quantities that cannot be used, either because the particular shade is no longer applicable for the next dyeing process, or because the dilution is too great to make recovery economically possible.

Like dyeing, printing is a type of colouration process, but it is a 'localised' colouration in which only a desired portion of the textile material, normally fabric, is being coloured. Printing is usually achieved by applying thickened pastes containing dyes or pigments onto a fabric surface according to the given colour design. In textile printing, the printing paste that is being used contains a thick substance such as starch, gum or resin that greatly increases the viscosity of the colouring agent applied. However, if an excessive amount of printing paste is used and discarded after the printing process, subsequent water pollution problems will be created. In order to reduce the risk of pollution, the printing paste often can be easily collected, with a lesser amount lost but that is also easily absorbed by the fibre. In textile printing, the printing paste is forced to colour the fabric; if the absorption of the printing paste can be increased by applying additional force, the wash-out process after printing can be minimised, thereby leading to a lower amount of discarded pollutants.

1.2.3 Finishing Process

The general aim of textile finishing is to improve the attractiveness and/or serviceability of the textile material, mostly in fabric form. There are many finishing processes, each producing a different effect, but they all share five specific objectives of finishing (Leung, Lo, and Yeung 1996):

1. Improve the dimensional stability of the fabric, e.g., by stentering, compressive shrinkage and heat setting
2. Modify the handle of the fabric, e.g., by softening, stiffening and resin finishing
3. Improve the appearance of the fabric, e.g., by calendering and pressing
4. Modify the serviceability of the fabric, e.g., by waterproofing and flame retardation
5. Improve the durability of the fabric, e.g., by mothproofing and mildew proofing

Although all fabrics do not need to undergo the same finishing processes, different types of chemicals and chemical reactions would be involved in achieving the final effect. The chemicals used in various finishing processes may be harmful when discharged into the wastewater stream, and the presence of treatment chemicals in the finished products can cause skin reactions in some people. Disposal of finished products after prolonged use will allow leaching of any residual finish to take place, with further harm resulting once it reaches the groundwater.

1.3 WATER REQUIREMENT FOR TEXTILE WET PROCESSING

In textile wet processing, water consumption is far greater than the amounts of fibres processed. It is the ubiquitous solvent for the solutions of chemicals used. Rinsing and washing operations alone consume enormous amounts of water. Steam is still the major heat-transfer medium for many processes, and the quality of water fed to boilers is often critical (Broadbent 2001).

Water for textile wet processing may come from a variety of sources. These include surface water from rivers and lakes as well as subterranean water from wells. The water may be obtained directly from the source or from the local municipality. Natural and pretreated water may contain a variety of chemical species that can influence textile wet processing. Different kinds of salts may be in the water, depending on the geological formations through which the water has flowed. These salts are mainly the carbonates (CO_3^{2-}), hydrocarbonates (HCO_3^- more commonly named bicarbonates), sulphates (SO_4^{2-}) and chlorides (Cl^-) of calcium (Ca^{2+}), magnesium (Mg^{2+}) and sodium (Na^+). Other than dissolved salts of natural origins, water may also contain a variety of other salts from human or industrial activities. These include nitrates (NO_3^-), phosphates (HPO_4^{2-} and $H_2PO_4^-$) and various kinds of metal ions. The ions of certain transition metals such as iron, copper and manganese can be easily found. The typical quality of water for textile wet processing is shown in Table 1.1 (Broadbent 2001).

After textile wet processing, a large quantity of effluent water will be discharged into public sewerage systems or as surface water on open land, where it is treated until it achieves a given tolerance limit. The characteristics of effluent water vary widely among various textile wet-processing methods, and the overall estimated range is pH value (6.7–9.5); total alkalinity (500–796 ppm); total dissolved

TABLE 1.1
Typical Water Quality for Textile Wet Processing

Characteristics	Acceptable Concentration (ppm)
Hardness	0–25 $CaCO_3$
Iron	0.02–0.1
Manganese	0.02
Silica	0.5–3.0
Alkalinity to pH 4	35–65 $CaCO_3$
Dissolved solids	65–150

Source: Broadbent 2001.

substances (2180–3600 ppm); suspended solids (80–720 ppm); biological oxygen demand (60–540 ppm); chemical oxygen demand (592–800 ppm); Chlorides (as Cl^-) (488–1390 ppm); sulphate (SO_4^{2-}) (47–500 ppm); calcium (Ca^{2+}) (8–76 ppm); magnesium (Mg^{2+}) and sodium (as Na^+) (610–2175 ppm).

1.4 SUSTAINABILITY CONSIDERATION OF CONVENTIONAL TEXTILE WET PROCESSING

For sustainable development in textile wet processing, Easton (2009) reviewed comprehensively that the industry is now evaluating the following considerations.

1.4.1 Accurate Colour Communication

A dyer is required to match the colour of the client's 'standard' or reference shade on a particular quality of fabric and with the equipment available in the factory. The target may be electronic, in the form of reflectance data, or it may be a physical sample of coloured material or a combination of both. In the case of a physical target shade, this can be anything from a scrap of coloured paper or fabric to a fully engineered colour standard. The retailer or buyer usually requires the dyer to match the 'standard' under several light sources to produce a nonmetameric match. In order to avoid gross changes of colour, the colour standard should be as colour constant as possible under different illuminants (Collishaw, Weide, and Bradbury 2004).

Achieving a particular colour typically involves a mixture of three dyes, or trichromatic, of yellow, red and blue. A dyer will often have a preferred set of primaries that have good dyeing behaviour and from which the widest range of shades can be economically achieved, along with additional dyes for specific requirements of shade, fastness or metamerism (Bide 2007).

By communicating the dye combination used to make the colour standard, it is possible for the dyer to save time by having a 'flying start' for a laboratory formulation of the shade, even if later recipe correction is required to suit the particular substrate being processed and the dyeing machinery available in the factory. The recipe used in bulk should consist of dyes that have compatible dyeing profiles so that level and reproducible dyeings are more likely to be achieved. If environmental factors limit the choice of dyes to be used (e.g., as with the restrictions on dyes and chemicals included in some organic textile standards), then it can become more difficult to find a dye combination that performs well under bulk application conditions and produces a non-metameric match with acceptable fastness performance (Easton 2009).

Specific fastness requirements for particular articles, such as multiple wash fastness or perspiration light fastness, are important considerations. Designers and colour specifiers must understand the implications of their selection of seasonal palette shades further up the supply chain. Some shade and substrate combinations are more difficult to dye than others, and the dyehouse may consume large quantities of dyes, chemicals, water and energy trying to hit a difficult shade at the limits of achievability when a slightly different shade may have presented a more easily achievable target (Easton 2009).

1.4.2 Intelligent Dye Selection for Product Durability

Fastness is the resistance of a dyed textile to colour removal or modification of shade under the action of a range of agencies, including light, water, washing, perspiration, environmental contaminants, physical abrasion, etc. Standard test methods for assessing the fastness of dyed textiles are available from the International Organisation for Standardisation (ISO) or from the American Association of Textile Chemists and Colorists (AATCC). Meeting the fastness requirements of the customer is mainly achieved by (Easton 2009):

1. Intelligent dye selection
2. Efficient washing-off processes in which loose or unfixed dye is removed from the fibre after dyeing

As with issues of shade and levelness, restriction of dye choice for environmental reasons can limit the fastness achievable. This is particularly an issue for high-performance textiles such as golfwear, where the dye recipe selected to meet the demanding conditions of combined light and perspiration fastness testing has to take into account the very different responses of the different dye chromophores available (Imada, Harada, and Takagishi 1994).

As noted previously, colour fastness also depends on the removal of unfixed dye from the fibre at the end of the dyeing process. The use of rinsing processes that are efficient in water and energy use can significantly reduce the impact of these rinses. Intelligent selection of dyes with good wash-off performance, e.g., alkali-clearing disperse dyes for polyester fibres, can simplify processes, thereby leading to savings in chemicals, water and energy without compromising on fastness performance (Leadbetter and Leaver 1989).

Colour fastness is a key element of product durability and is sometimes at odds with the current ethos of the fashion industry, which has increasingly promoted rapid change and lowered expectations of durability for fast fashion items, particularly in womenswear. Peers warned that cheap, fast fashion encourages consumers to dispose of clothes which have been worn only a few times in favour of new, cheap garments which themselves will also go out of fashion and be discarded within a matter of months. However, there are emerging signs that greater product durability (or 'slow fashion') is being recognised as a key component of a more sustainable clothing industry (Easton 2009).

Most of the major dye manufacturing companies have ranges of dyes for all fibres that can deliver high fastness and enhanced product durability at relatively low cost compared with the retail price of the garment. The challenge is to get the designers and buyers to specify fastness standards for durable clothing and then stick to them in the face of pressure from vendors to accept something cheaper that does not quite meet the standard. With the pressure to get clothes into stores as fast as possible, cutting corners on quality can have significant hidden environmental costs (Easton 2009).

1.4.3 Intelligent Dye Selection for Chemical Compliance

Factories producing fabrics use a wide range of chemicals, some of which have the potential to harm workers and cause irreversible damage if allowed to enter the environment untreated. Small quantities (residues) of some harmful chemicals on clothing can also pose a risk to consumers and reputational damage for the retailer or brand. Most responsible dye manufacturers abandoned the production of carcinogenic benzidine dyes many years ago in the light of evidence of increased levels of bladder cancer among their own workers. However, many of these dyes are still available today in major textile manufacturing locations. In situations where there is little or no regard for health and safety in the workplace, this can have tragic consequences for those involved in handling these chemicals (Easton 2009).

New, more-demanding safety regulations covering all products containing chemicals manufactured and imported into the European Union will be introduced over the next few years under the Registration, Evaluation, Authorisation and Restriction of Chemicals (REACH) regulation, which came into force in 2007 but which will not be fully implemented until 2018 at the earliest. REACH creates a legal framework for the evaluation of the risks associated with the use of chemicals and requires greater transparency and communication on hazardous chemicals in the supply chain (Regulation No. 1907/2006 of European Parliament and of the Council, OJEC L396, 31.12.03, pp. 1–849).

1.4.4 Intelligent Process Selection for Improved Resource Efficiency

Freshwater is an increasingly scarce resource as the demands of an ever-growing world population and the agricultural activity needed to support it consume a steadily rising proportion of global freshwater resources. The textile industry generally needs to find ways to reduce its water consumption, and as a major user, and potential polluter, of water, the textile wet-processing industry is under particular pressure to reduce water consumption on both environmental and economic grounds. Securing a reliable and economic supply of water is now a strategic imperative for textile operation (Easton 2005).

We have therefore witnessed a recent upsurge of interest in the so-called 'water footprint' of products, in particular cotton textiles with their associated issues of irrigation and pesticide use (Chapagain et al. 2006). The preparation, colouration and finishing stages of fabric manufacture are significant contributors to the overall water footprint of textiles and clothing, and so there is a renewed interest in optimised water use and investigation of the possibilities of water re-use in dyeing.

As so many textile processing steps require the use of hot water, minimising the water consumption per unit of production also has a concomitant benefit in energy consumption. Given the levels of public and corporate concern about global warming and climate change, the textile industry cannot afford to ignore the pressure to reduce the amount of energy embedded in its products—in other words, to reduce their 'carbon footprint'. In the Best Available Techniques (BAT) reference document for the textile sector (European Commission, IPPC BATREF Guide for Textile Sector, Brussels, 2003), the following measures as BAT for water and energy management are identified (Easton 2009):

- Monitor water and energy consumption in the various processes
- Install flow-control devices and automatic stop valves on continuous machinery

- Install automatic controllers for control of fill volume and liquor temperature in batchwise dyeing machinery
- Establish standard operating procedures in order to avoid wastage of resources
- Optimise production scheduling
- Investigate possibilities for combining process steps, e.g., scour/dye, dye/finish
- Install low-liquor-ratio machinery for batch processing
- Install low add-on equipment for continuous application
- Improve washing efficiency in batch and continuous processing
- Re-use cooling water as process water
- Install heat-recovery systems to win back thermal energy from dropped dye baths and wash baths

The importance of right-first-time dyeing to minimise waste during the dyeing process has long been emphasised by leading dye manufacturers, and much of their innovation in terms of both new dyes and new application processes over the past 20 years has been directed towards reducing the demand for both water and energy.

1.4.5 Waste Minimisation and Pollution Control

Basically there are two approaches to reducing pollution arising from the textile wet-processing sector (Easton 2009):

1. Effluent treatment or end-of-pipe solutions
2. Waste minimisation or source-reduction solutions

The first approach has no financial payback and is literally money down the drain but provides the dyeing facility with its 'licence to operate'. In contrast, the waste minimisation approach not only reduces environmental impact, but also delivers reduced costs—a situation that has been referred to as a 'win-win scenario'. For a given dyeing process and given fabric using a common preparation and dyeing procedure, it has been shown by modelling studies that waste minimisation can best be achieved by operating at the lowest liquor ratio possible and maximising right-first-time performance (Glover and Hill 1993).

The Controlled Colouration™ concept describes textile colouration processes carried out in a way that minimises the impact on the environment by achieving high levels of right-first-time production (Collishaw, Glover, and Bradbury 1992; Collishaw, Philips, and Bradbury 1993; Cunningham 1995). The controls that the dyestuff manufacturer can exert are (Easton 2009)

- Control of dyeing behaviour
- Control of product quality
- Control of application processes
- Control of environmental impact

For example, some of the factors that must be taken into account when designing reactive dyes for reduced environmental impact are (Easton 2004)

- *Careful choice of intermediates*: no banned amines, minimum adsorbable organic halogen (AOX)
- *High colour yield*: high-fixation, multifunctional dyes leading to reduced levels of colour in effluent
- *Suitability for ultra-low liquor ratio dyeing machinery*: to minimise energy, water and chemicals consumption
- *Right-first-time dyeing through dyestuff compatibility*: to minimise wasteful shading additions or reprocessing

The environmental problems facing the textile wet-processing industry cannot be solved using outdated products, processes or machinery. Innovation is required to address the environmental issues facing the supply chain, and dyestuff manufacturers have a key role to play. As well as product innovation, the other major contribution that the innovative dyestuff company can make to cleaner textile production is application process innovation. This combination of novel dyestuffs and optimised application processes leads to (Easton 2009):

- Minimised resource consumption for lower environmental impact
- Maximised productivity by achieving higher throughput from available assets

There are many ways of treating dyehouse effluent in order to reduce its impact when it is eventually discharged to surface water. Some of these treatments are best carried out immediately as the effluent is produced, i.e., before it is mixed with other types of effluent that may interfere with the efficiency of the chosen treatment technology. This is known as the partial (or segregated) stream approach.

Another alternative is to create a mixed or balanced effluent to smooth out the peaks and troughs in flow or composition typical of a dyehouse effluent, and to then treat this in a multi-stage wastewater treatment plant with a large hydraulic capacity. Depending on the availability of public sewerage systems, this can either be carried out on-site at the dyehouse or off-site in a centralised third-party treatment facility, be this privately or publicly owned and operated.

Either of these strategies may be acceptable options when properly evaluated and applied to the actual wastewater situation. Well-accepted general principles for wastewater management and treatment include (Glover and Hill 1993):

- Characterisation of the different waste streams arising from the processes carried out
- Segregation of effluents at source according to their contaminant load and type
- Allocating contaminated wastewater streams to the most appropriate treatment

- Avoiding the introduction of wastewater components into biological treatment systems that could cause the system to malfunction
- Treating waste streams containing a relevant non-biological fraction by appropriate techniques before, or instead of, final biological treatment

Technology options for all of these treatment scenarios have been extensively reviewed in the literature (Dos Santos, Cervantes, and Van Lier 2007; Shukla 2007).

1.5 DEVELOPMENT OF NON-AQUEOUS GREEN TREATMENT

Growing demands regarding the environmental friendliness of finishing processes as well as the functionality of textiles have increased the interest in physically induced techniques for surface modification and coating of textiles. In general, after the application of water-based finishing systems, the textile needs to be dried. The removal of water is energy intensive and therefore environmentally harmful and expensive. Therefore, non-aqueous treatment should be a green solution to conventional textile wet processing (Stegmaier et al. 2009).

Plasma treatment, being a dry process, represents an economical alternative. A plasma can be described as a mixture of partially ionised gases, where the constituents are activated by addition of external energy, thereby producing atoms, radicals and electrons in a dynamic equilibrium. If the energy input is steered in such a way that the gas temperature remains in the range of room temperature, the process is referred to as 'cold' or 'low-temperature' plasma (Stegmaier et al. 2009). The main advantages of such plasma treatments are:

- The electrons in low-temperature plasmas are able to cleave covalent chemical bonds, thereby producing physical and chemical modifications of the surface of the treated substrate without changing the fibre properties.
- There is a minimal consumption of chemicals and no drying process is required.
- The processes have a high level of environmental compatibility.
- The processes can be applied to almost all kinds of fibres.

In principle, all polymeric and natural fibres can be plasma treated to achieve the following effects (Stegmaier et al. 2009):

- Wool degreasing
- Desizing
- Change in fibre wettability (hydrophilic and hydrophobic properties)
- Increase in dyestuff affinity
- Improved dye-levelling properties
- Anti-felt finishing in wool
- Sterilisation (bactericidal treatment), etc.

During plasma treatment, the textile stays dry and, accordingly, drying processes can be avoided, no wastewater is generated, and no (or less) chemicals are required. Further advantages of plasma technology include the extremely short treatment time and the low application temperature. Therefore, plasma treatment represents an energy-efficient and economic alternative to classical textile finishing processes. The following chapters discuss the application of plasma treatment as an alternative to different textile wet-processing methods (Stegmaier et al. 2009).

1.6 CONCLUSIONS

This chapter reviewed recent sustainable considerations of textile wet processing and introduced the concept of plasma treatment to replace conventional textile wet processing. The remaining chapters discuss the application of plasma treatment as an alternative to different methods of textile wet processing.

REFERENCES

Anastas, P. T., and J. C. Warner. 1998. *Green chemistry: Theory and practice*. New York: Oxford University Press.

Bide, M. 2007. Environmentally responsible application of textile dyes. In *Environmental aspects of textile dyeing*, ed. R. M. Christie, 74–92. Cambridge, UK: Woodhead Publishing.

Broadbent, A. D. 2001. *Basic principles of textile colouration*. Margate, UK: Society of Dyers and Colourists/Thanet Press.

Chapagain, A. K., A. Y. Hoekstra, H. H. G. Savenije, and R. Gautam. 2006. The water footprint of cotton consumption: An assessment of the impact of worldwide consumption of cotton products on the water resources in the cotton producing countries. *Ecological Economics* 60(1): 186–203.

Collishaw, P. S., B. Glover, and M. J. Bradbury. 1992. Achieving right-first-time production through control. *Journal of the Society of Dyers and Colourists* 108: 13–17.

Collishaw, P. S., D. A. S. Philips, and M. J. Bradbury. 1993. Controlled colouration: A success strategy for the dyeing of cellulosic fibres with reactive dyes. *Journal of the Society of Dyers and Colourists* 109: 284–92.

Collishaw, P. S., S. Weide, and M. J. Bradbury. 2004. Colour (in) constancy: What is achievable on retailer standards? *AATCC Review* 4(9): 16–18.

Cunningham, A. D. 1995. The controlled colouration approach for right first time dyeing of polyester. *AATCC International Conference Book of Papers* 1995: 424–36.

Dos Santos, A. B., F. J. Cervantes, and J. B. Van Lier. 2007. Review paper on current technologies for decolourisation of textile wastewater: Perspectives for anaerobic biotechnology. *Bioresource Technology* 98: 2369–85.

Easton, J. R. 2004. Supply chain partnerships for sustainable textile production. In *Eco-textiles '04*, 50–58. Cambridge, UK: Woodhead Publishing.

Easton, J. R. 2005. General considerations in reuse of water: Reuse from colouration processes. In *Water recycling in wet processing*, ed. J. K. Skelly, 3–15. Bradford, UK: Society of Dyers and Colourist.

Easton, J. R. 2009. Key sustainability issues in textile dyeing. In *Sustainable textiles: Life cycles and environmental impact*, ed. R. S. Blackburn, 139–54. Cambridge, UK: Woodhead Publishing.

Glover, B., and L. Hill. 1993. Waste minimisation in the dyehouse. *Textile Chemist & Colourist* 25(6): 15–20.

Imada, K., N. Harada, and T. Takagishi. 1994. Fading of azo reactive dyes by perspiration and light. *Journal of the Society of Dyers and Colourists* 110: 231–34.

Leadbetter, P. W., and A. T. Leaver. 1989. Disperse dyes: The challenge of the 1990s (meeting demands for increasingly higher levels of wash fastness in the exhaust dyeing of polyester/cellulosic blends). *Review of Progress in Colouration and Related Topics* 19: 33–39.

Leung, K. T., M. T. Lo, and K. W. Yeung. 1996. *Knowledge of materials II*. Hong Kong: Institute of Textiles and Clothing, The Hong Kong Polytechnic University.

Roy Choudhury, A. K. 2006. *Textile preparation and dyeing*. Enfield, NH: Science Publishers.

Shukla, S. R. 2007. Pollution abatement and waste minimization in textile dyeing. In *Environmental aspects of textile dyeing*, ed. R. M. Christie, 116–48. Cambridge, UK: Woodhead Publishing.

Stegmaier, T., M. Linke, A. Dinkelmann, V. Von Arnim, and H. Planck. 2009. Environmentally friendly plasma technologies for textiles. In *Sustainable textiles: Life cycles and environmental impact*, ed. R. S. Blackburn, 155–78. Cambridge, UK: Woodhead Publishing.

2 Textile Materials

2.1 TEXTILE FIBRES

Textile fibres had been defined differently by various organizations and researchers (Hui, Leung, and Lo 1996; The Textile Institute 1995; The Hong Kong Cotton Spinners Association 2007; Roy Choudhury 2006). Generally speaking, textile fibres are defined as units of matter characterised by flexibility, fineness and a high ratio of length to thickness. They should have sufficient strength to resist breakage due to stress applied during manufacture and use. They should also possess enough thermal and chemical stability to withstand the environment to which the fibres are exposed. Moreover, an extensibility of 5% to 50% is required, depending on the end use of the final product.

Until the introduction of man-made fibres, one had to rely on fibres from natural sources. Not all of these fibres were suitable for use as textile fibres, because they lack certain characteristics, for example, many were not long, flexible or strong enough. Soil, feed and other climatic and environmental conditions affect natural fibres, resulting in non-uniform properties of these fibres. Man-made fibres are not much influenced by these factors, and greater control can be exercised over their production. However, even with greater control, slight variations in the production of man-made fibres can give rise to significant variation in dyeability, strength and some other properties (Hui, Leung, and Lo 1996; The Textile Institute 1996; The Hong Kong Cotton Spinners Association 2007; Roy Choudhury 2006).

Each individual fibre is made of millions of individual long molecular chains of discrete chemical structure. The morphology, i.e., the arrangement and orientation, of these molecules within the individual fibre as well as the gross cross-section and shape of the fibre influence the fibre properties. However, the basic physical and chemical properties largely depend on the total number of units that repeat themselves in a chain, which can vary from a few units to several hundreds and is termed as the degree of polymerization (DP) for molecules within the fibre (Hui, Leung, and Lo 1996; The Textile Institute 1996; The Hong Kong Cotton Spinners Association 2007; Roy Choudhury 2006).

2.2 CLASSIFICATION OF TEXTILE FIBRES

A textile fibre is an individual, fine, hair-like substance, which forms the fundamental element of textile yarn and fabric. Fibres are either found in nature or made by man. Natural fibres are obtained from plants, animals and minerals, while man-made fibres are produced either purely chemically (synthetic fibres) or by modifying natural fibres by chemical means (regenerated fibres). The classification of textile fibres is presented in Figure 2.1

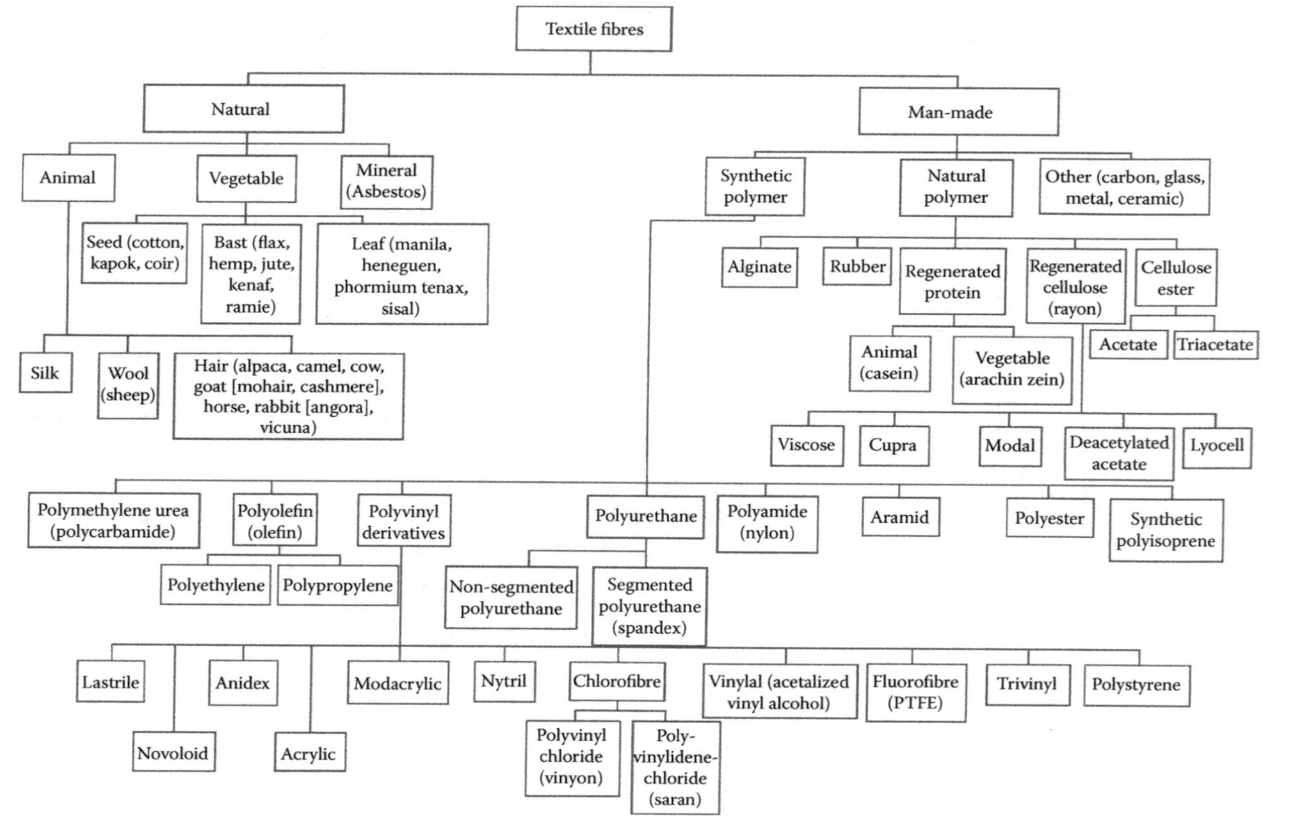

FIGURE 2.1 Classification of textile fibres. (Summarised from The Textile Institute 1995; and The Hong Kong Cotton Spinners Association 2007.)

(The Textile Institute 1995; The Hong Kong Cotton Spinners Association 2007). The characteristics of some important natural and man-made fibres are described in Sections 2.2.1, 2.2.2 and 2.2.3.

2.2.1 Natural Vegetable Fibres (Cellulosic Fibre)

Cellulose is the most abundant of all naturally organic polymers, with thousands of millions of tonnes being produced by photosynthesis annually throughout the world. Cellulosic fibres may be obtained naturally or man-made. However, natural cellulosic fibres are easily obtained from plants. Cotton is a fibre that grows on the seeds of cotton plants, and is therefore called seed fibre. Fibres obtained from the stems of plants are called bast fibres, examples of which are flax (linen fibre), ramie and jute. Leaf fibres are obtained from the leaves of plants, e.g., sisal and piña fibres. Some information about cellulosic fibres commonly used for textiles is given in the following subsections (The Hong Kong Cotton Spinners Association 2007).

2.2.1.1 Cotton

Cotton is the seed hair of the plant of the genus *Gossypium*. It is classified as a natural, cellulosic, monocellular, staple fibre. Different kinds and types of cotton are grown in various parts of the world. Variations occur among cotton fibres because of growth conditions, including such factors as soil, climate, fertilisers and pests. The quality of cotton fibre is based on its colour, staple, fineness and strength. Usually, longer fibres are finer and stronger. The particular kind of cotton is often identified by the name of the country or geographical area where it is produced (The Hong Kong Cotton Spinners Association 2007).

When dry, cotton fibre is almost entirely made up of cellulose (up to 96%). The other components, being regarded as impurities, include small amounts of protein, pectin, wax, ash, organic acids and pigment. The length of cotton is normally between ½ to 2½ in. (12.7–63.5 mm). When viewed under a microscope, cotton has a shape like a flat twisted tube (The Hong Kong Cotton Spinners Association 2007).

Cotton fibre has good strength and abrasion resistance. It is hydrophilic, absorbs moisture quickly and dries quickly, and has no static or pilling problems. However, cotton has little lustre and has poor elasticity and resiliency. It is attacked by mildew and weakened by resin chemicals used in finishing and by acids, but is highly resistant to alkalis (The Hong Kong Cotton Spinners Association 2007).

2.2.1.2 Flax (Linen)

Flax comes from the stem of the flax plant of the species *Linum usitatissimum*. It is classified as a natural cellulose, bast and multicellular fibre. When the fibre is processed into fabric, it is called linen. Flax fibre is the stem of flax plant, with a fibre length ranging from 2 to 36 in. (50.8–914.4 mm). The fibres are hackled (combed) to separate the long line and short tow fibres. The line fibres are generally drafted and doubled, and then lightly twisted before undergoing a wet spinning process. This produces strong, fine yarn. The short tow fibres are carded and drafted and then spun using a dry spinning method. Dry-spun yarns have a heavier count and are used for furnishing fabrics, heavy apparel and household textiles and knitwear (The Hong Kong Cotton Spinners Association 2007).

Flax is lint-free because of the absence of very short fibres. It has excellent strength, especially when wet. It has poorer drape and resiliency than cotton but is more hydrophilic, and therefore good for hot weather (The Hong Kong Cotton Spinners Association 2007).

2.2.1.3 Jute

Jute is the common name given to fibre extracted from the stems of plants belonging to the genus *Corchorus*. It is yellowish-brown in colour, and is classified as natural, bast and long staple fibre. People all over the world have used jute for most of their packaging requirements. Demand for jute goods reached a peak between the two World Wars, but since then the industry has experienced strong competition from bulk-handling paper sacks and more recently synthetic fibres like polypropylene and polyethylene (The Hong Kong Cotton Spinners Association 2007).

2.2.1.4 Ramie

Ramie is a bast fibre from a plant known as *reha* and *China grass*. The fibre is stiff, more brittle than linen, and highly lustrous. It can be bleached to extreme whiteness. Ramie lends itself to general processing for textile yarns, but its retting operation is difficult and costly, making the fibre unprofitable for general use. Ramie is seldom used for garments worn next to the skin because it can cause a prickly and itchy sensation to the human skin, particularly when the skin is moistened, such as during humid summers. It has poorer drape and resiliency than cotton. The prickly and itchy effect can be compensated by enzyme treatment. Since it is hydrophilic, similar to linen, it was used to produce a fabric called 'summer cloth', to be worn in summer in China (The Hong Kong Cotton Spinners Association 2007).

2.2.2 Natural Animal Fibres (Protein Fibres)

Protein fibres are obtained mainly from natural resources which are formed by animal sources through condensation of α-amino acids to produce repeating polyamide units with various substituents on the α-carbon atom. The sequence and type of amino acids making up individual protein chains contribute to the overall properties of the resultant fibre. Two major classes of natural protein fibres are keratin (hair or fur) and secreted (insect) fibres. The most important members of the two groups, respectively, are wool (derived from sheep) and silk (excreted by various moth larvae such as *Bombyx mori*). Wool is composed of an extremely complicated protein called keratin, which is highly cross-linked by disulphide bonds from cystine amino acid residues. However, the silk fibre is composed of much simpler secreted protein chains, which are arranged in a linear pleated structure with hydrogen bonds between amide groups on adjacent protein chains. Generally speaking, protein fibres are fibres of moderate strength, resiliency and elasticity. They have excellent moisture absorbency, and static charge is not built up on them. They are fairly resistant to acids, but are readily attacked by bases and oxidizing agents. The fibres have a tendency of yellowing in sunlight. They are comfortable under most environmental conditions, and they possess excellent aesthetic qualities. Some information about protein fibres commonly used for textiles is given in the following subsections (The Hong Kong Cotton Spinners Association 2007).

2.2.2.1 Wool

Wool is the fibre from the fleece of sheep. It is a natural, protein, multicellular, staple fibre. Early wool was a very coarse fibre. Its development into the soft, fleecy coat so familiar today is the result of long and continued selective breeding. The breeding of animals and the production of the wool fibre into fabric are more costly processes than the cultivation of plant fibre (e.g., cotton, linen) and their manufacture (The Hong Kong Cotton Spinners Association 2007).

Wool fibre is composed of proteins and organic substances with composition of carbon, hydrogen, oxygen, nitrogen and sulphur. All wool fibres have scales in the fibre surface (The Hong Kong Cotton Spinners Association 2007).

Wool fibre has good resiliency when dry, but poor when wet. It retains warmth because of slow moisture absorption and slow drying, which creates no cooling effect, and its natural crimp provides an air trap for insulation. The major disadvantages of wool are felting, and it is easily attacked by moths (The Hong Kong Cotton Spinners Association 2007).

There are about 40 breeds of sheep. Counting the crossbreeds, there are over 200 distinct grades of sheep. Those that produce wool may be classified into four groupings according to the quality of the wool produced (The Hong Kong Cotton Spinners Association 2007).

1. *Class-one or Merino wool*: Merino sheep produce the best wool, which is relatively short, ranging from 25 to 125 mm, but the fibre is strong, fine and elastic and has good working properties.
2. *Class-two wool*: These are not quite as good as the Merino wool, but this variety is nevertheless a very good quality wool. It is 50–200 mm in length, has a large number of scales, and has good working properties. This class of sheep originated in England, Scotland, Ireland and Wales.
3. *Class-three wool*: These fibres range in length from 100 to 455 mm, are coarser, and have fewer scales and less crimp than Merino and class-two wools. As a result, they are smoother and, therefore, they have more lustre. These wools are less elastic and resilient.
4. *Class-four wool*: These fibres are from 25 to 400 mm long, are coarse and hair-like, have relatively fewer scales and crimp, and therefore are smoother and more lustrous. This wool is less desirable, with the least elasticity and strength.

Wool can be classified by fleece. Wool shorn from young lambs differs in quality from that of older sheep. Also, fleeces differ according to whether they come from live or dead sheep, which necessitates standards for the classification of fleeces. The typical classes of fleece are lamb's wool, hogget wool, wether wool, pulled wool, dead wool and taglocks (The Hong Kong Cotton Spinners Association 2007).

Lamb's wool is the first fleece sheared from a lamb about six to eight months old; sometimes it is referred to as the first clip. This wool is of very fine quality. The fibres are tapered because the ends have never been clipped. Such fibres produce a softness of texture in fabric, and it is a popular material for sweaters (The Hong Kong Cotton Spinners Association 2007).

2.2.2.2 Hair

Hair fibres that have qualities of wool are obtained from certain kinds of animals. The hair of these animals has been so adapted by nature for the climate in which they live that the cloth produced from the fibres gives warmth with light weight. Hair fibres include camel hair, mohair, cashmere, llama, alpaca, vicuna and angora (rabbit hair). Camel hair is a fine and naturally water-repellent hair that is able to protect the body in both heat and cold. Camel hair fabrics are ideal for comfort, particularly when used for overcoating, as they are especially warm but light in weight. These fabrics are divided into three grades. Grade 1 is the soft and silky light-tan underhair with a length of 1¼ to 3½ in. (30–90 mm), and it is the choicest quality. Grade 2 consists of short hairs and partly of coarse outer hairs, ranging from 1½ to 5 in. (40–125 mm) in length. Grade 3 consists entirely of coarse outer hairs measuring up to 15 in. (380 mm) in length and varying in colour from brownish black to reddish brown (The Hong Kong Cotton Spinners Association 2007).

Mohair is the hair of the angora goat, which is a smooth, strong and resilient fibre. It does not attract or hold dirt particles. It absorbs dye evenly and permanently. Its fine silk-like lustre permits interesting decorative effects. It is more uniform in diameter than wool fibre and, therefore, does not shrink or felt as readily as wool (The Hong Kong Cotton Spinners Association 2007).

The underhair of the cashmere goat is made into luxuriously soft wool-like yarns with a characteristic highly napped finish. This fine cashmere fibre is soft, lighter in weight than wool, and quite warm; however, because it is a soft, delicate fibre, fabrics produced from cashmere are not as durable as wool (The Hong Kong Cotton Spinners Association 2007).

Alpaca fibre is valued for its silky beauty as well as for its strength. The hair of the alpaca is stronger than ordinary sheep's wool, is water-repellent, and has a high insulative quality. It is as delicate, soft and lustrous as the finest silk. The best selected type of alpaca is the suri, which is longer, silkier and finer and has curl throughout its length (The Hong Kong Cotton Spinners Association 2007).

The Angora rabbit produces long, fine, silky white hair, and comes mainly from France, Italy and Japan. Its smooth, silky texture makes it difficult to spin, and the fibres tend to slip out of the yarn and shed from the fabric (The Hong Kong Cotton Spinners Association 2007).

2.2.2.3 Silk

Silk is a continuous strand of protein filament cemented together forming the cocoon of the silkworm *Bombyx mori*. The silkworm forms silk by forcing two fine streams of thick liquid out of tiny openings in its head. In contact with air, the streams of liquid harden into filaments. Silk is classified as a natural, protein filament fibre (The Hong Kong Cotton Spinners Association 2007).

Silk classification depends whether it comes from cultivated or wild silkworms. The cultivated silkworms, which solely live on mulberry leaves, produce the finest lustrous fibres, and these fibres are known as real silk fibre. The wild silkworms, which are not cultivated, feed on the leaves of other trees such as oak and cherry, and produce brown, much thicker and less lustrous fibres. These fibres are known as tussah silk and are used for heavier, rough-textured fabrics.

Dupioni (also known as Douppioni) is made when two silkworms make their cocoons at the same time, thus joining together in one cocoon (The Hong Kong Cotton Spinners Association 2007).

Silk has excellent drape, lustre and luxurious hand. It is hydrophilic and with little problem in static, and its wet strength is 15% less than its dry strength. It has poor resistance to prolonged exposure to sunlight and can be attacked by moths. Under strong alkaline condition, silk will be weakened (The Hong Kong Cotton Spinners Association 2007).

2.2.3 Man-Made Fibres

Man-made fibres are fibres that are made by industrial processes, starting with materials which are completely different physically and chemically from the fibres made. For example, some man-made fibres are made from timber, while others are made from gas. They can be further divided into the following (Hui, Leung, and Lo 1996; The Hong Kong Cotton Spinners Association 2007):

1. *Regenerated man-made fibre (regenerated fibre)*: Regenerated man-made fibres (or regenerated fibres) are made from polymers. A polymer is a substance with a high molecular weight, formed by chemically joining numerous molecules of substances of low molecular weight. Certain polymers, such as cellulose and protein, occur naturally, and it is these natural polymers which are used to make regenerated man-made fibres. Examples of this type of fibre are viscose rayon, which is made from cellulose.
2. *Synthetic man-made fibre (synthetic fibre)*: Synthetic man-made fibres (or synthetic fibre) are also produced from polymers, but from man-made, not naturally occurring ones. Synthetic fibre can be further divided into subclasses, four of which are very important, namely, in the order of world production: polyester, nylon (or polyamide), polyvinyl derivatives, and polyolefins. There are many types of fibre within the polyvinyl derivatives group, the most important ones being acrylic and modacrylic. In the polyolefins subclass, there are two main fibres of commercial importance: polyethylene and polypropylene. It is thought that polypropylene will become increasingly important in the future, for both textile and non-textile uses.

2.2.3.1 Acetate

According to *Textile Terms and Definitions* (The Textile Institute 1995), acetate fibre, the common name for cellulose acetate, is defined as "manufactured fibre in which less than 92% but at least 74% of the hydroxyl groups (of cellulose) are acetylated". The "hydroxyl group is acetylated" means that the hydroxyl group of cellulose is changed to an acetyl group. Each cellulose molecule contains three hydroxyl groups, of which, on average, 2.3 hydroxyl groups are acetylated. This is an average figure and should not be rounded off to 2.0 (The Hong Kong Cotton Spinners Association 2007).

Acetate is a regenerated man-made fibre made from acetylation of cellulose by acetic acid. The cellulose source is similar to viscose rayon, but it differs greatly in chemical nature because the acetylation of the cellulose gives it a hydrophobic character. Different degrees of acetylation will result in different fibre properties and entitlement, i.e., secondary acetate and triacetate. It gives excellent drape and a luxurious hand, has no pilling problem and generates little static, and is also inexpensive. The characteristics of acetate are quite different from those of all the other man-made fibres. One of its unique characteristics is its thermoplasticity; that is, it can be softened by the application of heat and placed or pressed into a particular shape. Creases and pleats that are heat set into acetate fabrics are relatively durable and are retained better than in cotton, linen, wool, silk, or rayon. It is not very absorbent and, in fact, it is one of the weakest textile fibres, weaker than any rayon (The Hong Kong Cotton Spinners Association 2007).

2.2.3.2 Viscose Rayon

Viscose rayon, a man-made fibre composed of 100% regenerated cellulose, was discovered in 1891, and the first commercial production was undertaken in 1905 by Courtaulds. It is made from cotton linters or wood pulp, usually obtained from spruce and pine trees. Initially, viscose was called "artificial silk" and later named as rayon because of its brightness and similarities in structure to cotton (*rayon* is a combination of *sunray* and *cotton*). The other two regenerated cellulose fibres are cuprammonium (cupro) and polynosic (modal). In the United States, these three regenerated cellulose fibres are still referred to collectively by the generic term *rayon*. But the International Organisation for Standardisation (ISO) prefers the name *viscose* and defines viscose as regenerated cellulose obtained by the viscose process. The name *viscose* was derived from the word *viscous*, which describe the liquid state of the spinning solution (The Hong Kong Cotton Spinners Association 2007).

Viscose rayon is weak, with high elongation at break and a low modulus. It loses 30% to 50% of its strength when wet, and needs careful laundering. It also shrinks appreciably from washing. Viscose rayon is one of the most absorbent of all textiles. It is more absorbent than cotton or linen and is exceeded in absorbency only by wool and silk. A variation of rayon is classified as high wet modulus (HWM) rayon or polynosic rayon. This type of rayon is launderable (The Hong Kong Cotton Spinners Association 2007).

2.2.3.3 Lyocell (Trade Name 'Tencel')

Lyocell is a regenerated cellulose fibre made by a newly invented process which causes less effluent problems and less pollution than the process of making viscose rayon. Lyocell has very good properties, such as high tenacity and low elongation. It has a high moisture regain of 11.5% and at the same time retains tenacity well when wet. The drawback of lyocell is that the fibre can split quite easily. Lyocell, modal and polynosic rayon have very similar properties, so their end uses are also similar (The Hong Kong Cotton Spinners Association 2007).

2.2.3.4 Polyester

Polyester fibre is a man-made synthetic fibre which is a long-chain polymer produced from elements derived from coal, air, water and petroleum. Polyester fibre is chemically composed of at least 85% by weight of an ester of a substituted aromatic carboxylic acid, including but not restricted to substituted terephthalic units and *para*-substituted hydroxybenzoate units (The Hong Kong Cotton Spinners Association 2007).

The polyester fibres may be primarily divided into two varieties, i.e., PET (polyethylene terephthalate) and PCDT (poly-1,4-cyclohexylene dimethylene terephthalate). Most of the production is PET. Modification of each of these varieties is engineered to provide specific properties. It is also possible to produce other variants of polyester. PBT (polybutylene terephthalate) is the other polyester made by condensing terephthalic acid with butane diol and is melt-spun to get the filaments (The Hong Kong Cotton Spinners Association 2007).

Fabrics made of PET polyester yarn should be given compressive shrinking and heat setting to obtain dimensional stability to subsequent finishing processes and washing. PCDT polyester fabrics need not be initially heat-set because they are inherently stable. Both forms of polyester fabrics can be permanently pressed, since they are thermoplastic and hold their shape exceedingly well. Polyester fibres are subject to the accumulation of static electricity. In general, PET polyester filament yarns used for tires and industrial purposes are extremely strong. The abrasion resistance of polyester fibre is exceptionally good, being exceeded only by nylon among all of the commonly used fibres. However, the abrasion resistance of low-pilling types, including those of PCDT polyester, is of a lower order that is generally similar to wool. As polyester fibres do not have a high degree of elasticity, their strength, abrasion resistance, and stability make them very suitable for sewing thread (The Hong Kong Cotton Spinners Association 2007).

Polyester fibres are dyed almost exclusively with disperse dyes. Because of its rigid structure, well-developed crystallinity and lack of reactive dye sites, PET absorbs very little dye in conventional dye systems. Research work has been done to improve the dyeability of PET fibres. Polymerizing a third monomer, such as dimethyl ester, has successfully produced a cationic dyeable polyester fibre into the macromolecular chain. This third monomer has introduced functional groups as the sites to which the cationic dyes can be attached. The third monomer also contributes to disturbing the regularity of PET polymer chains, so as to make the structure of cationic dyeable polyester less compact than that of normal PET fibres. The disturbed structure is good for the penetration of dyes into the fibre. The disadvantage of adding a third monomer is the decrease in tensile strength (The Hong Kong Cotton Spinners Association 2007).

2.2.3.5 Nylon (Polyamide)

Nylon is a man-made synthetic polymer, polyamide filament or staple fibre. It is a long-chain synthetic polyamide in which less than 85% of amide linkages are attached to two aromatic rings (The Hong Kong Cotton Spinners Association 2007).

The first nylon produced by Du Pont was nylon 6.6, so called because its chemical components contain six carbon atoms per molecule. Nylon 6 was produced from a polyamide called caprolactam, which contain six carbon atoms. Some other forms of nylon were also developed, known as nylon 7, nylon 11, nylon 6.12 and nylon 4, 8, 10, and 6.10 (The Hong Kong Cotton Spinners Association 2007).

Nylon is produced in both regular and high-tenacity strengths. Although it is one of the lightest textile fibres, it is also one of the strongest. The strength of nylon will not deteriorate with age. It has the highest resistance to abrasion of any fibre. It can take a tremendous amount of rubbing, scraping, bending, and twisting without breaking down. Nylon is one of the elastic fibres; however, such stretch nylon yarns as Helanca and Agilon have exceptional elasticity. It has excellent resilience and draping qualities. Nylon does not absorb much moisture; therefore the nylon fabric feels clammy and uncomfortable in warm, humid weather (The Hong Kong Cotton Spinners Association 2007).

2.2.3.6 Acrylic and Modacrylic

According to *Textile Terms and Definitions* (The Textile Institute 1995), acrylic is defined as 'fibre made of co-polymer composed of 85% to 90% acrylonitrile and 10% to 15% other basic constituent which has an affinity with acid dyes'. Modacrylic (or modified acrylic) is defined as 'fibre made of copolymer composed of 35% to 85% acrylonitrile and 15% to 65% other basic constituent'. If the amount of acrylonitrile used is more than 90%, the fibre produced is called polyacrylonitrile. The other basic constituent may be vinyl chloride, vinyl acetate, vinyl alcohol or some other suitable substance (The Hong Kong Cotton Spinners Association 2007).

Looking at some of these terms more closely, acrylonitrile is a poisonous clear liquid, soluble in water. It can be easily polymerised, at about 40°C in the presence of catalysts, to form polyacrylonitrile (PAN) (The Hong Kong Cotton Spinners Association 2007).

Polyacrylonitrile is a polymer which can be made into fibre. It is now more common, however, to make fibre from copolymer based on acrylonitrile and other monomers. A copolymer is a type of polymer formed when two or more types of monomers, each of which can be polymerised on its own, are polymerised together. Both acrylic and modacrylic are copolymers (The Hong Kong Cotton Spinners Association 2007).

Acrylonitrile alone can, for example, be polymerised to polyacrylonitrile (PAN), while another monomer, say, vinyl chloride, can also be polymerised on its own to polyvinyl chloride (PVC). If acrylonitrile and vinyl chloride are put together to carry out polymerization, a copolymer is formed; the process is usually called *copolymerisation.* It is possible to have many kinds of acrylic and modacrylic, depending on the percentage of acrylonitrile used and the nature of the other basic constituents. Even so, all types of acrylic and modacrylic are very similar in their manufacture, properties and end uses (The Hong Kong Cotton Spinners Association 2007).

2.2.3.7 Polyolefins

Polyolefin is a manufactured fibre in which the fibre-forming substance is any long-chain synthetic polyethylene, polypropylene, or other olefin unit, except amorphous (noncrystalline) polyolefins. It is a very lightweight fibre that possesses very good strength and abrasion resistance. It possesses a unique combination of low moisture absorbency and exceptional wicking of water, which are advantages in providing comfortable apparel in certain circumstances. Polyolefin is almost completely hydrophobic. Such moisture properties are sought for active sportswear fabrics, particularly sweatshirts, socks and warm-up suits, because water is moved away from the skin. This fibre also has excellent sunlight resistance and weatherability, and can be washed or dry-cleaned easily. However, because olefin is sensitive to perchloroethylene, the most frequently used dry-cleaning solvent, petroleum solvent generally should be specified if necessary. Polyolefins are used for making indoor/outdoor carpeting and bathroom floor covering because of their low specific gravity. Important apparel end uses are athletic clothes, exercise suits and underwear because of its excellent wicking action (The Hong Kong Cotton Spinners Association 2007).

Polypropylene, or PP, is the most popular types of polyolefin, which is the fourth largest synthetic man-made fibre in terms of world production, after polyester, nylon and acrylic. Polypropylene is currently gaining in importance, and its share of the market is growing, so that in the future an increasing number of textile articles will be made of polypropylene. Since the raw material for making polypropylene, propylene, is abundant and cheap, the fibre produced is also very cheap (The Hong Kong Cotton Spinners Association 2007).

2.3 ESSENTIAL PROPERTIES OF TEXTILE FIBRES

Fibres usually are grouped and twisted together into yarn, and the yarns are then used to make woven and knitted fabrics. A fibre's structure contributes to the performance characteristics of a fabric made from it. The properties are determined by a fibre's physical attributes, chemical composition and molecular formation. This section summarises different essential properties of textile fibres (Hui, Leung, and Lo 1996; The Hong Kong Cotton Spinners Association 2007; Morton and Hearle 2008; Gordon Cook 1984a, 1984b).

1. *Fibre length*: Fibres which can be measured in length are called staple fibres, while fibres of infinite length are called filament fibres. For staple fibres, the longer the fibre length, the better is the quality of yarn being spun.
2. *Cross-sectional shape and surface*: These determine the bulk, texture, lustre and hand feel of the fibre. For example, round-shape fibres do not pack as well as flat fibres (which results in bulkier products), and they have a smoother and more slippery hand.
3. *Straightness*: Fibres may be straight, twisted, coiled or crimped, and this affects the performance properties such as resiliency, elasticity and abrasion resistance, e.g., crimp on fibre, bulk, warmth, elongation and absorbency.
4. *Strength*: Fibre strength contributes to yarn strength and eventually affects fabric durability. It is usually expressed as tenacity (grams per tex).
5. *Extensibility and elasticity*: This is the ability to increase in length when under tension and then return to the original length when released. Fibres that can elongate at least 100% are called elastomeric fibres.
6. *Hand feel*: This is affected by its shape, surface and configuration. Terms such as *soft*, *crisp*, *dry*, *silky* or *harsh* are used to describe the hand feel.
7. *Plasticity*: A thermoplastic fibre melts or softens when heat is applied. Thus, permanent creased and pleats can be made on fabrics containing thermoplastic fibres.
8. *Absorbency*: This is the ability to take in moisture. It is expressed as a percentage of moisture regain (MR). MR is the amount of water a bone-dry fibre will absorb from the air under 20°C and 65% relative humidity.

Fibres which absorb water easily are called hydrophilic, while those which only absorb a small amount are called hydrophobic.

9. *Abrasion resistance*: This is the ability to resist wear form rubbing; fibres that have high breaking strength and abrasion resistance are more durable.
10. *Resiliency*: This is the ability of a material to spring back to shape after being distorted. It is closely related to wrinkle recovery.
11. *Lustre*: This refers to the light reflected from a surface. Various characteristics of a fibre will affect the amount of lustre produced. The cross-sectional shape, lengthwise appearance, yarn type, weave and finish will affect the amount of lustre.
12. *Density*: A fibre of low density can produce a thick and lofty but still relatively lightweight fabric.
13. *Wicking*: This is the ability of a fibre to transfer moisture from one section to another. Usually the moisture is along the fibre surface. The wicking propensity of a fibre is usually based on the chemical and physical composition of the outer surface, where a smooth surface reduces wicking action. A hydrophilic fibre such as cotton possesses good wicking action, while a hydrophobic fibre such as polypropylene also possesses good wicking action when it is in an extra-thin filament form. This property is especially desirable for sportswear; body perspiration is transported by wicking action along the fibre surface to the outer surface of the cloth, thus providing improved comfort.

2.4 CONCLUSIONS

This chapter reviewed the classification of the fibres commonly used in the textile industry. The chemical and physical nature of these textile fibres affect the subsequent textile wet processing to be applied. The chapter concluded with a discussion of general fibre properties.

REFERENCES

Gordon Cook, J. 1984a. *Textile fibres*. Vol. 1, *Natural fibres*. Darlington, UK: Merrow Publishing.

Gordon Cook, J. 1984b. *Textile fibres*. Vol. 2, *Man-made fibres*. Darlington, UK: Merrow Publishing.

Hui, H. W., K. T. Leung, and M. T. Lo. 1996. *Knowledge of materials*. Vol. 1. Hong Kong: Institute of Textiles and Clothing, Hong Kong Polytechnic University.

Morton, W. E., and J. W. S. Hearle. 2008. *Physical properties of textile fibres*. 4th ed. Cambridge, UK: Woodhead Publishing.

Roy Choudhury, A. K. 2006. *Textile preparation and dyeing*. Enfield, NH: Science Publishers, Edenbridge.

The Hong Kong Cotton Spinners Association. 2007. *Textile handbook 2007*. Hong Kong: Hong Kong Cotton Spinners Association.

The Textile Institute. 1995. *Textile terms and definitions*. 10th ed. Manchester, UK: The Textile Institute.

3 Processes for Treating Textile Fibres

After textile materials have been made, by being spun into yarn or woven into fabric, they still contain impurities which make them undesirable for immediate use. Such textiles—usually referred to as 'grey goods'—are unattractive to consumers because of their appearance, handle and lack of serviceability and durability. *Textile wet processing* is the collective term for the processes that are used to improve the textiles in terms of these desirable properties. The most common way to examine textile wet processing is to split it into the following three stages (Roy Choudhury 2006; The Hong Kong Cotton Spinners Association 2007; Leung, Lo, and Yeung 1996):

Pretreatment (preparation)
Colouration (dyeing and printing)
Finishing

3.1 PRETREATMENT (PREPARATION) PROCESS

Different textile materials possess different inherent impurities. The pretreatment processes, therefore, vary for textile substrates. However, they can be generalised into a few types as follows (Roy Choudhury 2006; The Hong Kong Cotton Spinners Association 2007; Leung, Lo, and Yeung 1996; Vigo 1994):

1. Removal of loose fibres or yarns projecting on the surface. In a singeing process, the removal is done by burning, while in shearing and cropping processes, the removal is done by cutting (most projecting threads, which hamper uniform printing) with the help of an extruder type of blade.
2. Removal of starch or other sizing materials, which are applied on yarn before weaving, for better weavability. The process is known as *desizing*.
3. Removal of most of the water-soluble and water-insoluble impurities. This is the most important preparatory step and is known as *scouring*.
4. Removal of colouring substances present in the textile materials during bleaching, which is the last pretreatment step. The material becomes whiter in appearance; consequently, the dyed or printed materials become brighter and purer in colour. If the final product is to be sold as white, the bleached material may be subsequently treated with a blue pigment or an optical whitening agent to make the appearance dazzling white.
5. For synthetic textile materials, an additional pretreatment step called *heat-setting* is done before or after scouring to achieve dimensional stability of the material during subsequent treatments at high temperature.

3.1.1 Pretreatment (Preparation) for Cellulosic Fibres

A typical continuous processing sequence for the pretreatment of cotton cellulosic fabric is given as follows (The Hong Kong Cotton Spinners Association 2007; Leung, Lo, and Yeung 1996):

Grey inspection → Singeing → Desizing → Scouring → Bleaching → Mercerization

3.1.1.1 Grey Inspection

This is to identify any fabric faults that exist in the grey state. The inspection is carried out visually, either on a traditional long inspection table or by means of a specially designed inspection machine. Faults are usually marked with coloured threads or other means of markings on the fabric selvedges for identification. If the number of faults exceeds the inspection requirement and the faults cannot be mended, the goods will be rejected at this stage. After inspection, the goods will be made up into a continuous form by sewing the short pieces together (The Hong Kong Cotton Spinners Association 2007; Leung, Lo, and Yeung 1996).

3.1.1.2 Singeing

Singeing of cotton goods is necessary to remove the unwanted surface loose hair so as to give a uniform surface. The process is done by passing the fabric rapidly over a singeing machine with gas burners. The burners are specifically arranged such that both sides of the fabric are singed. The speed at which the fabric travels must be carefully adjusted so that it will not catch fire or be scorched. After singeing, cooling of the fabric, by either cooling cylinders or water bath, is needed to recover the handle (The Hong Kong Cotton Spinners Association 2007; Leung, Lo, and Yeung 1996).

3.1.1.3 Desizing

Desizing is carried out to remove the sizes that have been applied to the warp yarns for reinforcement during the weaving process. Sizes are normally composed of starches or other synthetic chemical agents, like polyvinyl alcohol (PVA). Removal of sizes is traditionally done by some chemical treatments, such as acid steeping, alkali steeping and oxidative treatment, etc., which are quite time consuming and polluting. Nowadays, a more practical and faster method is to make use of biochemical enzymes which decompose and digest the size, especially starch-based ones. The enzymes, together with the decomposed size residues, are then removed by successive hot- and cold-water rinsing (The Hong Kong Cotton Spinners Association 2007; Leung, Lo, and Yeung 1996).

3.1.1.4 Scouring

Scouring is carried out to remove the natural impurities and dirts of grey textiles so as to produce a clean white ground and improved water-absorbing property, which is suitable for later colour processing. The main types of impurities associated with the cellulosic cotton are seed fragments, waxes, oils, dirts and natural colouring materials. The most common method used for scouring is the treatment using strong alkali (sodium hydroxide, i.e., caustic soda) with detergent at a temperature near the boiling point for an optimum period of time. Excessive scouring will result in oxidation of cellulose, which will weaken the fabric strength (The Hong Kong Cotton Spinners Association 2007; Leung, Lo, and Yeung 1996).

3.1.1.5 Bleaching

Bleaching enhances the whiteness of the scoured textiles to produce a brighter white ground suitable for a fluorescent whitening effect as well as for coloration in brilliant shades. The three common kinds of bleaching agents used are sodium hypochlorite (NaClO), sodium chlorite ($NaClO_2$) and hydrogen peroxide (H_2O_2) (The Hong Kong Cotton Spinners Association 2007; Leung, Lo, and Yeung 1996).

3.1.1.6 Mercerization

Mercerization is the chemical treatment of the cotton fabric in a cold bath of concentrated caustic soda under controlled tension in a specialised mercerizing machine. The effect of the strong alkali causes the flat, twisted, ribbon-like cotton fibre to swell into a straight and round-shaped appearance, which results in a silk-like outlook with better lustre, dimensional stability, fabric strength and chemical reactivity. Double mercerization is a process whereby yarns are mercerised and knitted into a fabric, and the fabric is subsequently mercerised. Another chemical treatment, called the 'liquid ammonia' process, is especially used for yarn and knitted fabric, and it can also produce the same effect as mercerization. However, it is not as popular as mercerization because of the complicated application (operation temperature at –30°C in a vacuum chamber) (The Hong Kong Cotton Spinners Association 2007; Leung, Lo, and Yeung 1996).

3.1.1.7 Fluorescent Brightening

Fluorescent brightening, also called fluorescent whitening or optical brightening/whitening, is a process to further enhance the bleached textiles that are used for white goods. Due to its complex organic structure, the fluorescent brightening agent (FBA) possesses the ability to absorb light energy in the ultraviolet zone and re-emit the energy in the visible blue or violet light region. This results in a fluorescent whitening effect that makes the treated material look whiter. In some cases, a blue or violet tint may be incorporated together with the FBA, which results in an even whiter appearance (The Hong Kong Cotton Spinners Association 2007; Leung, Lo, and Yeung 1996).

3.1.2 Pretreatment for Protein Fibres

The pretreatment of two types of protein fibre, i.e., wool and silk, is discussed in the following subsections (Leung, Lo, and Yeung 1996).

3.1.2.1 Pretreatment of Wool

The pretreatment of wool involves the following processes (Leung, Lo, and Yeung 1996):

Setting → Scouring → Milling → Carbonising → Bleaching

3.1.2.1.1 Setting

Fabrics sometimes have an unsatisfactory appearance owing to cockling (waviness of the selvedges) and distortion of the design and weave structure. The main cause of this defect is the stretching of wool fibres during the spinning and fabric-manufacturing processes, which exert tension on the material. Wool setting is required to avoid this defect and also avoid the formation of permanent creases in fabrics. There are two techniques for setting of wool, namely decatising and crabbing, both of which impart dimensional stability to woollen fabrics. Decatising is a setting process in which the fabric is wound around a perforated drum, through which steam is passed in alternate directions. It preshrinks the fabric, and improves its appearance, lustre and handle. Crabbing performs virtually the same functions as decatising, but uses hot water instead of steam (Leung, Lo, and Yeung 1996).

3.1.2.1.2 Scouring

Raw wool is a very impure substance. The amount of impurities is greatest in the finer wools, ranging from 20% to 60%. The following three kinds of impurity are present (Leung, Lo, and Yeung 1996):

1. Wool grease and wax, which consist largely of sterols in association with fatty acids and can be removed either by emulsification or by means of organic solvents
2. Suint, i.e., the dried sweat of sheep, which is a mixture containing high levels of potassium salts of fatty acids and is water soluble
3. Mechanically adhering impurities, e.g., dirt, sand and vegetable matter

All these impurities should be removed before the wool is processed. Consequently, the greasy raw wool is scoured in its loose state. Wool yarn contains about 3%–15% of spinning oil and some dirt from processing. These impurities must be removed from woollen and worsted cloths by a scouring process.

The scouring methods for wool fibre include (Leung, Lo, and Yeung 1996):

Emulsion scouring: Wool is scoured by means of soap and soda ash. The scouring liquor should be carefully controlled to provide efficient scouring with minimum damage to the wool.

Solvent scouring: Grease and fats can be removed by means of solvents, e.g., perchloroethylene, trichloroethylene or carbon tetrachloride, etc. This method is less harmful to wool, and the grease can be recovered during solvent recovery.

Freeze scouring: Wool fibres are frozen in a cooling chamber at very low temperature (e.g., −40°C). The heavy dirt and grease are then removed by mechanical beating, followed by a modified emulsion-scouring process.

3.1.2.1.3 Milling

The aim of milling is to felt the fabrics so that they become fuller and denser, with the pattern of the weave structure disappearing to some extent during this process.

Milling can be carried out with one of the following processes: soap milling, acid milling or grease milling (Leung, Lo, and Yeung 1996).

Soap milling: A hard soap is used, which gives a gelatinous solution at the temperature employed (40°C) and provides a milling medium possessing a lubricating action. It gives the softest and most attractive handle to the wool goods.

Acid milling: This is quicker and, in the case of dyed goods, carries less danger of bleeding (loss of colour in the solution). The washing process is easier than for the other two milling processes (soap milling, grease milling), so lower water consumption is expected, but the handle is not as good as that obtained when mild soap is used. The material is typically treated at temperatures up to 70°C–80°C.

Grease milling: This is seldom used. It is a cheap and dirty milling method that is performed in the presence of dirt during the scouring process.

3.1.2.1.4 Carbonising

The objective of carbonising is the removal of cellulosic impurities such as burrs and vegetable fibres from wool. The main carbonising agent used is sulphuric acid. The scoured and milled pieces are immersed in 5% (wt/vol) sulphuric acid until they are thoroughly saturated. They are then squeezed, dried slowly and evenly in open width at 60°C–70°C, and finally baked at 100°C–110°C for about 5 min. This causes the vegetable matter to blacken. The charred cellulosic material is then easily removed from the fabric in a crushing machine resembling a milling machine, but possessing iron rollers, which shake the dust out. Carbonised pieces contain about 6% of sulphuric acid and are usually dyed with equalizing (levelling) acid dyes without neutralising (Leung, Lo, and Yeung 1996).

3.1.2.1.5 Bleaching

The purpose of wool bleaching is to remove, as far as possible, the natural cream colour of wool, either for the production of white goods or, when it is desired, to produce bright pale shades. The most common bleaching methods for wool involve the use of oxidising agents such as hydrogen peroxide and reducing agents such as sulphur dioxide or sodium hydrosulphite. If a good and permanent whiteness is required, both agents may be used. The wool is usually treated with an oxidising agent first and then with a reducing agent. Note that sodium hypochlorite, a common bleaching agent for cotton, is not used because it discolours and damages woollen fibre. Of the methods for bleaching wool, hydrogen peroxide treatment is the most common (Leung, Lo, and Yeung 1996).

3.1.2.2 Pretreatment of Silk

Silk usually undergoes three preparatory processes: degumming, bleaching and weighting (Leung, Lo, and Yeung 1996).

3.1.2.2.1 *Degumming*

In the degumming process, silk yarn or fabric is treated in a boiling soap solution for up to 2 h. This removes the outer layer of sericin and brings out silk's most valuable qualities, namely its lustre and softness. The raw silk contains around 15%–25% sericin. The quality of a particular silk depends on how much sericin is boiled off in the degumming process. The more sericin that is boiled off, the better is the silk quality. Complete removal of sericin results in weight loss of 25% (Leung, Lo, and Yeung 1996).

3.1.2.2.2 *Bleaching*

The bleaching agents used for silk are similar to those used for wool, with hydrogen peroxide again being the most popular bleaching agent. The bleaching of silk is carried out at 70°C–90°C, which is higher than the bleaching temperature for wool, since silk has better stability under alkaline conditions than wool (Leung, Lo, and Yeung 1996).

3.1.2.2.3 *Weighting*

Since silk fibre is very expensive, any weight loss caused by degumming may make it even more expensive. The process of weighting, by which metal atoms are incorporated into the silk fibre, is sometimes used to offset this weight loss. Tin salt such as stannic chloride is generally used as the weighting agent. In this process, the silk fibre is immersed alternately in solutions of stannic chloride and sodium phosphate until the desired increase in weight is obtained. The silk is then given a final treatment in a bath containing sodium silicate. Weighting may be regarded as a finishing process which gives a fuller and richer handle to the silk fibre and keeps the weighting within reasonable limits. The improvement is not due solely to the increase in weight, but also to the added smoothness of the fibre, which is caused by swelling (Leung, Lo, and Yeung 1996).

3.1.3 Pretreatment for Man-Made (Synthetic) Fibres

The amount of preparation required on a man-made fibre product is usually considerably less than necessary for the natural fibres such as cotton or wool. The main reason is that they contain far less impurity. A man-made fibre may be of high purity when it first enters the textile processing industry, but it then begins to acquire impurities. The fibre may have lubricants or antistatic agents added to assist in spinning the yarn or knitting a fabric (Leung, Lo, and Yeung 1996; The Society of Dyers and Colourists 2005).

- Residual dirt, oils and grease need to be removed before a final setting of the fabric, particularly relevant in setting of polyester and polyamide. It may be beneficial to pre-set at a low temperature, to impart a degree of stability, whilst ensuring contaminants are not set into the fibre during the process. In this case, the fabric will be post-scoured and then rinsed and dried before giving it a final heat-set prior to dyeing.
- Size, lubricating or antistatic agents, which are usually water soluble and readily cleared in a washing process, should be removed before dyeing.

- Sighting colours can often be removed during the normal scouring process, but in the case of specific fibres, it may be advisable to use a suitable reducing agent.
- Certain high-bulk fabrics which tend to relax when treated in hot water will bulk during the scouring process.

The preparatory processes for synthetic fibres are not as fixed as those for the natural fibres. The preparation required may differ for each particular synthetic fibre. In addition to the preparatory processes that are discussed here, note that desizing is sometimes necessary for woven synthetic goods. For example, the pretreatment route for synthetic knitted fabric is as follows (Leung, Lo, and Yeung 1996; The Society of Dyers and Colourists 2005):

Grey Inspection (& Mending) → Slitting → Joining → Desizing → Scouring and Relaxing → Heat Setting → Bleaching

3.1.3.1 Desizing

If a starch size is present, an enzyme desizing process will be satisfactory and will be performed in much the same way as that for a cotton fabric. Sizes on synthetics include polyvinyl alcohol, carboxylcellulose, polyacrylic acids, etc. They are normally removed during aqueous scouring at a temperature of about 70°C (Leung, Lo, and Yeung 1996; The Society of Dyers and Colourists 2005).

3.1.3.2 Scouring and Relaxation

Synthetic fibres are generally free from natural oils, waxes and impurities. The only possible sources of contamination are the spinning oils and lubricating agents that are added to assist the spinning, weaving or knitting processes. Since any contamination is likely to impair the levelness of dyeing, they must still be cleaned before being processed further. As the amount of the impurities is usually very small, and they can be easily removed, only a mild scouring (sometimes called pre-cleansing) is required. Another function of this process is to relax material so as to develop the bulkiness, handle and stretchability of the knitted fabric (Leung, Lo, and Yeung 1996; The Society of Dyers and Colourists 2005).

For the pre-cleansing of polyester and polyamide, detergent and alkali (soda ash or ammonia) are the usual scouring agents. The fibres are treated at 60°C–70°C for 30 min. For the pre-cleansing of acrylic, an acid wash with detergent and acetic acid at 50°C–60°C for 30 min can be used as an alternative to alkali treatment to produce a good result (Leung, Lo, and Yeung 1996; The Society of Dyers and Colourists 2005).

3.1.3.3 Heat Setting

Fabrics made of polyester and polyamide can be heat-treated to eliminate the internal tensions within the fibres. These tensions are generally formed during manufacture and further processing such as weaving and knitting. The fibres' new relaxed state is fixed (or set) by rapid cooling after the heat treatment. Without this setting, the fabrics might shrink and crease during later washing, dyeing and drying processes (Leung, Lo, and Yeung 1996; The Society of Dyers and Colourists 2005).

A fabric can be exposed to dry heat or steam at a temperature above the softening temperature while being held flat and to a particular width. If the width is held while the fabric cools, the fabric will be very resistant to stretch or shrink in the width at any subsequent stage of processing so long as the setting temperature is not reached or closely approached. Acrylic fibres cannot be heat-set because of the very loose structure of acrylic fibre molecules, and it discolours at temperature above 120°C (Leung, Lo, and Yeung 1996; The Society of Dyers and Colourists 2005).

Heat-setting can be carried out by hot-air-setting, hydro-setting or steam-setting. Hot-air-setting is by far the most popular heat-setting method, especially for polyester fabric. The process is usually carried out in a pin stenter, which holds the fabric at a pre-determined width by two rows of moving pins and feeds it into a heated chamber at 180°C–220°C for 30–60 s, where it is set. Hydro-setting is a treatment which uses water at 125°C–135°C for 20–30 min. Both this method and steam-setting at 130°C–135°C for 2–3 min are popular in the setting of polyamide fabric (Leung, Lo, and Yeung 1996; The Society of Dyers and Colourists 2005).

If the setting temperature is too high or the setting time is too long, yellowing and harsh handle of fabric can occur. If the setting temperature is too low, poor dimensional stability of fabric can occur. Uneven setting temperature during the process will give unlevel dyeing (Leung, Lo, and Yeung 1996; The Society of Dyers and Colourists 2005).

3.1.3.4 Bleaching

Synthetic fibres are sufficiently white for many purposes, without bleaching. However, if they are to be used as white goods or need to be used for brilliant shades, or for those materials that have yellowed after heat-setting, such fibres can be bleached in a similar manner to natural ones (Leung, Lo, and Yeung 1996; The Society of Dyers and Colourists 2005).

In the case of polyamide, both a reducing action with sodium hydrosulphite and an oxidizing action with hydrogen peroxide can be used to bleach the fibre. The bleaching methods suitable for wool can also be used (Leung, Lo, and Yeung 1996; The Society of Dyers and Colourists 2005).

Polyester, in contrast, can only be bleached by sodium chlorite, since no other bleach produces a satisfactory white effect. The bleaching is done in a slightly acidic solution at boiling point (Leung, Lo, and Yeung 1996; The Society of Dyers and Colourists 2005).

Acrylic can also be bleached by sodium chlorite in an acidic solution at boiling point. Care must be taken in cooling down the bleached material, as any sudden cooling may permanently impair the fabric handle (Leung, Lo, and Yeung 1996; The Society of Dyers and Colourists 2005).

3.2 DYEING PROCESS

Textile dyeing together with printing is commonly called the textile coloration process. In both cases, either dyes or pigments are used as the colorants (Leung, Lo, and Yeung 1996; The Society of Dyers and Colourists 2005; Roy Choudhury 2006).

Dyeing essentially involves the use of highly complex organic synthetic (but occasionally natural) dyestuffs, which under proper conditions (such as presence of salts, acids/alkalis, temperature and pressure) actually combine with the textile fibre molecules. Coloration by pigments, however, differs from dyeing in the sense that pigments do not combine with the textile fibre molecules as dyes do. They are only physically held and glued onto the fibre surface by means of resin binders.

The differences between dyes and pigments are that dyes are small in particle size, soluble in the application medium (such as water) and have affinity to fibres; pigments are large in particle size, insoluble in the application medium (such as water) and without affinity to fibres.

3.2.1 Factors that Affect Dyeing

Generally speaking, the dyeing process is a very complex chemical reaction. It involves the use of dyes as well as other assisting chemicals, known as auxiliaries, and the process is carried out under its optimum conditions. Different classes of dye applied on different types of fibre require a careful selection of such dyeing auxiliaries and processing conditions.

The major factors that affect dyeing are listed as follows (The Hong Kong Cotton Spinners Association 2007):

- Quantities of dyeing auxiliaries, such as salts, levelling agent, etc.
- pH, to be adjusted by acids and alkalis
- Liquid ratio
- Temperature and rate of heating
- Treatment time
- Agitation
- Any after-treatment to be given, such as dye fixation, oxidation, etc.

3.2.2 Classification of Dyes

There is no class of dye that is capable of dyeing all textile fibres. A specific class of dye can only be applied to a given type of textile fibre. The classification of dyes and their general description are summarised in Table 3.1 (The Hong Kong Cotton Spinners Association 2007; Leung, Lo, and Yeung 1996).

The choice of a specific colour for a particular material is the responsibility of the textile designer or colourist who perceives the colour to be in conformity with the fashion requirement. It is the job of the textile dyer to match the designer's colour with the proper dyes or pigments as well as to meet the colour-fastness requirements for the specific end use of the material. In brief, the designer's role is part of the world of artistry and creativity, while the dyer's role is in the world of science and technology.

Matching of colour shades by the dyer requires the skillful blending and formulation of different dyes and pigments, as well as an understanding of the nature of fibres and the numerous chemicals needed to carry the dyeing process. Colour-match recipes are first developed on a small laboratory basis. Once the dyer has formulated a colour match and achieved a satisfactory sampling (often known as the labdip),

TABLE 3.1
Classification of Dyes for Textile Fibres

Dye Class	General Description	Main Application
Direct	Simple application; cheap; complete colour range; moderate colour fastness but can be improved by after-treatment with copper salts and cationic fixing agents	Mainly used for cellulosic fibres, but can also be applied on rayon, silk and wool
Azoic	Complicated application; limited colour range (red, orange, navy among the best); bright shade at moderate cost; generally good wet fastness but moderate to poor dry cleaning and rubbing fastness; also called *naphthol dye* due to the use of naphthol or *ice colour* because of the usage of ice during application	Mainly applied on cellulosic fibres, especially on brilliant red shades
Vat	Difficult to apply (requires reduction treatment to make soluble in water and oxidation to resume insoluble state after dyeing); most expensive; incomplete colour range (strong in blue and green but weak in brilliant red); good all-around fastness except indigo and sulphurised vat species; tending to decrease in popularity due to increasing use of reactive dyes	Commonly used for high-quality cotton goods (e.g., towel), especially in the dyeing of denim fabric
Sulphur	Difficult to apply (application similar to vat dyes); cheap, particularly for dark shade; incomplete colour range (strong in black, navy, khaki and brown but no bright shade); poor washing and rubbing fastness and sensitive to chlorine; may cause fabric rendering of cellulose upon storage (aging)	Mostly used for heavy cellulosic goods in dark shades
Reactive	Easy application; moderate price; complete colour range; good fastness due to direct reaction with fibres	Commonly used for all cellulosic goods, especially in knitted fabric batchwise dyeing; selective dyes can also be applied on wool, silk and rayon; increasingly used in printing due to good fastness
Acid	Easy application; complete colour range with very good bright shades; fastness properties may vary among individual dyes	Commonly used for wool, silk and nylon
Metal-complex	Relatively difficult to apply; expensive; complete colour range but duller shade than acid dyes; good fastness due to high molecular size and metal complex structure	Mainly used for wool and nylon
Chrome	Complicated application; expensive; complete colour range but very dull shade; good all-around fastness	Mainly used for wool products, especially for the end use of carpet

TABLE 3.1 (*Continued*)
Classification of Dyes for Textile Fibres

Dye Class	General Description	Main Application
Disperse	Require skill in application (either by carrier or under high temperature); moderate price; complete colour range; limited solubility in water (normally dispersed in water for application); good fastness after reduction clearing treatment; sublimation property	Mostly used for polyester and acetate; can also be applied on nylon and acrylic
Basic (Cationic)	Careful application required to prevent unlevel dyeing and adverse effect in hand-feel; complete colour range with very good brilliant shades	Mainly used for acrylic

Source: Summarised from The Hong Kong Cotton Spinners Association (2007); and Leung, Lo, and Yeung (1996).

this becomes the standard which all future dye lots or batches must follow. In actual production, however, each dye lot is more or less different in shade from all other lots. This lot-to-lot shade variation is caused by several factors such as differences in dyes/auxiliaries concentration, fabric lots and different dyeing machine settings, etc.

A good dye must possess an acceptable degree of retention by the textile fibre so as to withstand the subsequent treatment (e.g., laundering, dry cleaning, etc.) or environmental wearing (e.g., rubbing, light exposure, etc.). The degree to which a dyed material can withstand such treatments and wearing is called colour fastness. No dye or pigment used for textiles is absolutely fast to all colour-fastness conditions. Only a careful selection and formulation of dyes and auxiliaries can result in a desirable dyeing, and conform with the colour-fastness requirements. The most common colour-fastness items for textile material are laundering (washing), light exposure, dry cleaning, perspiration and rubbing (crocking). The common fastness properties of different dye classes are summarised in Table 3.2 (Leung, Lo, and Yeung 1996; The Hong Kong Cotton Spinners Association 2007).

3.2.3 Application of Pigments

Pigments are very popular in textile coloration, especially in printing. The advantages of using pigments are given as follows (Leung, Lo, and Yeung 1996; The Hong Kong Cotton Spinners Association 2007):

- Easy to apply with good shade matching from lot to lot
- Full colour range
- Can be applied on all textile fibres and their blends

The application of pigments involves the use of adhesives (commonly known as binders) to bond the pigments to the surface of textile fibres. Pigment-dyed materials generally have inferior washing, dry cleaning and rubbing fastness properties. The handle of the pigmented dyed materials are also adversely affected due to the binder

TABLE 3.2
Common Colour-Fastness Properties of Different Dye Classes

	Colour Fastness to				
Dye Class	**Washing**	**Light**	**Dry Cleaning**	**Perspiration**	**Rubbing**
Direct	Moderate (can be improved after proper after-treatment)	Moderate (can be improved after proper after-treatment)	Good	Good	Good
Azoic	Good	Good	Moderate	Good	Moderate
Vat (except indigo)	Excellent	Excellent	Good	Excellent	Good
Sulphur	Moderate (sensitive to chlorine)	Good	Good	Good	Moderate (poor on dark shade)
Reactive	Good	Good	Excellent	Excellent	Good
Acid	Moderate to poor	Good	Good	Moderate	Good
Metal complex	Good	Excellent	Good	Good	Good
Chrome	Excellent	Excellent	Good	Good	Good
Disperse	Good	Good	Good	Good	Good
Basic (cationic)	Good	Moderate to poor	Good	Good	Good

Source: Summarised from Leung, Lo, and Yeung (1996); The Hong Kong Cotton Spinners Association (2007).

film that is formed on the fibre surface. Nowadays, a fashion trend of 'garment wash' or 'pigment wash' has become increasingly popular. The basic principle is to make use of the properties of the poor washing and rubbing fastness of pigments to deliberately produce a colour-fading effect after a severe machine washing (The Hong Kong Cotton Spinners Association 2007).

3.2.4 Methods of Dyeing

Dyeing can be done during any stage, from the raw fibre right through to the finished garment. The following are the four stages at which dyeing can be carried out (Leung, Lo, and Yeung 1996; The Hong Kong Cotton Spinners Association 2007):

1. Fibre dyeing (also called stock dyeing)
2. Yarn dyeing
3. Fabric dyeing (also piece dyeing)
4. Garment dyeing

The considerations that govern the choice of the dyeing stage are mainly technical and economical. Dyeing at the fibre stage is much more expensive than at other stages, but it will normally give the best fastness result as well as enabling production of a fancy colour effect such as mottle, heather, etc., after a fibre blending process. Yarn dyeing is mainly used to produce a checks or stripe effect with different coloured yarns. The details of these dyeing stages are discussed in the following subsections.

3.2.4.1 Fibre Dyeing

Fibre dyeing is carried out on loose stock before the fibre is spun into yarn. It is done by putting packed loose fibres into the container of the loose-stock dyeing machine in which the dyeing will take place. The process is mostly used in the production of woolen materials, especially when mottled and heather-like colour effects are desired. Fibre dyeing usually results in even dyeing and excellent penetration of dyes into the fibres, and hence good fastness properties. However, it is the most costly method for dyeing because the production rate is relatively low and there is 10%–15% waste due to loss of fibres during the process (The Hong Kong Cotton Spinners Association 2007; Leung, Lo, and Yeung 1996).

3.2.4.2 Yarn Dyeing

Yarn dyeing is carried out after fibre has been spun into yarn. The process is mainly used to produce various effects of stripes, checks, plaids or other multicolour designs with different coloured yarns in the weaving and knitting process. Moreover, threads for sewing and embroidery purposes are all dyed at the yarn stage. Yarns may be dyed in different forms (Leung, Lo, and Yeung 1996; The Hong Kong Cotton Spinners Association 2007):

- Hank dyeing
- Package dyeing (cone dyeing)
- Slasher dyeing
- Rope dyeing

3.2.4.3 Fabric Dyeing

Fabric dyeing, also called piece dyeing, is the most popular production method for solid colour, as it gives the greatest flexibility to the manufacturer in terms of inventory as well as production capacity. Large orders of woven fabric can be dyed in a continuous process, i.e., pad dyeing, while the batchwise jig dyeing is suitable for small batch production. For tension-sensitive materials, such as knitted fabrics or some thin woven fabrics, the batchwise process of winch dyeing and jet dyeing are the most appropriate dyeing methods. The description of each dyeing method is discussed in the following list (Leung, Lo, and Yeung 1996; The Hong Kong Cotton Spinners Association 2007):

1. *Pad dyeing* is a continuous process used to dye fabric in open-width form. The fabric is passed through a trough of dye liquid and then squeezed evenly by a pair of pad mangles to impregnate dye deep into the fibres. After padding, a series of operations like intermediate drying, chemical fixing of the dye, steaming, successive washing off and final drying will be followed (The Hong Kong Cotton Spinners Association 2007; Leung, Lo, and Yeung 1996).

2. *Jig dyeing* is commonly used in the production of tightly woven fabric of medium to heavy weight, e.g., corduroy, canvas, etc. In the jig-dyeing machine (also called jigger), the fabric is held on rollers in open-width form and transferred repeatedly from one roller to another through a trough of dye liquor (The Hong Kong Cotton Spinners Association 2007; Leung, Lo, and Yeung 1996).
3. *Winch dyeing* is the traditional dyeing method in the production of knitted fabrics. Pieces of fabric (to make up appropriate weight) sewn in rope form are loaded into the winch which essentially consists of a dye vessel fitted with a motor-driven wheel (also named *winch*). During operation, the winch will rotate and initiate an endless fabric movement with minimised tension through the dye liquid. The operation of winch dyeing is among the simplest, and the dyed materials retain much of their original fullness and softness (The Hong Kong Cotton Spinners Association 2007; Leung, Lo, and Yeung 1996).
4. *Jet dyeing* is somewhat similar to winch dyeing in the sense that fabric in rope form is circulated through a dye liquid bath. However, in a jet-dyeing machine fabric movement is generated by a jet flow of dye solution instead of mechanical pull as in winch dyeing. As such, the dyed fabrics are almost free from creases and rope marks. Pressurised dyeing up to 130°C is also possible for synthetic fibres. Recent developments in jet-dyeing machines are moving towards process automation and cost reduction (The Hong Kong Cotton Spinners Association 2007; Leung, Lo, and Yeung 1996).

3.2.4.4 Garment Dyeing

Garment dyeing is the dyeing of completed garments. It is primarily an economical method used for non-tailored garments such as sweaters, hosiery and pantyhose, etc. Nowadays fashion trends also require some cotton jeans and shirts to be dyed in completed garment form. During processing, an appropriate number of garments are loaded into the specific dyeing machine, either a paddle type or a rotating-drum type, which is more or less similar to that of a domestic washing machine, and the garments are agitated in the dye bath during colour buildup (The Hong Kong Cotton Spinners Association 2007; Leung, Lo, and Yeung 1996).

3.2.5 Dyeing of Cellulosic Fibres

Cellulose is the most important structural material in nature and the most abundant of the natural polymers. Cellulose is a condensation polymer of β-D-glucopyranose with 1,4-glycosidic bonds between successive pyranose rings. All the substituents, including glycosidic links, project from the pyranose ring in the same plane, and only the four C–H bonds are perpendicular to this plane. The essential features of the polymer chain are the main sequence of main-chain units, the non-reducing end group and the reducing end group. The reducing group is a cyclic hemiacetal that exhibits the characteristics of both a secondary alcohol and an aldehyde under appropriate conditions. The degree of polymerization (DP) of cellulose varies with the source and is usually expressed as an average, since a wide distribution is found.

In raw cellulosic material, it can be as high as 14,000, but purification treatments involving alkali usually lower this to about 1000 to 2000. Furthermore, the hemi-acetal end groups are oxidised to carboxylic acid groups under these conditions, so that purified cotton normally has no reducing power if degradation has not taken place (Broadbent 2001; Shore 1995; Roy Choudhury 2006). This section reviews the selection of dye for cellulosic fibres based on application method. The dyes used for dyeing cellulosic fibres include direct dyes, sulphur dyes, reactive dyes, azoic dyes and vat dyes.

3.2.5.1 Direct Dyes

This class includes the first synthetic dyes to be manufactured which had affinity for cellulose. They are mainly high-molecular-weight azo compounds containing sulphonic acid groups which make the dye water soluble. A wide range of colours is available varying in cost and fastness properties. Generally speaking, direct dyes are classified into three categories (The Society of Dyers and Colourists 2005):

Class A: *Self-levelling dyes* have good migrating power and thus very good levelling properties.
Class B: *Salt-controllable dyes* generally migrate less well. Their uptake can be controlled with gradual additions of salt (electrolyte).
Class C: *Temperature controllable dyes* have poor migrating properties, and once they are taken up unevenly are difficult to level. To achieve level dyeing, it is necessary to use a gradual increase in temperature to help control dye uptake.

3.2.5.1.1 Dyeing Method for Direct Dyes on Cellulosic Fibre

The dye, usually in powder form, is pasted and then dissolved in hot water before adding to the dye bath. The dye bath will contain dyeing assistants and possibly a sequestering agent. The temperature is set at around 30°C–40°C, and the pre-wet material is added before raising the temperature to 85°C–95°C. The material is dyed for 30–45 min and, where necessary, electrolyte is added to increase the exhaustion (The Society of Dyers and Colourists 2005).

After dyeing, the material is rinsed and then dried. It is important not to allow the bath to become too strongly alkaline, as the dye may be partially destroyed by reduction under alkaline conditions (The Society of Dyers and Colourists 2005).

Direct dyes will also dye wool, silk and polyamide, but as higher fastness properties can be achieved on such fabrics by using other dyes, the use of direct dyes is largely confined to cellulosic fibres (The Society of Dyers and Colourists 2005).

3.2.5.1.2 After-Treatment of Direct Dyes on Cellulosic Fibres

1. *Diazotisation and coupling*: The dye capable of being after-treated in this way is dyed using the method appropriate to its dye class. The fabric is then treated in a solution of sodium nitrile and hydrochloric acid. After rinsing, it is then treated in a second bath containing a coupling component such

as a naphthol. An after-soaping is essential if good wash fastness is to be obtained. The after-soaping process can result in considerable shade change (The Society of Dyers and Colourists 2005).

2. *Coupling with diazotised amines*: The dyed fabric is treated in a bath containing a diazotised amine, resulting in the formation of a new coloured product in the fibre (The Society of Dyers and Colourists 2005).
3. *Treatment with formaldehyde*: Azo dyes which contain two adjacent hydroxyl or amine groups are able to form complexes with formaldehyde by after-treatment in a bath containing formaldehyde at around 60°C for 20 min. This generally improves wet fastness; however, light fastness is often impaired (The Society of Dyers and Colourists 2005).
4. *After-treatment with metal salts*: Chromium and copper salts, e.g., chromium fluoride and copper sulphate, can be used for this purpose (The Society of Dyers and Colourists 2005).
5. *After-treatment with synthetic resins and cationic agents*: Cross-linking finishes such as urea-formaldehyde are applied to the dyed materials followed by a curing process. Whilst fixing the dye in the fibre improves wet durability, the light fastness may be impaired and the shade may change (The Society of Dyers and Colourists 2005).

3.2.5.2 Sulphur Dyes

These dyes are still widely used on cellulose to produce cheap dull shades which are generally fast to washing and light but not fast to hypochlorite bleach liquors. Sulphur dyes have gained increased popularity for dyeing of fashion jeans. Such dyes have an uncertain constitution but are usually produced by heating together sulphur and organic materials containing nitrogen with alkali. The dyes are commercially available in liquid and powder form, some as pre-reduced solutions, and modern systems have concentrated on using less-hazardous reducing agents than the commonly used sodium sulphide systems (The Society of Dyers and Colourists 2005).

3.2.5.2.1 Dyeing Method of Sulphur Dyes

After the dye is added to water, a solution is obtained by adding the necessary amounts of alkali and reducing agent, e.g., sodium sulphide. The mix is heated until a solution is obtained. This is then added to the dye bath and the wet-out fabric is introduced before raising the temperature to a boil. Dyeing is continued for 1–1½ h, and salt is usually added during the dyeing cycle to increase dye uptake. After dyeing, the fabric is rinsed, oxidised, soaped, rinsed in cold water and dried. No sulphur must be left in the fabric after dyeing, and a thorough oxidation and rinsing is essential (The Society of Dyers and Colourists 2005).

3.2.5.2.2 After-Treatments for Sulphur Dyes

One of the major problems associated with deep-shade sulphur dyeing is the relatively poor wet rub fastness. This may be improved by padding with a synthetic cationic acrylic or polyurethane binder (The Society of Dyers and Colourists 2005).

3.2.5.3 Reactive Dyes

This range of dyes contains a reactive group which will react with hydroxyl groups in cellulose under alkaline conditions to form a covalent bond. When using reactive dyes, care must be exercised, as reaction with water will also take place giving a hydrolysed dye. Table 3.3 provides a list of reactive dyes by chemical type, by their reactivity characteristics and by their respective fixation temperature (The Society of Dyers and Colourists 2005).

3.2.5.3.1 Dyeing Method of Reactive Dyes

Reactive dyes are applied batchwise in the following general manner (The Society of Dyers and Colourists 2005).

- Dissolve dye in water
- Set bath temperature relative to dye type, e.g., between 20°C–50°C
- Add the wet-out material and continue dyeing for approximately 15 min
- Add common salt to the bath at intervals, then add alkali to fix the dye, continuing dyeing for 30 min
- After dyeing: rinse, soap, rinse and dry

Cold-dyeing types are most reactive, e.g., dichlorotriazine. Hot-dyeing types were developed in which the dyeing method is similar to the previous method, but the alkali fixation stage of the process is carried out at a higher temperature, often between 70°C–90°C. The addition of resist salt is advisable at these high temperatures to prevent reduction of the dye. Reactive dyes are also very suitable for pad application (The Society of Dyers and Colourists 2005).

3.2.5.3.2 After-Treatment of Reactive Dyes

As with direct dyes, it is possible to after-treat reactive dye to improve fastness properties using certain cationic agents. Synthetic cationic acrylic binders may also be used for this purpose. Careful selection of a suitable fixing agent must be carried out

TABLE 3.3
Properties of Major Types of Reactive Dyes

Reactive Group	Fixation Temperature (°C)	Relative Reactivity[a]
Dichlorotriazine	30	1
Difluorochloropyrimidine	40	2
Dichloroquinoxaline	50	3
Monofluorotriazine	50	3
Vinylsulphone	60	4
Monochlorotriazine	80	5
Dichloro- and trichloro-pyrimidine	95	6

Source: The Society of Dyers and Colourists (2005).
[a] 1 = most reactive; 6 = least reactive.

to ensure that shade is not adversely affected and that wet and dry rubbing plus wash fastness are all improved without adversely affecting light fastness (The Society of Dyers and Colourists 2005).

3.2.5.4 Azoic Dyes

These are formed in the cellulose fibre by reacting a coupling component, phenolic type with a diazotised base or stabilised salt. They produce extremely bright hues on cellulose with good light and wash durability and high chlorine fastness. However, they show poor solvent fastness, and rubbing fastness can also be a problem (The Society of Dyers and Colourists 2005).

The application comprises four stages:

1. *Impregnation*: The naphthol is first dissolved in sodium hydroxide and then diluted with cold water. This is then applied to the fabric by padding or by exhaustion dyeing. The ultimate hue will depend on the uptake of naphthol, which is affected by the following factors:
   - The substantivity of the naphthol for the cellulose (This can be improved by the use of common salt.)
   - The concentration of the solution applied
   - Temperature of the solution, which may be 25°C–30°C (Levelling is often best achieved at higher temperatures; however, in this case, a lower level of exhaustion is obtained. The time of impregnation for piece-goods is in the region of 25–30 min, whereas longer times are required for package dyeing.)

     Once applied, it is extremely important to protect the naphthol-treated fabric from acid or chlorine fumes and from direct sunlight.
2. *Removal of excess naphthol*: After impregnation with naphthol, the fabric should be mangled or hydroextracted and may be dried before subsequent processing.
3. *Development*: At this stage, the development is achieved by treating the naphtholated material in a solution of the diazotised base or fast salt. The latter is more expensive but easier to apply.
4. *Soaping*: After rinsing, the fabric is soaped, which performs a number of functions: removing mechanically held pigment, developing the true shade and giving the end fastness.

3.2.5.5 Vat Dyes

Vat dyes are used where high light and wash fastness is required. Such dyes are largely based on the anthraquinone and indigo types. The method of dyeing involves reducing the insoluble dye to its soluble leuco form by treatment with reducing agent and alkali. In this state, the dye has affinity for cellulose. After dyeing in this reduced state, the dye within the fibre is oxidised back to its insoluble form, where it remains fixed in the fibre. Batchwise application to cellulose involves four stages (The Society of Dyers and Colourists 2005):

Vatting → Dyeing → Oxidation → Soaping

1. *Vatting*: The dye is pre-dispersed in water, and the required amounts of caustic soda (alkali) and sodium hydrosulphite (hydros) are added. The solution is allowed to stand for around 10–15 min at the required temperature until reduction (vatting) is complete.
2. *Dyeing*: The vatted dye is then added to the dyeing machine containing water at the desired temperature. The pre-wet-out material is added to the dye vessel and dyeing continues for approximately 30 min. It is essential to monitor the alkali and hydros concentrations during processing.
3. *Oxidation*: After rinsing, the dyed material is oxidised in water or with a suitable oxidizing agent.
4. *Soaping*: After rinsing and neutralizing, the material is soaped at or as near to the boil as possible before rinsing and drying.

3.2.6 Dyeing of Protein Fibres

Proteins are natural polymers of high relative molecular mass formed by condensation of α-amino acids through their carboxyl and amino groups. Proteins are widespread in nature, being essential components of animal and plant tissue, and they have thus been the subject of much research. Wool belongs to a class of proteins known as keratins. The α-amino acids have the general formula and are distinguished from one another by the R-group, which can be basic, acidic or nonpolar. Examples include the amino acids proline and cystine. Cystine plays a key role in the properties of wool but is absent from the structure of silk. Each protein chain contains a primary amine ($-NH_2$) group at one end of the protein chain and a carboxylic acid ($-COOH$) group at the other end. Wool is by far the most important of the animal fleece fibres in the protein fibre category, but several others are of interest. These include the mohair and cashmere from species of goat, alpaca and vicuna from camel species, angora fur from rabbits, wool from the llama that is native to the Andean plateau, and yak hair from the humped Tibetan ox. Most of these speciality fibres, notably cashmere, angora and mohair, are relatively scarce and costly, but they may be blended with high-quality wool to increase lustre and give a distinctive appearance. Demand for such blends is largely subject to the dictates of fashion. Silk fibroin (silk fibre), like wool keratin, is a protein fibre, but there are major differences in composition, most notably the virtual absence of cystine cross-links. Basic and acidic side chains occur much less frequently along the silk fibroin peptide chains, and therefore electrostatic links between them make a much lower contribution to the internal structure of the fibre (Broadbent 2001; Lewis 1992; Roy Choudhury 2006; The Society of Dyers and Colourists 2005).

This section reviews the selection of dye for protein fibres based on application method. The dyes used for dyeing protein fibres include acid dyes, chrome dyes, metal-complex dyes and reactive dyes.

3.2.6.1 Acid Dyes

Acid dyes are anionic and will dye any fibre which contains positively charged ions. Such dyes are used for dyeing wool, silk and polyamide. They ionise in solution to form sodium ions and large negatively charged coloured ions. Wool contains more potential dye sites than polyamide and, as a result, it is possible to achieve heavier depths of shade. Acid dyes for wool may be classified into a number of distinct groups. Traditionally, they were mainly of the type: (a) acid-levelling dyes and (b) acid-milling dyes (The Society of Dyers and Colourists 2005).

Acid levelling dyes have a relatively low affinity for wool under neutral conditions, whereas under acid conditions the tendency is to give unlevel dyeing due to the high rate of strike. In practice, it is common to dye in the presence of an acid such as sulphuric acid (H_2SO_4), which increases the affinity of the dye for the fibre, plus sodium sulphate (Glauber's salt), which slows down the rate of dyeing. Acid-levelling dyes tend to have poor durability to washing and alkaline treatments, yet they are often preferred for their good levelling characteristics. Acid-milling types produce better wash durability. The classification into acid-levelling and -milling types is not sharply defined, and the Society of Dyers and Colourists has proposed a method of classifying in terms of the dye-bath acidity (The Society of Dyers and Colourists 2005).

Acid-milling types possess higher affinity for wool and are therefore applied from a much weaker acid bath, for example, using ammonium sulphate or acetate. A typical dyeing method would be as for levelling types but replacing the acid and omitting Glauber's salt (The Society of Dyers and Colourists 2005).

3.2.6.2 Chrome Dyes

Chrome dyes are less popular today largely due to health and environmental constraints. They may be considered as acid dyes which contain groups capable of forming a stable complex with chromium. There are three main methods of application on wool: (a) chrome mordant, (b) after-chrome (or top-chrome) and (c) meta-chrome (The Society of Dyers and Colourists 2005). It has already been stated that chromium is a problem to the dyer, particularly as consent limits for discharge are becoming tighter. Even with the strictest effluent control and 'optimised dyeing methods', the future for chrome dyes may be questionable.

3.2.6.2.1 Chrome Mordant Method

This is the oldest technique but not the most important. It involves mordanting the wool using 2%–4% potassium dichromate ($K_2Cr_2O_7$), the bath being made up at around 50°C with a small amount of an acid. The wool is then added, the temperature raised to the boil, and the process continued for 1–1½ h. After pre-mordanting, the fabric is dyed in the manner described previously (The Society of Dyers and Colourists 2005).

3.2.6.2.2 After-Chrome Process

The wool is first dyed with dye plus around 10% Glauber's salt for 1 h at the boil. Any dye remaining in the bath is then exhausted with an addition of acetic acid. Then 1%–2% potassium dichromate ($K_2Cr_2O_7$) is added and the process continued for 30 min. For deep shades, e.g., black, it is common to add an acid such as sulphuric acid. Varying the concentration of the acid will give different depths of black (The Society of Dyers and Colourists 2005).

3.2.6.2.3 *Meta-Chrome Method*

Certain chrome dyes can be dyed and mordanted simultaneously in the same bath, but it is very important that (a) the dyeing take place well under neutral conditions, (b) the dye must not be affected by salt in the bath and (c) the dichromate or chromate present in the bath needs to be reduced on the fibre; otherwise the dye will be precipitated. In this method the metachrome mordant is first produced, probably comprising one part sodium chromate to two parts ammonium sulphate. The dye bath is set at around 50°C with water, dye, and 2%–8% of the metachrome mordant. Dyeing takes place in the usual manner. The level of the mordant will vary depending on the desired shade (The Society of Dyers and Colourists 2005).

3.2.6.3 Metal-Complex Dyes for Wool

Metal-complex dyes consist of an acid dye–type structure in a complex with a metal ion (The Society of Dyers and Colourists 2005; Lewis 1992). In addition to natural fibres such as wool and silk, they can be applied to polyamide. Two key categories exist: (a) 2:1 metal-complex dyes which consist of two parts dye to one part metal and (b) 1:1 metal-complex dyes which consist of one part dye to one part metal.

The 2:1 metal-complex dyes are neutral dyeing, mainly complexes of chromium with water-soluble azo dyes, while 1:1 metal-complex dyes are seldom used on polyamide, mainly on wool. On wool they are applied from a low-pH (2) dye bath using up to 8% sulphuric acid. As dyeing is carried out at the boil at low pH, care must be exercised due to the possibility of damage to the wool. Under such conditions there may also be noticeable shade changes of the dyes (The Society of Dyers and Colourists 2005).

3.2.6.4 Reactive Dyes for Wool

Reactive dyes are dyes which form a covalent bond with the fibre; hence, the durability of reactive dyes, particularly to washing, is generally good. Reaction takes place on wool, as on cotton, with hydroxyl groups present in the fibre. However, on wool, both thiol and amino groups are also available for bonding with the dye. The application of reactive dyes on wool requires the addition of auxiliaries to prevent skitter and grossly unlevel dyeing. Such auxiliaries include surface-active agents which are capable of forming dye-surfactant complexes which aid the dyeing operation (The Society of Dyers and Colourists 2005; Lewis 1992).

3.2.7 Dyeing of Silk

The dyeing behaviour of silk is similar to that of wool, but silk typically requires between two and four times as much dye as wool to achieve a similar visual depth because of the fineness of the silk filament. Acid-milling, 2:1 metal-complex, and direct and reactive dyes are the most important classes for silk dyeing. Traditionally, the concentrated soap solution remaining after degumming was used to provide an anionic levelling system for the application of direct and acid dyes. This approach has been superseded by dyeing with 2:1 metal complex and acid-milling dyes in the presence of levelling and penetrating agents. Typically, dyeing is commenced at 30°C in the presence of 1%–2% of a weakly cationic levelling agent at pH 4.5–5.0.

The temperature is raised at 1°C per minute to a top temperature in the range 70°C–85°C, depending on dye exhaustion and the type of equipment available. Dyeing is continued for 45–60 min at top temperature before cooling and rinsing (The Society of Dyers and Colourists 2005).

3.2.8 Dyeing of Hair Fibres

Other than wool fibre, animal hair fibres are protein fibres in nature, and all dyes used for dyeing wool fibre can be used for dyeing animal hair fibres. However, the properties of animal hair fibre should be considered before undertaking the dyeing process. For example, angora is only processed in blends with wool, sometimes with the addition of a small proportion of nylon to improve the durability. For economic and technical reasons, 2:1 metal-complex and acid-milling dyes are preferred. Chrome dyes and 1:1 metal-complex types are seldom used because strongly acidic dye baths may damage the angora. Major outlets for wool/mohair blends are worsted outerwear and fabrics for suiting. Such fabrics may be made from intimate blends for both warp and weft, but they often consist of a mohair warp with a botany wool weft. Wool/mohair fabrics may be piece-dyed with 2:1 metal-complex dyes or more economically with levelling-acid dyes (The Society of Dyers and Colourists 2005).

3.2.9 Dyeing of Synthetic Fibres

Disperse dyes, basic dyes and acid dyes can be used for dyeing synthetic fibres, depending on the physical and chemical natures of the synthetic fibres.

3.2.9.1 Disperse Dyes

Early disperse dyes were developed for dyeing cellulose 'diacetate'. The dyes had a small molecular size, and although the fibre could be dyed quite easily, fastness properties were limited. With the introduction of polyester fibre, similar dyes gave much better wet fastness. Disperse dyes are only marginally water soluble, and they form a dispersion in water. The marginal water solubility is essential for their application by exhaustion methods. The rate of dyeing is determined by the speed at which the dye is transferred from the aqueous dispersion to the solution and into the fibre (Hawkyard 2004; Nunn 1979; The Society of Dyers and Colourists 2005). Disperse dyes are normally classified according to their degree of sublimation. Class A dyes are of small molecular size and readily sublime, and therefore show low heat fastness, whereas Class B, C and D dyes have increasingly better sublimation properties. In the case of dyeing with disperse dye, a reduction clearing process should be conducted to remove any dyestuff which remains unfixed plus any dyeing aids, including carrier residues remaining on the surface of the fibre (The Society of Dyers and Colourists 2005).

3.2.9.1.1 Application of Disperse Dyes to Cellulose Acetate/Triacetate

This fabric has traditionally been dyed using batchwise methods, often jig dyed. In the case of jig dyeing, the fabric is passed end to end through a trough containing the dye. The temperature for the fabric on the rollers should be around 80°C, and dyeing is continued for up to 1 h. The conditions for dyeing cellulose triacetate are very similar, but a higher temperature of around 120°C is needed. If this is not possible, then dyeing should be at the boil with the aid of a carrier. A number of carriers may be used, including diethyl phthalate and butyl benzoate. Normally, carriers are effective in low concentrations, but in the case of jig dyeing at low liquor ratios, higher levels can be effective at up to 20 g/L. Additional auxiliaries used in dyeing of disperse dyes on triacetate include buffers that are required to control pH and sequestering agents to reduce any problems associated with water hardness. In the case of certain blue and violet shades, it is sometimes necessary to use gas-fume fading inhibitors (Hawkyard 2004; Nunn 1979; The Society of Dyers and Colourists 2005).

3.2.9.1.2 Application of Disperse Dyes to Polyamide

Disperse dyes can be used for dyeing polyamide if levelness of shade is more critical than fastness. Compared to acid dyes on polyamide, the hues produced are often duller. Application conditions are similar to those used on triacetate. Temperatures of 95°C–100°C or up to 120°C are employed at times of up to 45 min. The pH of the dye bath is buffered to 5.5–6.5 (Hawkyard 2004; Nunn 1979; The Society of Dyers and Colourists 2005).

3.2.9.1.3 Application of Disperse Dyes to Polyester

For polyester, the less-volatile dyes are most suitable, particularly for dyeing fabrics which have to withstand high-temperature processing. There are two conventional batchwise methods for dyeing polyester with disperse dyes: (a) at the boil with the aid of a carrier and (b) at 130°C under pressure (Hawkyard 2004; Nunn 1979; The Society of Dyers and Colourists 2005).

The dye bath must be buffered to maintain a slightly acid pH. The dye bath is set with water, disperse dye, dispersing agent and buffers. If a carrier is to be used, this is added as required. The temperature is raised to the required level, and dyeing is continued for 30–60 min, after which the bath is cooled and drained. The material is then rinsed, soaped, reduction cleared, rinsed again and then dried. Cooling the bath is essential before discharging when temperatures above 100°C are used (The Society of Dyers and Colourists 2005).

3.2.9.2 Basic Dyes

Basic dyes are amongst the oldest known dyes, and some are still known by their original names such as methyl violet, malachite green and rhodamine magenta. Dyes of this type dissolve in water to give positively charged coloured ions which can dye any fibre that contains a negatively charged group. Thus in addition to being used on natural fibres such as cotton, wool and silk, they may also be used for dyeing polyamide, although this is not a popular commercial choice (The Society of Dyers and Colourists 2005). Their main application for synthetic fibres comes in dyeing modified acrylic fibres, which give better fastness than other dye classes. Basic dyes are adsorbed rapidly; thus it is necessary to control carefully their dyeing rate with the aid of retarding agents, which can be negatively or positively charged. It is their respective charge which determines their mode of action. A positively charged retarding agent will complex with the fibre, slowing down the dye uptake, whereas a negatively charged product reacts with the dye to control dye uptake. During heating in the dye bath, this association breaks down to allow dyeing of the fibre to take place (Hawkyard 2004; Nunn 1979; The Society of Dyers and Colourists 2005).

3.2.9.3 Acid Dyes

Acid dyes can be used for dyeing natural fibres such as wool and silk. They are also used for dyeing polyamide. Acid dyes are anionic, and in solution they will ionise into sodium ions plus large negatively charged coloured ions. For this reason, they will dye any fibres which have positively charged sites, In comparison to wool, polyamide has fewer sites available for dyeing, and thus the shades produced on polyamide are generally weaker than on wool but of higher fastness (The Society of Dyers and Colourists 2005).

3.2.10 Pigment Dyeing

The main technique used for pigment 'dyeing' is that of padding from a short liquor ratio, the degree of pickup being largely determined by the substrate and application conditions. Pigment 'dyeing' differs from conventional dyeing processes in that inorganic and organic insoluble pigments, with little or no affinity for the fibre, are applied to the substrate in conjunction with an adhesive, normally termed a *binder* (The Society of Dyers and Colourists 2005).

Conventional pigment dispersions cannot make a direct bond with the fibre and thus require the use of a binder to fix the pigments to the substrate. The binder must form a film, enclosing the pigment particles, with adequate adhesion to the fibre surfaces. The pigment 'dyeing' process is relatively simple compared to conventional dye systems, requiring only a Pad → Dry → Cure process. Fixation may be achieved in a hot-air baker or during stenter finishing, either alone or in combination with other finishing auxiliaries. No wash-off process is normally required with pigments (The Society of Dyers and Colourists 2005).

3.3 PRINTING PROCESS

Textile printing has often been described as a localised dyeing process to produce single- or multi-coloured patterns on fabrics. It utilises the same dyes or pigments of similar principles of application and has the same colour fastness properties as that of textile dyeing. Among all the colorants available in dyeing, the most commonly used is the pigment. The advantages of using pigment are summarised as follows (The Hong Kong Cotton Spinners Association 2007):

- Economical in terms of lower machine investment, good reproducibility, less difficulty in washing off, quick sampling and high production rate
- Can be applied on all fibres and their blends
- Complete colour range
- Especially suitable for colour-resistant and discharge printing due to inert chemical nature

However, the disadvantages of pigment printing include inferior washing, rubbing and dry-cleaning fastness properties as well as inevitably harsh (hardened) hand-feel. Therefore, other dye classes may sometimes be used to overcome the weakness of pigment. Not all dye classes can be used in printing because of such reasons as solubility difficulty, low colour yield and poor print-paste stability. The appropriate dye classes applied on different fibre types are (1) reactive dyes for cotton and other cellulosic fibres, rayon, wool and other animal fibres, silk; (2) acid dyes for wool and other animal fibres, silk; (3) disperse dyes for polyester and acetate and (4) basic dyes for acrylic (The Hong Kong Cotton Spinners Association 2007).

In spite of the various methods and styles of printing, the typical textile printing procedures include the following steps (The Hong Kong Cotton Spinners Association 2007):

- Preparation of ground colour of the fabric
- Preparation of print paste
- Printing
- Drying
- Fixation
- Washing off

1. *Preparation of ground colour*: The printing process can be either done on a bleached white ground or over-printed on a coloured ground. In the former case, the preparation of fabric involves a series of wet-processing steps such as desizing, scouring and bleaching of the grey fabric, the details of which will be discussed later on. In the latter case, the coloured ground may be produced by either a dyeing or printing process (The Hong Kong Cotton Spinners Association 2007).
2. *Preparation of print paste*: Before a printing process can take place, print paste of the required colour must be prepared. The basic components of print paste are dye or pigment, water, thickening agent (thickener) and suitable chemicals. Water is used to dissolve or disperse the dye, pigments and chemicals. It also serves to regulate the overall viscosity of the print paste. The thickening agent is used to give the print paste the desired viscosity so that the printed patterns are of good sharpness and definition. The most common thickening agents are starch and gum products which, however, tend to form a hard gel after drying and should be removed afterwards to avoid stiff handle. Emulsion thickener, a cream-like product produced from a high-speed stirring of mixture of water and petroleum oil, may be used as a substitute, or to mix with starch and gum, to give a softer handle. Other types of thickening agents that may sometimes be used in particular occasions include starch derivatives, cellulose derivatives, alginates and synthetic polymer thickeners. The functions of adding chemicals are to assist dye solubility or pigment dispersion, dye fixation and print paste stability, etc. (The Hong Kong Cotton Spinners Association 2007).
3. *Printing*: The methods of printing can mainly be classified as roller printing, rotary-screen printing and flatbed screen printing. For each method of

production, various techniques or types of prints can be applied (The Hong Kong Cotton Spinners Association 2007).

4. *Drying*: After printing, the fabric may be required to dry to prevent smudging (marking off) of the wet print. Drying may not be necessary if a dry-heat (hot-air baking) fixation is applied just after printing (The Hong Kong Cotton Spinners Association 2007).
5. *Fixation*: Fixation is normally done by means of either dry heat (over 100°C) or steaming. It is an important step to enable the absorbed dye or pigment molecules to penetrate and fix to the fibres (The Hong Kong Cotton Spinners Association 2007).
6. *Washing off*: A washing-off process, with soaping and sufficient numbers of cold and hot water rinsing, is used to remove the unfixed dye, thickening agent and residue chemicals. Particular care, such as control of washing temperature and addition of special detergent, will be taken during the process to prevent the staining of the unprinted areas by the unfixed dye in the wash liquid (The Hong Kong Cotton Spinners Association 2007).

3.3.1 Methods of Printing

Roller printing, flatbed screen printing and rotary-screen printing are widely used in commercial production. However, transfer printing and digital printing are entering the market. Details of each method are discussed in the following subsections.

3.3.1.1 Roller Printing

Roller printing is a continuous automatic production in which the process is carried out with the aid of engraved copper rollers. A separate engraved roller is used for each colour. The size of the print repeat is governed by the printing machine and the size of the roller (The Hong Kong Cotton Spinners Association 2007; Leung, Lo, and Yeung 1996).

The design was originally engraved on the roller by hand but this was found to be unreliable and time-consuming. Nowadays, a photographic engraving method which involves the etching of the copper rollers with chemicals (strong acids) is used to produce the engravings on the rollers.

In the operation of roller printing, firstly, the engraved copper rollers are arranged around a large impression cylinder. Next, print paste is fed to the printing roller by a furnishing roller. Surplus paste is scraped off the roller surface with a very sharp stainless steel blade known as a doctor blade, so that only the engraved lines are filled with colour. The fabric then passes between the roller and the impression cylinder, taking up the colour from the engravings, to complete the printing process. Several rollers can be fitted onto the machine if a multi-coloured pattern is desired.

3.3.1.2 Flatbed Screen Printing

This printing is so called because the process makes use of a meshed screen for transferring the print paste to the fabric. In the past, the screen was made of silk, and thus the process was also called silk-screen printing. Nowadays, due to the high cost of silk fabric as well as its inferior durability, strong synthetic fibres such as nylon and polyester are commonly used as the screen material. In this method, the design is transferred to the screen, either by manual manipulation or by photochemical means. The unprinted areas are blocked, whereas the areas to be printed are left open. The screen is then mounted onto a wooden or metal frame. One screen is needed for each colour in the design. During printing, the fabric to be printed is spread smoothly onto a table whose surface has been coated with a light semi-permanent adhesive, such that the fabric is guaranteed not to move during printing. The mounted screen is positioned accurately, and the print paste is then poured onto the screen. The print paste is forced through the open areas of the screen with a flexible, synthetic rubber blade (known as the squeegee). The squeegee is drawn steadily across the screen at a constant angle under pressure to ensure a sharp and uniform printing quality. The steps are repeated when multicolour is required in the design. The printed fabric is then taken away from the table for successive drying, fixation and washing off (The Hong Kong Cotton Spinners Association 2007).

The quality of flatbed screen printing is normally governed by several variables, such as mesh size of the screen, the fraction of the open area (i.e., the style of the design), the squeegee angle and pressure, and the number and speed of squeegee strokes. To control and standardise such variables, the commercial practice is shifting from hand screen printing towards fully automatic production (The Hong Kong Cotton Spinners Association 2007).

Screen printing was originally done by hand. All operations, such as precise positioning of the screen frame, operating the squeegee and handling of the fabric, etc., are manipulated by workers. Although the method is of less commercial interest in the sense of long production runs, it is used for small-sized production of garment parts or very large designs which cannot be achieved by other means (The Hong Kong Cotton Spinners Association 2007).

In fully automatic production, the fabric to be printed is gummed onto an endless conveyor belt–type blanket which moves and stops at intermittent fashion, one screen-repeat distance at a time. All the screens equipped with automatically operated squeegees are positioned precisely over the blanket. Printing is done simultaneously among all screens while the fabric is stationary. After one operation step, the screens are lifted up and the blanket moves to carry the fabric to a next stop for printing (The Hong Kong Cotton Spinners Association 2007).

3.3.1.3 Rotary-Screen Printing

This printing method, which utilises seamless cylindrical screens made of nickel foil, combines the advantages of the high-production roller printing and flexible flatbed screen printing. It has proved very successful, especially in terms of large production at relatively lower cost than roller printing. The basic operation of rotary-screen printing is very similar to that of the fully automatic flatbed screen machine. The fabric is gummed onto an endless conveyor belt blanket which moves continuously, in contrast to the intermittent action of the automatic flatbed screen machine. The rotary screens, which are equipped with special types of print-paste-feeding devices and squeegee inside the screens are continuously rotating at a relatively high speed to force the print paste through the screens to the fabric. The printed fabric successively undergoes drying, fixation and washing off (The Hong Kong Cotton Spinners Association 2007).

3.3.1.4 Transfer Printing

By a simple heat process, a design printed on a piece of paper is transferred to the fabric. The dyes used are capable of vaporizing under the heat conditions of the process, and therefore they have a high affinity for the fibres of the fabric. Transfer printing has long been used on polyester and polyamide.

3.3.1.5 Digital Printing

This is a process of creating prints generated and designed on a computer and then printing the design on textile substrates using ink-jet technology. The use of digital technology means that a digital file can be communicated worldwide, and fabrics of identical colour patterns and quality can be produced anywhere in the world. This technology reduces traditional order-to-delivery lead times. Ink-jet printing also makes it possible for firms to print images which cannot normally be produced by rotary-screen or flatbed screen printing. For instance, very fine lines and photographic imagery can be printed by ink jet, as well as an infinite number of colours and sizes (The Hong Kong Cotton Spinners Association 2007).

3.3.2 Printing of Cellulosic Fibres with Reactive Dyes

Reactive dyes are widely used for printing cellulosic fabrics. Not all reactive dyes are suitable for use in printing, and the choice of dye depends on a number of factors such as fixation method, style being printed and fabric type. High- and low-reactive dyes can be used and are available as powders or liquids. It is best to use the recommendations of the supplier to select the correct dye for a given set of conditions. An extensive colour range is available due to the wide variety of chromophores possible (The Society of Dyers and Colourists 2005; Miles 2003).

3.3.3 Printing of Cellulosic Fibres with Vat Dyes

Vat dyes are used in textile printing on cellulosic goods which require good wash and light fastness; for example, they are used in printing drapes and upholstery fabrics. Vat dyes used in modern printing systems are supplied in liquid or paste form as aqueous dispersions (The Society of Dyers and Colourists 2005; Miles 2003).

3.3.4 Printing of Protein Fibres

Acid and metal-complex dyes are the most popular dye classes for wool fibre. Acid dyes offer a wide shade range varying in brightness and fastness while generally offering good light and wet fastness. Acid dyes are the main dye class for printing silk, although their fastness properties vary with the type of dye chosen (The Society of Dyers and Colourists 2005; Miles 2003).

3.3.5 Printing of Polyester Fibre

Polyester is a key synthetic fibre whether used alone or in combination with other fibres. Pigments can be used for printing polyester, particularly for cheaper articles and speciality print effects, such as metallics. The main limitation of pigments in the past has been their reduced fastness and harsher handle on polyester. However, with the modern printing binders, these restrictions may be reduced. In general, high wet fastness properties can be achieved, but problems in rub fastness can result from poorly prepared fabric, in particular if non-ionic surfactants are left in the fabric from previous processes. For direct printing of polyester, it is necessary to select dyes which have high sublimation fastness, as this helps prevent dye migration to adjacent white or coloured areas. Choosing low-affinity dyes will reduce staining of white areas during fixation and wash off (The Society of Dyers and Colourists 2005; Miles 2003).

Disperse dye is selected for printing polyester. It should be mentioned that the diffusion of disperse dye into polyester is relatively slow, and it is necessary to use high temperatures for fixation or lower temperatures in conjunction with a carrier. The choice of dye is also dependent on the fixation conditions. Thickeners used in disperse printing are chosen relative to the desired end result and the proposed fixation conditions. They must not have a tendency to crack during processing and must be good film formers to prevent mark-off during processing (The Society of Dyers and Colourists 2005; Miles 2003).

3.3.6 Printing of Acetate Fibres

Cellulose acetate is produced from an ester of cellulose and has some affinity for disperse dyes, but its wet durability is limited. Various methods of improving the performance have been tried, including the use of swelling agents in combination with acid, basic and vat dyes. Basic dyes may also be used for printing cellulose acetate where bright hues are required, but light fastness properties may be limited, restricting applications where the dyes can be used (The Society of Dyers and Colourists 2005; Miles 2003).

Cellulose triacetate has a higher degree of esterification than acetate. It is less hydrophilic, less sensitive to alkali and offers greater mechanical and chemical stability. Disperse dyes are used due to the large colour range and their ease of application, and the printing method is the same as for cellulose acetate. Whilst disperse dyes are mainly used, it is possible to print cellulose triacetate with acid dyes. These offer high colour yield with bright hues and good wet fastness properties. The main advantage would be the improved fastness to dye sublimation, which can be a problem with some disperse dyes. Moreover, if extremely bright prints are required, cellulose triacetate can be printed with basic dye, but the overall fastness properties are inferior (The Society of Dyers and Colourists 2005; Miles 2003).

3.3.7 Printing of Acrylic Fibres

Acrylic is the general name for a series of addition copolymers produced from vinyl compounds, at least 85% being acrylonitrile. Commercial fibres vary in their characteristics due to the fact that other groups are incorporated into the polymer. On its own, acrylonitrile has little affinity for dyes. It is necessary to incorporate other products to make the polymer dyeable. Acrylic fibres are widely used for knitwear and in blends with other fibres in dresswear and carpets. They have a tendency to attract dirt due to their relatively low moisture regain and are susceptible to yellowing at heat treatments above 110°C. Drying and fixation conditions are thus critical (The Society of Dyers and Colourists 2005; Miles 2003).

3.3.8 Printing of Polyamide Fibres

Other than acid dye, metal-complex dyes can also be used, particularly where very good fastness to wet treatments and high light fastness is desired. By careful selection of the dyes, it is possible to achieve good fixation under neutral conditions; hence no acid-liberating catalyst needs to be incorporated in the previous recipe. It may also be possible to fix without using a fibre-swelling agent (The Society of Dyers and Colourists 2005; Miles 2005; The Hong Kong Cotton Spinners Association 2007).

3.3.9 Pigment Printing

Pigment printing and dyeing differs from conventional dyeing and printing processes. Inorganic and organic insoluble pigments with little or no affinity for the fibre are applied to the material by the aid of an adhesive, normally termed a *binder*. Conventional pigment dispersions cannot make a direct bond with the fibre and thus require the use of a binder to adhere the pigments to the substrate. The nature of the binding force is largely determined by the binder type and its ionic charge. Fixation of the pigment is time/temperature dependant and is normally achieved by a dry-heat cure at temperatures of between 130°C for 5 min to 180°C for 1 min. The specific conditions are determined by the nature of the binder, pigment, substrate and application conditions. The pigment printing (and pigment dyeing) process is relatively simple compared to conventional dye systems, only requiring a print (pad in the case of dyeing) → dry → cure process. Fixation may be achieved during stenter finishing, although a pre-baking process is to be preferred in the case of heavy shade prints. No wash-off process is normally required (The Society of Dyers and Colourists 2005; Miles 2003; The Hong Kong Cotton Spinners Association 2007).

3.4 FINISHING PROCESS

Textile finishing in the narrow sense is the final step in the fabric manufacturing process, the last chance to provide the properties that customers will value. Finishing completes the fabric's performance and gives it special functional properties, including the final 'touch'. However, the term *finishing* is also used in its broad sense: 'Any operation for improving the appearance or usefulness of a fabric after it leaves the loom or knitting machine can be considered a finishing step' (Tomasino 1992). This broad definition includes pretreatments such as washing, bleaching and coloration. In this book, the term *finishing* is used in the narrow definition to include all those processes that usually follow coloration and that add useful qualities to the fabric, ranging from interesting appearance and fashion aspects to high-performance properties for industrial needs (Schindler and Hauser 2004).

Most finishes are applied to fabrics such as wovens, knitwear or nonwovens. But there are also other finishing processes such as yarn finishing, for example sewing yarn with silicones and garment finishing. Textile finishing can be subdivided into two distinctly different areas: chemical finishing and mechanical finishing. Chemical finishing '(wet finishing)' involves the addition of chemicals to textiles to achieve a desired result. Physical properties such as dimensional stability and chemical properties such as flame retardancy can both be improved with chemical finishing. Typically, the appearance of the textile is unchanged after chemical finishing. Mechanical finishing '(dry finishing)' uses mainly physical (especially mechanical) means to change fabric properties and usually alters the fabric appearance as well. Mechanical finishing also encompasses thermal processes such as heat setting (thermal finishing). Typical mechanical finishes include calendering, emerising, compressive shrinkage, raising, brushing, and shearing or cropping, and especially for wool fabrics, it involves milling, pressing and setting with crabbing and decatising (Heywood 2003). Mechanical and chemical finishing often overlap. Some mechanical finishes need chemicals, e.g., the use of milling agents for the fulling process or use of reductive and fixation agents for the decatering of wool fabrics. On the other hand, chemical finishing is impossible without mechanical assistance, such as fabric transport and product application. The assignment to mechanical or chemical finishing depends on the circumstance—whether the major component of the fabric's improvement step is more mechanically or chemically based (Schindler and Hauser 2004). The principal chemical and mechanical finishing processes are shown in Table 3.4 (The Society of Dyers and Colourists 2005).

As highlighted in Table 3.4, the various finishing processes are reviewed in the following subsections.

3.4.1 Mechanical Finishing

3.4.1.1 Calendering

Calendering is essentially an ironing process which adds sheen to a fabric. After going through the preparation, colouration and finishing processes, the surface of the fabric is generally distorted, resulting in a loss of lustre. The calendering process can compensate for this and improve the fabric's lustre. This involves passing the fabric between hollow heated or unheated rollers under great pressure, with the lustre produced being directly proportional to the amount of heat and pressure applied (Rouette 2001). After calendering, the fabric has a smooth and flat surface. Calendering is a relatively cheap process to obtain lustre. It is not permanent to washing; better durability can be obtained by following calendering with a resin treatment.

3.4.1.2 Friction Calendering

Friction calendering gives a higher gloss and a greater closing of the yarns; it is produced by bringing the cloth into contact with a three-bowl process consisting of a heated bowl, a polishing bowl and a chilled-iron bowl which is travelling at a faster speed than the cloth itself. Three-bowl heavy-friction calenders are suitable for the finishing of highly glazed materials. The bottom bowl is usually made of cast iron; the middle bowl consists of cotton and is of greater diameter than the others to

TABLE 3.4
Principal Chemical and Mechanical Finishes

Fibre Type	Mechanical Finish	Chemical Finish
Cellulosic: natural	Calender	Crease resist
	Friction	Minimum iron
	Schreiner	Permanent press
	Emboss	Water and oil repellent
	Chintz	Flame retardant
	Chasing	Handle modification
	Swiss	Soften
	Compressive shrink	Stiffen
	Brush/raise	Soil release
		Antimicrobial
Cellulosic: regenerated	Calender	Crease resist
	Friction	Water repellent
	Schreiner	Flameproof
	Emboss	
	Chintz	
	Chasing	
	Swiss	
	Compressive shrink	
	Brush/raise	
Protein	Press	Shrink resist
	Raise	Water repellent
	Crop	Insect repellent
	Decatise	Antimicrobial
		Mothproofing
Cellulose diacetate	Calender	Water repellent
	Schriener	Antistatic
	Emboss	
	Moiré	
Cellulose triacetate	Calender	Lamination
	Schriener	S finish
	Emboss	
	Moiré	
	Heat set	
Polyamide	Friction	Water repellent
	Schreiner	Antistatic
	Emboss	Handle modification
	Bulk	Soften
	Heat set	Stiffen
Polyester	Friction	Water repellent
	Schreiner	Antistatic
	Emboss	Handle modification
	Bulk	Soften
	Heat set	Stiffen
		S finish

TABLE 3.4 (*Continued*)
Principal Chemical and Mechanical Finishes

Fibre Type	Mechanical Finish	Chemical Finish
Polyacrylics	Stenter at low temperatures	Antistatic Soften
All fibres		Antimicrobial/fungal

Source: The Society of Dyers and Colourists (2005).

allow for wear; the top or glazing bowl is made of highly polished chilled iron and is heated by steam or gas. An arrangement of spur wheels enables the top bowl to produce a surface speed 1.5 to 2 times that of the lower bowls. The cloth is passed into the bottom nip and around the middle bowl, which is revolving at the same surface speed as the bottom bowl; the top bowl, with its higher surface speed, produces the friction effect by polishing the cloth. In certain cases, where a very high gloss is required, the fabric is often pre-impregnated with a wax emulsion which further enhances the polished effect (The Society of Dyers and Colourists 2005).

3.4.1.3 Schreinering

Schreinering is an inexpensive way of producing a very high degree of lustre in cotton fabrics. If cotton is subjected to slight pressure, a low degree of lustre is obtained, whereas if the pressure is great, the numerous small surfaces are merged and give no satisfactory lustre, but only a specular reflection, as in a mirror. In order to obtain a silky lustre, it is necessary to produce on the fabric a very large number of small reflecting surfaces distributed in several plains. Engravings of 125 to 500 lines per inch were made on various metallic calender bowls and applied to different fabrics. The operating principle is similar to that of calendering, except that the upper heated roller is engraved with very fine lines, which are set at a slight angle to the warp or weft of the fabric. These lines are usually at an angle of about 30° to the warp threads. After the schreinering process, the reflection of light from the ridges impressed on the fabrics by the engraved roller will give it a lustre that is similar to that obtained through mercerization. If the fabric has already been mercerised, the additional process produces a lustre which simulates that of silk. Therefore, schreinering is also called silk finishing (Leung, Lo, and Yeung 1996; The Society of Dyers and Colourists 2005).

3.4.1.4 Embossing

Embossing produces a raised relief design which is permanent on thermoplastic synthetic fibres but only temporary on cotton. This finishing effect can, however, be made permanent with the aid of chemical resins, which have a binding action on fibres. A fabric is embossed by passing it between heated, engraved rollers that imprint or emboss the design on the fabric surface (Rouette 2001). Plain woven fabrics may, after embossing, resemble fabric such as pique, which is generally more expensive. This process is popular with woollen fabrics (Leung, Lo, and Yeung 1996; The Society of Dyers and Colourists 2005).

3.4.1.5 Chintz Finishing

Chintz finishing is a high-glaze finish produced by friction calendering cellulosic fabrics previously impregnated with wax or paraffin emulsions, dried, and adjusted to an appropriate residual moisture content. Permanent chintz effects are produced on fabrics to which resin finishes (mainly melamine-formaldehyde compounds) have been applied prior to calendering, followed by subsequent curing. The finishing procedure is similar to embossing. Nowadays, depending on market requirements, higher demands are placed on the wash fastness of chintz finishes; in these cases, reactant resins are applied together with fluorocarbon or silicone derivatives (The Society of Dyers and Colourists 2005).

3.4.1.6 Chasing

During the chasing process, the fabric is fed in a spiral through the calender by a roller device. The goods are arranged in seven layers, one above the other in the roller system, and are led over a board into the middle of the calender. The goods run spirally towards the outside, so that the last outer layer leaves the upper nip and is led to the winder (Rouette 2001). The goods are compressed in the circulation through the chasing calender. The aim of the treatment is to get the required density and air permeability at the same time. Because there are multiple layers of fabric in the press, the fibres remain round and bulky (Rouette 2001).

3.4.1.7 Swissing Finishes (Normal Gloss)

Swissing finishes are obtained merely by passing the cloth, suitably conditioned, through the nips of a calender in which the surface speed of all the bowls is the same; the cloth is then batched or plaited as required. A smooth appearance is thus obtained according to the number and composition of the bowls (The Society of Dyers and Colourists 2005).

3.4.1.8 Compressive Shrinkage

Processing stresses are introduced during the bleaching, dyeing and finishing of a fabric by pulling the fabric in the warp direction. This tends to remove the warp crimp from the fabric. In order to replace the warp crimp and so minimise warp shrinkage, a process known as compressive shrinkage is carried out on the fabric to replace the crimp which has been pulled out in preparation and coloration processes. This process is normally used for woven cotton fabrics (The Society of Dyers and Colourists 2005).

3.4.1.9 Raising

The process of raising is also known as 'napping'. It produces a fuzzy or hairy surface on a fabric by abrading (scarping) it and pulling the fibre ends to the surface. The degree of hairiness on the surface can be adjusted according to techniques used. The fabric to be raised should be made from soft twisted yarns; it is it is processed by the action of revolving cylinders which are covered by fine wires with small hooks on the ends. The hooks scrape the surface of the fabric, pulling fibre ends up out of the yarn. The fuzzy and soft surface of the raised fabric makes it warm, owing to the insulating air cells in the naps. The greater the extent of raising (i.e., the more numerous the air cells), the greater is the warmth. Raising may also serve to cover any weaving imperfections, although too much raising will weaken the finished fabric and increase any tendency of the fibre to pill (to ball or roll up). Napped fabrics should not be confused with pile constructions that have been woven or knitted. In a pile construction, the thickness represents a true three-dimensional effect, which is produced by additional fabric pile (Leung, Lo, and Yeung 1996).

3.4.1.10 Brushing

Brushing is an operation in dry finishing which serves particularly the cleaning of woven wool fabrics, but also those produced from other fibres, especially when fibre fly, dust or threads have to be removed. Furthermore, the raising effect can be affected by brushing in such a way that the raised hairs point in a specific direction, or become entangled with each other. Brushing is carried out on so-called brushing machines, before or after shearing and before pressing, mostly simultaneous with steaming. Brushing units are also employed between singeing and desizing or dyeing and printing. In the first case, severe contamination of the scouring liquor is prevented, while, in the second case, fibre deposition on the roller-printing machine roller is avoided. Brushing is by far of greatest importance in the production of woven pile fabrics, e.g., cord and velvet (Rouette 2001).

3.4.1.11 Pressing

Pressing is usually performed prior to decatising and before cutting and also immediately before completion. The flat press, one of the oldest devices in the cloth industry, is typically used for wool and wool-mix fabrics, which are folded edge to edge with intermediate layers of polished pressboard inserted; electrically heated boards are inserted between each pack, and every 3–5 cm this is replaced by an electrically heated pressboard. The entire pack, consisting of different pieces, one on top of the other, is enclosed at the top and bottom by two thick sheets of cardboard (fire-resistant cover) and is usually left overnight in a hydraulic press at a pressure of 100–600 bar. The most common procedure is re-cutting, with the break points of the first pressing in the middle of the pack, and then pressing again. Achieving the particular pressing effect by a flat press is labour intensive and thus costly, so the flat press is rarely used. The rotary cloth press or cylinder press works quicker, is cheaper to run, and guarantees fold-free material. However, the effect is not always as good and as permanent and, furthermore, a slight loss of width and longitudinal stretch can occur (Rouette 2001).

3.4.1.12 Shearing (Cropping)

Fabrics that have been napped usually undergo shearing to give them an attractive, smooth and level surface. This process is very popular with woollen materials as well as certain cotton fabrics. This process is therefore also called *cutting* or *cropping*. The shearing device has revolving blades similar to those of a lawn mower, with revolving shearing blade, ledger blade and shearing bed (Leung, Lo, and Yeung 1996; Rouette 2001). During the operation, the spiral shearing blade revolves and is in contact with the ledger blade. Fabric passes over the shearing bed in front of the ledger blade, and the raised fibres fall against the ledger blade and are cut by the rotary shearing blade. The blades can be adjusted to give different levels of cutting. All fibres longer than the setting are cut off, leaving an attractive surface resembling a pile effect. Shearing can also be used to produce a pattern: This is done by using a special blade which cuts into the fabric, leaving high and low surface levels. Following shearing, any loose-cut fibres can be removed by passing the sheared fabric through brush cylinders or a vacuum extraction system (Rouette 2001).

3.4.1.13 Emerising (Sanding, Sueding)

Emerising—also called *'sueding'*, *sand wash* or *peach skin finish*—sands off uneven and protruding fibres on fabrics to produce a soft, supple and velvety or suede-like handle. The emerised fabric is sometimes known as imitation suede. A fine pile is formed without destroying the fabric's structure. The process involves the sanding of fabrics by passing them through a series of emery-covered rollers. The sandpaper or emery paper cuts the fibres that are on the surface of the fabric and produces a fuzzy surface. The main difference between emerising and raising is that, in raising, the fibre ends are plucked out of the fabric, whereas in sueding they are cut (Leung, Lo, and Yeung 1996).

3.4.1.14 Decatising

Decatising is a finishing process mainly used to improve the characteristics of wool and wool-blend piece goods with regard to appearance, shape, handle, lustre and smoothness, thereby producing a fabric in the desired finished state. Decatising is accomplished by steaming under pressure, i.e., by subjecting the well-packed or firmly wound fabric to a steam treatment. Since this treatment is usually carried out as the final operation after the rotary press, the pressed state of the fabric is retained by the decatising treatment. The purpose of decatising is therefore to improve handle by the action of moisture, to moderate and set the lustre imparted by pressing or calendering (lustre effects resistant to steam pressing), to achieve a finish fast to water spotting and to set the fabric in length and width (Rouette 2001).

3.4.1.15 Moiré

Moiré aims at production of wave-shaped moiré effects (so-called soaking), which occurs due to partial even printing of weft ribs on viscose and silk fabrics. Real moiré without repeat of figures is produced by calendering with ribbed rollers and possibly following outside passage of an irregularly perforated edge, whereby weft threads are shifted from their position and the moiré effect is increased. Work also takes place on the normal calender and/or the plate press by straining two material lengths lying on top of each other or a piece folded lengthwise. Designs running symmetrically to the centre of the material are produced in the latter case by unfolding (Rouette 2001).

3.4.1.16 Heat Setting

An essential process in the pretreatment of man-made fibres, heat setting is also of importance in subsequent stages as intermediate setting or post-setting. The effect of heat setting is greater in fibres with an increasingly hydrophobic character and is carried out on polyolefin, polyester, polyurethane, polyacrylonitrile, polyamide, triacetate, acetate and viscose fibres (where the fibres are listed in order of decreasing hydrophobic character). The objectives of heat-setting processes include structure homogenization and the elimination of internal tensions within the fibre, resulting in reduced shrinkage, improved dimensional stability, reduced creasing propensity and reduced edge-curl in woven and knitted fabrics. To this extent, the process may be better described as thermal relaxation. Heat setting changes not only the mechanical, but also the dyeing properties of man-made fibres. The principle is based on heating the fibre within a fibre-specific temperature range which is limited at the upper end by the melting point (softening range) and the respective glass-transition temperature (necessary to break the secondary bonds) at the lower end (Rouette 2001).

3.4.1.17 Stentering

The process of drying textile fabrics is carried out in a stenter (Rouette 2001; Heywod 2003), which has units for thermal treatment of textile fabrics that set the fabric width.

3.4.2 Chemical Finishing

3.4.2.1 Resin Finishing (Crease Resist, Permanent Press)

Resin finishing is a general finishing process that gives each fibre additional properties depending on requirements (soil release, moth repellent, antifelting, creasing resistance). This finishing method has a significant practical value and causes permanent improvement in wear resistance (wash and dry-cleaning resistant), and particularly in shrinkage stability and crease recovery, of textiles made out of cellulose or cellulose compounds, by means of intercalation and/or modification of the cellulose with certain finishing products. This is also known as permanent-press process; wash-and-wear finishing; anticrease finish; non-shrink finish; swelling-resistant finish; easy-care finish; and no-iron, non-iron, durable-press, minimum-iron, and rapid-iron finishes (Rouette 2001). The behaviour of components, which are primarily cellulosic, in chemical reactions can be explained from knowledge of the structure and properties of cotton. It is assumed that the conversion of cellulose fibres is a case of reactions in a heterogeneous system. These can progress to a greater or lesser degree, depending on accessibility of the different cellulose areas. Three different situations are possible in the reaction (Rouette 2001; Heywod 2003; Schindler and Hauser 2004):

1. Formation of a covalent bond between a monofunctional or polyfunctional reagent and a cellulose chain
2. Formation of at least two covalent bonds between a polyfunctional reagent and a cellulose chain
3. Formation of at least two covalent bonds between a polyfunctional reagent and two cellulose chains

All of these reactions, known as cross-linking, that lead to covalent bonds have the effect of improving crease properties. The cross-linking of the cellulose limits the possibility of chain-molecule displacement, which means that crease recovery and dimensional stability improve correspondingly with the level of cross-linking. However, at the same time, the elasticity of the fibre is reduced, which results in brittleness and therefore a reduction in abrasion resistance, breaking strength and resistance to tear propagation. These depletions in strength are particularly common in native cellulose fibres.

There are different types of resin finishing processes (Leung, Lo, and Yeung 1996):

- *Wrinkle-resistant finish (crease-resistant finishes)*
- *Shrink-resistant finish*: mainly to maintain the dimensional stability of the textiles
- *Wash-and-wear finish (drip-dry finish)*: for textiles requiring only minimal ironing or no ironing after washing
- *Permanent-press finishing (durable-press finish)*: to impart permanent pleats for pleated skirts

1. *Wrinkle-resistant finish*: This finish is sometimes referred to as a *wrinkle-free finish.* Unlike wool, silk and thermoplastic synthetic fibres, cotton fibres do not have good resilience. This means that cotton fibres lack the ability to retain their shape, as they crease and crush easily. To make them competitive with other fibres in a market that demands easy-care garments, cotton and its blends must be chemically treated with resins to make them wrinkle resistant. The wrinkle-resistant finish is based on resins or reactants that combine chemically with the cotton fibre through a cross-linking process. The cross-linkings join the molecular chains of cotton together to provide greater rigidity and to prevent molecular slippage, thereby producing a fibre capable of returning to its original position after blending. The resin is usually applied to the fabric by padding, followed by a curing process at high temperature. This allows the resins to form the cross-linking on a stenter machine.
2. *Wash-and-wear finish*: This finish is also known as a *drip-dry finish.* It is like the wrinkle-resistant finish, but it usually has a higher resin content. The wash-and-wear finish is normally described as a pre-cure process, which means that the fabric is cured before being cut and sewn into garments. A common manufacturing route for wash-and-wear garments is as follows (The Hong Kong Cotton Spinners Association 2007):
   - The fabric is padded with the resin cross-linking solution and dried.
   - The fabric is cured in a curing oven.
   - The garment is cut out and sewn.

   When garments made of fabrics with a wash-and-wear finish can be washed in a home laundering machine, they will dry smoothly and require only very little ironing. This finish is very popular with lightweight cotton shirting materials. Not all the fabrics which have drip-dry properties are wash-and-wear finished. This is particularly so in the case of synthetic fabrics, which are thermoplastic and naturally keep their shape well.

3. *Permanent-press finish*: This finish is also commonly known as the *pp* or *durable-press finish.* It is similar in the concept to the wash-and-wear finish, but, since the curing step is carried out only after a garment has been sewn, it is described as a post-cure process. Another difference is that the resin content of the permanent-press finish is one of the highest of all resin finishes. The operational steps of the process are as follows (The Hong Kong Cotton Spinners Association 2007):
   - The fabric is padded with a resin cross-linking solution and dried.
   - The garment is cut and sewn, and then pressed into the required shape with a hot press.
   - The pressed garment is cured.

   As a post-cure process, the permanent-press finish is useful in the manufacture of garments such as trousers and pleated skirts, since it enables permanent pleats to be incorporated into a garment after it has been madeup.

3.4.2.2 Resin Finishes for Wool

Wool has a natural 'memory'. It is resilient, elastic and has a strong tendency to return to its original shape after compression. Hence, the only resin finish that is commonly used on woollen fabric is the permanent-press process, by which long-lasting pleats or creases can be incorporated into woollen garments. There are a number of permanent-press processes for woollen garments, developed by different research organizations. The resin used for wool is different than that used for cotton, so that the problems associated with cotton resin treatment do not necessarily apply in this case (Leung, Lo, and Yeung 1996).

3.4.2.3 Water-Repellent Finishing

Undesirable substances which become deposited on the surface of a textile during use or cleaning represent foreign matter. Textile finishes capable of repelling foreign matter directly from the surface of a textile fabric include (Rouette 2001):

- Soil-release finishes (effective dry)
- Oil-repellent finishes (effective dry)
- Water-repellent finishes (effective wet)
- Anti–soil redeposition in washing (effective wet)

A distinction has therefore to be made between those finishes that are effective at the solid/gaseous (fabric/air) interface in the dry state and those which are effective at the solid/liquid (fabric/liquor) interface in the wet state (e.g., during washing). Products with intermediate properties between these extremes, i.e., the surfactants, also have a role to play in water-repellent and oil-repellent finishes as well as hydrophilic and soil-release finishes. Anti-soil finishes are applied to reduce or prevent the soiling of textiles in use.

Fluorocarbon compounds and organosilicates have the unique property of reducing the critical surface tension of textile fibres (Rouette 2001). If the surface of a textile is wetted (soiled) by a liquid, then the liquid must have a lower surface tension than the finish. Consequently, polymers with a high fluorocarbon content are used for oil- and water-repellent effects, and organosilicates are used for dry soil repellency.

Depending on the particular end use of a textile fabric, it may be given either a fully water-impermeable finish (i.e., resistant to water pressure) or a water-repellent finish only. In the latter case, the aim is to achieve a substantial water-repellent effect together with permeability to air and water vapour. In principle, all these processes are based on the deposition of water-repellent and water-insoluble substances on the fibre (Rouette 2001).

3.4.2.4 Oil-Repellent Finishing

The textile auxiliaries used for water-repellent finishing are not sufficient to protect textiles against grease and oil stains. For this, special products are used, e.g., fluorocarbon polymers, which are used in the form of emulsion, sometimes in padding and sometimes in the exhaustion method. However, these products in turn do not provide a good water-repellent effect, which is why in practice oil-repellents and water-repellents are always used together. Some of the products available on the market provide effects which are resistant to washing and dry cleaning. The synthesis technique of telomerisation provides access to perfluoropolymers. Sterically small base groups permit the arrangement of the fluoroalkyl chains close and parallel to one another, so that the result is oil repellence and water repellence (Rouette 2001).

3.4.2.5 Soil-Release Finish

Applications of soil-release finishing processes are specifically designed to ensure more efficient laundering of soil and stains. Soil-release products currently in use can be based on the following chemicals, for example: silicium compounds, carboxymethylcellulose, ethoxylated compounds, polyglycol ester of terephthalic acid, acrylic acid polymers, and fluorochemicals. These are frequently utilised in combination with resin finishing agents under the conditions specified for cross-linkers. Generally speaking, there are no conditions of application specific to soil-release finishing. The level of permanence of the soil-release effects achieved is dependent on the product used. Good soil-release effects are achieved through application of the dual-action principle, using fluorochemicals that are oleophobic and hydrophilic (Rouette 2001).

3.4.2.6 Soil Repellency

Soil repellency refers to the resistance to soiling as a finishing effect, which prevents soil penetration or makes it difficult. Examples of soiling include dry soil (dust), wet soil (fruit juice, ink), and oils and fats (engine oil and skin grease). *Soil-repellent finish* is an alternative term for *antisoiling finish*. It prevents dry soil deposits on synthetic fibre textiles and should not be confused with a soil-release finish (Rouette 2001).

3.4.2.7 Flame-Retardant Finishing

The expression of 'flameproof' is incorrect, since only certain fibre materials can be said to be flameproof, mainly those of inorganic origin. Polyacryl nitrile fibres and cellulose fibres are the easiest to ignite and burn. The flame-proofing procedure therefore mainly concentrates on these textiles. For decades, the cotton materials used for interior decoration in public buildings, for example, have been given flame-retardant finishing with relatively few problems. However, problems started when certain countries (initially the United States) started to issue legal regulations for clothing materials and banned the sale of easily flammable articles. The processes that had been previously used were no longer viable, since there was inadequate wash-fastness and the handle had been too greatly affected. It soon emerged that state regulations had exceeded technical feasibility. It was difficult to find a flame-retardant finish of adequate wash-fastness that did not irritate the skin and did not adversely affect the handle of the material (Rouette 2001).

Known flame-retardant concepts can be grouped as follows (Rouette 2001):

1. *Dense material structure*
2. *Fibre-spinning mass modification*: (a) copolymerization of modacrylics (instead of polyacryl nitrile); (b) additives (viscose, acetate, polyacryl nitrile); (c) aramids, polyvinylchloride
3. *Chemical modification*: (a) grafting (vinyl chloride to polyacryl nitrile); (b) reaction between cellulose and phosphor-nitrogen compounds
4. *Finishing*: (a) low-molecular-weight compounds (tris-(2,3 dibromopropyl) nitrile chloride); (b) polymers (polyvinylchloride–Sb_2O_3 mixture); (c) in situ polymerization (tetrakis (hydroxymethyl) phosphonium chloride); (d) salts.

Finishes containing antimony have a tendency to after-glow. This process also causes the handle of the material to be adversely affected. The permanently flame-retardant finishes are therefore of primary interest, some of which can be used for synthetic fibres. The products that are used must not be toxic or carcinogenic, which also applies to the vapours and gases that are produced during pyrolysis. They must also be skin compatible. Flame-retardant finishes free from antimony and chlorine are therefore required for environmental reasons. There is also the additional problem of effluent, since most phosphor nitrogen compounds are not biodegradable.

The most important phosphorus-based product groups are (Rouette 2001; Schindler and Hauser 2004):

1. Tetrakis (hydroxymethyl) phosphonium chloride (THPC)
2. Tris-aziridinyl phosphine oxide (APO)
3. Phosphor nitrile chloride polymers (PNC)
4. N-methylol dialkyl phosphonium carbonic acid amide

3.4.2.8 Handle-Modifying Finishes

Handle-modifying finishes are applied to textile fabrics after bleaching, dyeing or printing for the purpose of achieving a particular handle which differs according to end use. The following subjective descriptions are in common usage: full handle, wool-like handle, silky handle, soft handle, hard handle, stiff handle, fleshy handle, firm handle, scroopy handle, etc. Filling finishes, which involve the application of cheap substances to increase the weight of a textile fabric (weighting), may also be included in this category (Rouette 2001; Schindler and Hauser 2004).

3.4.2.9 Soft-Handle Treatment

The final required handle for textiles is also linked to ease of processing and enhanced ease of use. A soft, smooth and supple fabric handle is a market demand for many textiles. Fibres of all types go through many different process stages and will lose the greater part of any contained fats, oils and preparations on the way. Modification of fabric handle is usually from an additional process and should also contribute to avoiding any tendency to build electrostatic charges (Rouette 2001; Schindler and Hauser 2004). Starting materials for classical fabric softeners include natural oils and fats that can be chemically modified. Selected silicones used as softeners in microemulsion form are considered to be 'super softeners'.

3.4.2.10 Stiffening

It is well known that the application of some starch during ironing makes the task easier and produces a smooth finish. This can be described as a stiffening process. The action of the stiffening agent (starch in this case) gives the cotton stiffness, smoothness, weight and strength.

As well as starch, other substances such as flour, dextrin, glue and gum can be used to stiffen fabrics. Note that all of these agents have only a temporary effect, because they are not fast to washing. A more permanent stiffening effect can be gained by the use of newer types of synthetic resin polymers, but the cost is greater. Wool and silk fabrics are not usually stiffened, as the process does not suit their natural fibre properties (Leung, Lo, and Yeung 1996).

3.4.2.11 Shrink Resist

Shrink-resist finishes involve the use of chemical and/or mechanical processes for the dimensional and form stabilization of textiles, thereby improving a fabric's ability to retain its size and shape (shrinkage) under conditions of heat, dampness and wetness. Repeated swelling and drying leads to shrinkage. To a lesser extent, this also applies to natural cellulose products. Fabrics and knitted goods of synthetic fibres are given a non-shrink finish by heat setting (thermosetting). Natural silk woven fabrics (with the exception of crêpes) do not tend to shrink due to the low swellability of the silk. Woollen woven fabrics are stabilised by crabbing, i.e., treatment with boiling water or steam and subsequent rapid cooling with cold water. The non-shrink finishing processes are principally divided into mechanical and chemical processes, with the combined process also of significance. These processes not only prevent shrinking, but also give other useful usage and care properties (Rouette 2001).

1. Chemical non-shrink finishes use chemical processes to limit or remove the fibre-specific causes of shrinkage (Rouette 2001):
   - In the case of cellulose textiles, shrinkage is most closely linked to the extreme swelling behaviour of the fibres. Reduction of swelling is done by resin finishing.
   - In the case of wool textiles, the shrinkage is linked among other aspects to the felting of the wool. Reduction of the felting tendency of the wool hair is achieved by using enzymes, plasma, synthetic resins, and oxidizing agents in an antifelting finish.

2. Mechanical non-shrink finishing uses physical measures in the final finishing to equalise the tensions in the fibre material caused by the manufacturing process, and to remove the foreseeable normal shrinkage which may occur during making up or use (Rouette 2001).
   - For cellulose textiles, dryers are generally used which, to a great extent, enable pre-shrinkage by tensions in the fibres caused during manufacture (width shrinkage, overfeed). Compressive shrinkage is largely limited to cotton. Size changes of ±1% after boil washing can be achieved under optimum conditions.
   - For wool textiles, compressive shrinkage processes are less common, usually only used as measures for tension compensation.
   - For synthetic fibres, a heat-setting process is used.

3.4.2.12 Antifelting Finish

The typical scale structure of the wool fibre can be affected by means of different processes so that it is no longer detrimental in terms of felting. A distinction is drawn between various processes (Rouette 2001):

1. *Subtractive process*: The exocuticle of wool scales is hydrophobic, whilst the endocuticle is hydrophilic. Hydrophilisation of the exocuticle by oxidation of the 35% cystine there with hypochlorite (chlorination) results in the oxidised outer fibre areas containing cysteic acid being covered by a film of water when they are washed. This water film neutralises the ratchet effect (Heywood 2003; Leung, Lo, and Yeung 1996).
2. *Additive process*: Masking the scales with polymers (forming a film-like cover), the additive process enables woven or warp-knitted fabrics to be finished from an aqueous or organic medium. Synthetic resins are applied to the fibre by padding, and their action is mainly based on the so-called spot-weld effect, i.e., the polymers, in condensing, glue the individual fibres together at their contact points (unsuitable for tops), providing good shrink-proof effects at low cost (Heywood 2003; Leung, Lo, and Yeung 1996).
3. *Combined process*: Like the Hercosett process, for example (Heywood 2003; Leung, Lo, and Yeung 1996), the action of chemical antifelt finishing is no longer explained merely by the morphological degradation of the wool scales, but also by the introduction of ionised groups. The surface of natural wool contains only a very few ionised groups and is hydrophobic, which produces very close fibre contact in the aqueous felting medium. The formation of ionised and hydrated groups in antifelt finishing makes the fibre surface hydrophilic; at the same time, an electric double layer is produced. Both factors together reduce fibre contact in the aqueous medium and prevent the longitudinal anisotropy of the frictional characteristics responsible for felting from being effective on wool fibres.

3.4.2.13 Antistatic Finishing

Antistatic finishing involves the treatment of textiles with special chemicals to increase surface conductivity in order to prevent the buildup of electrostatic charges (especially at relative air humidity levels below 30%) during spinning, combing, sizing, weaving, knitting and also for finished goods. These finishes cause a reduction in friction associated with increased softness and smoothness. The antistatic finishing of clothing materials for persons working in situations involving the risk of explosion is an area of increasing importance. The prescribed maximum concentration for each product must on no account be exceeded in any process application (problem of adhesion to machine parts). A permanent antistatic finish for polyamide still awaits development (Rouette 2001).

Textile auxiliaries (anionic, cationic or non-ionic) are used to prevent the development of electrostatic charges during the processing and use of synthetic fibres and yarns (which also includes some natural fibres such as wool). In the case of anionic and cationic antistatic agents, the antistatic effect becomes greater with increased chain length of the fatty acid residue. Presumably this is because a marked molecular adsorption, perpendicular to the fibre surface, becomes possible with longer chains. Antistatic action is essentially due to the combined effects of increased ionic conductivity, increased water-absorbing capacity and, possibly, a fibre lubricating effect as well. Antistatic agents have only a very limited effect on soil repellency. Wash-resistant antistatic agents are based on the principle of applying, for example, polymer compounds to fibres whose water solubility is due to the presence of hydrophilic side groups, after which the water-solubilising groups are blocked by salt formation or esterification (Rouette 2001).

3.4.2.14 Coating and Laminating

Both coating and laminating require a textile substrate to be treated. The substrate plays a major role in establishing the final properties of the finished article (Heywood 2003). In addition to the chemical and physical properties of the fibres themselves, yarn construction and fabric formation are significant factors. Yarns made from staple fibres provide rough surfaces that enhance adhesion to chemical coatings. Filament yarns generally must be pretreated with chemicals to generate a more reactive surface prior to coating or laminating. Fabric structure determines the extent of textile-finish interbonding while also influencing the final mechanical properties of the treated material. Knitted and nonwoven structures are especially useful for coating and laminating, but when strength and dimensional stability are required, wovens are preferred (Schindler and Hauser 2004).

The chemicals used for coating and laminating are polymeric materials, either naturally occurring or produced synthetically. These include natural and synthetic rubbers, polyvinyl chloride, polyvinyl alcohol, acrylic, phenolic resins, polyurethanes, silicones, fluorochemicals, epoxy resins and polyesters (Heywood 2003). Coating formulations typically include auxiliaries such as plasticisers, adhesion promoters, viscosity regulators, pigments, fillers, flame retardants, catalysts and the like.

3.4.2.15 S-Finish

The term *S-finish* originally applied to the partial surface saponification of cellulose ester fibres (acetate and triacetate) by alkaline hydrolysis to reduce static charge and improve fabric handle. Nowadays the term has also been adopted for an analogous treatment of polyester fibres in which a controlled partial saponification with alkali (sodium hydroxide) is carried out to achieve a silk-like lustre and handle, reduce the buildup of static charges and improve anti-soil properties. Degradation of the fibre surface layer typically results in a weight loss of 12%–14%. Precise control of the process is necessary in order to prevent excessive loss of tensile strength. Quaternary ammonium compounds are used as accelerators in the alkali treatment (Rouette 2001).

3.4.2.16 Antimicrobial/Antifungal Finishes

Antimicrobial finishes are effectively produced on textiles by (Rouette 2001):

1. Addition of microbicidal substances to the spinning solution in fibre manufacture
2. Modifications involving grafting or other chemical reactions
3. The finishing of textiles with suitable active substances

Such substances are fixed on textile materials after a thermal treatment (drying, curing) by incorporation into polymers and resin finishing agents. Antimicrobial effects, resistant to washing and dry-cleaning, are obtained, for example, by the incorporation of microbicides into spinning solutions as well as by chemical modification of the fibre itself. As a result, the textile material is protected from microbial attack and can no longer serve as a culture medium. It is, however, also necessary for the active constituent to be carried to the microorganism cells being targeted, either by water, e.g., after hydrolytic breakdown, or by leeching out of the textile material. This is an important prerequisite for an effective antimicrobial effect (Rouette 2001).

Many active substances suffer reduced effectiveness or even inactivation as a result of chemical reaction with, for example, the fibre. For this reason, finishing processes, which apply substances that can be incorporated into textile auxiliaries, and which do not cross-link with, but rather exhaust onto the fibre from where they are slowly released during use, have gained increasing importance. In this case, of course, resistance to washing and dry-cleaning is limited. This limited resistance is actually desirable in terms of effective germ resistance (Rouette 2001).

3.4.2.17 Insect-Repellent Finishes

In contrast to the use of insecticide (oral or contact poisons, etc.), special odoriferous substances (repellents) are employed for the purpose of repelling insects. These substances are unpleasant or unbearable to insects and therefore have a repellent effect as far as insect bites are concerned whilst, for humans, they have only a slight or even a pleasant odour. Repellents of this kind find widespread use as skin creams, body oils, etc. Experience has revealed that such insect repellency is rather non-specific, and a relationship between chemical constitution, physical data and insect repellency appears questionable. Useable repellents are, as a rule, neutral, viscous oils of low volatility or crystals with low melting points and, almost without exception, a bitter taste. In order to maintain the longest possible activity, such substances must, in addition, not give rise to skin irritations or cause damage to textiles. Suitable products for textile impregnations include, e.g., indalone, undecenoic acid, mandelic acid hexyl ester, N-cyclohexyl-2-(butoxyethoxy) acetamide, etc. A patent for the production of insect-repellent hosiery recommends saturation with quaternary ammonium compounds followed by subsequent drying and heating to approximately 100°C–150°C (Rouette 2001; Schindler and Hauser 2004).

3.4.2.18 Mothproofing

Mothproof finishing is directed towards the garment to prevent damage from the fur moth (and/or its larvae), the *Anthrenus* and *Attagenus* beetles, etc. (The Hong Kong Spinners Association 2007), which live as textile parasites on keratin-containing substances (wool amongst other protein fibres, fur, duvet feathers, etc.) (Rouette 2001; Schindler and Hauser 2004). Mothproofing agents should also protect against carpet beetles amongst other harmful insects. Insecticides are more or less poisonous. This applies in particular to dieldrin, which is banned in many countries. There is an intensive search for new, non-poisonous products. The synthetic pyrethroids also seem to be problematic. They are fish poisons as well, but can at least be removed from wastewater. Pyrethroids have a satisfactory protective effect against moths, but they are effective against carpet beetles only in high concentrations (Rouette 2001; Schindler and Hauser 2004).

3.5 CONCLUSIONS

This chapter reviewed the common textile wet processes involved in the preparation, coloration and finishing of common textile fibres. Depending on the physical and chemical natures of the textile fibres, the processes involved are different, but there are common features in that the processes typically involve the use of water, energy and chemicals. Thus, the review in this chapter suggests that there is room for the existing textile wet processes to have 'sustainable', 'greener' and 'cleaner' improvement.

REFERENCES

Broadbent, A. D. 2001. *Basic principles of textile coloration*. Bradford, UK: Society of Dyers and Colourists.

Hawkyard, C. 2004. *Synthetic fibre dyeing*. Bradford, UK: Society of Dyers and Colourists.

Heywood, D. 2003. *Textile finishing*. Bradford, UK: Society of Dyers and Colourists.

Leung, K. T., M. T. Lo, and K. W. Yeung. 1996. *Knowledge of materials II*. Hong Kong: Institute of Textiles and Clothing, Hong Kong Polytechnic University.

Lewis, D. M. 1992. *Wool dyeing*. Bradford, UK: Society of Dyers and Colourists.

Miles, L. W. C. 2003. *Textile printing*. Rev. 2nd ed. Bradford, UK: Society of Dyers and Colourists.

Nunn, D. M. 1979. *The dyeing of synthetic-polymer and acetate fibres*. Bradford, UK: Society of Dyers and Colourists.

Rouette, H. K. 2001. *Encyclopedia of textile finishing* (English ed.). Cambridge, UK: Woodhead Publishing.

Roy Choudhury, A. K. 2006. *Textile preparation and dyeing*. Enfield, NH: Science Publishers, Edenbridge.

Schindler, W. D., and P. J. Hauser. 2004. *Chemical finishing of textiles*. Cambridge, UK: Woodhead Publishing.

Shore, J. 1995. *Cellulosic dyeing*. Bradford, UK: Society of Dyers and Colourists.

The Hong Kong Cotton Spinners Association. 2007. *Textile handbook 2007*. Hong Kong: Hong Kong Cotton Spinners Association.

The Society of Dyers and Colourists. 2005. *SDC e-learning: Module package*. Bradford, UK: Society of Dyers and Colourists.

Tomasino, C. 1992. *Chemistry and technology of fabric preparation and finishing*. Raleigh, NC: College of Textiles, North Carolina State University.

Vigo, T. L. 1994. *Textile processing and properties: Preparation, dyeing, finishing and performance*. Amsterdam, the Netherlands: Elsevier Science BV.

4 What Is Plasma?

4.1 INTRODUCTION

The concept of plasma was introduced by Tonks and Langmuir (1929a, 1929b, 1929c). Plasma is defined as a state where a significant number of atoms and/or molecules are either electrically, thermally or magnetically charged or ionised. Plasma in general refers to the excited gaseous state consisting of atoms, molecules, ions, metastables, and excited state of these as well as electrons, such that the concentration of positively and negatively charged species is roughly the same. The ionised gas system displays significantly different physical and chemical properties when compared with its neutral condition. Theoretically, plasma is referred to as a "fourth state of matter" and is characterised in terms of the average electron temperature and the charge density within the system (Clark, Dilks, and Shuttleworth 1978; Chan 1994; Kan, Chan, and Yuen 2000).

The physical phenomena called plasma can be divided into hot plasma (equilibrium) and cold plasma (low temperature, non-equilibrium). The low-temperature plasma is commonly used in material modification. In low-temperature plasma, the electron temperature is 10 to 100 times higher than the gas temperature (Luo and van Ooij 2002). However, due to the very low density and very low heat capacity of the electrons, the very high temperature of electrons does not imply that the plasma is hot. This means that although the electron temperature rises over several ten thousand K, the gas temperature remains at 100K. Therefore, this explains why plasma is termed as low-temperature plasma (hereinafter called *plasma*) and used for the modification of polymer surfaces.

Depending on the gas pressure, two different forms of electrical discharge in gases are known which are often referred to as *plasma treatment* (Rakowski 1997):

1. *Corona discharge*: This is generated at gas pressures equal to or near to the atmospheric pressure with an electromagnetic field at high voltage (>15 kV) and frequency in the 20–40-kHz range for most practical applications.
2. *Glow discharge*: This is generated at gas pressures in the 0.1–10 MPa range with an electromagnetic field in a lower voltage range (0.4–8.0 kV) and a very broad frequency range (0–2.45 GHz).

4.1.1 Production of Plasma Species

The production of plasma species is briefly shown as follows, with oxygen gas being used as an example (Manos and Flamm 1989).

Ion and electron formation

$$e + O_2 \rightarrow O_2^+ + 2e$$

Atom and radical formation

$$e + O_2 \rightarrow O + O$$

Generation of heat and light

$$e + O_2 \rightarrow O_2^*$$

$$O_2^* \rightarrow hv$$

$$e + O \rightarrow O^*$$

$$O^* \rightarrow hv \text{ (represents heat and light)}$$

where O_2^* and O^* are excited states of O_2 and O; h is Planck constant; v is frequency of electromagnetic radiation.

In the plasma, each formation step balances various loss processes. The equilibrium between formation and loss determines the steady concentration of species in a discharge. For the charged species in plasma, these formation and loss processes may be grouped into a few categories, as shown in the following subsections.

4.1.1.1 Ionisation and Detachment

Ionisation reactions are the main source of ions and electrons. The general form of these reactions is:

$$e + M \rightarrow A^+ + 2e \ (+B)$$

where M is either a molecule (AB) or an atom. If the molecule dissociates in this process to yield the neutral fragment B, it is called *dissociative ionisation*. When the species M is a negative ion, the process is called *detachment*, since the negatively charged electron is said to be attached when a negative ion is formed in the first place. The detachment process, $e + A^- \rightarrow A + 2e$, is no less an ionisation and similarly creates a free electron. Less frequently, another process called *Penning ionisation* can make an important contribution. In Penning ionisation, an excited metastable state is formed by electron impact, $e + C \rightarrow C^* + e$, and the excitation energy of the metastable state is enough to ionise a second species via, $C^* + M \rightarrow C + M^+ + e$, or (where again M = AB) $C^* + AB \rightarrow C + A^+ + B + e$ or $C^* + M \rightarrow CM^+ + e$. Penning ionisation has been found to be significant in mixtures where C is a rare gas, e.g., neon, with a metastable state excited energy that is just above the ionisation energy for M (Kan and Yuen 2007).

4.1.1.2 Recombination, Detachment and Diffusion

A series of loss processes balance the formation steps outlined previously. Some of the most important loss changes in these mechanisms include electron-ion recombination, $e + M^+ \rightarrow A + B$; attachment, $e + M \rightarrow A^- + B$; and diffusion of ions and electrons to the walls of the reaction vessels. These reactions take place in a variety of ways, depending on the species involved. In the case of oxygen plasma, the dissociative recombination will be the most rapid ion-electron recombination process, e + $O_2^+ \rightarrow O + O$; in a pure argon discharge, only simple electron-ion recombination is possible, e + $A^+ \rightarrow A$ (Kan and Yuen 2007).

In highly exothermic reactions, reaction channels that form two or more product fragments with comparable mass are generally favoured because they make it easier to conserve both energy and momentum. In the plasma, reactive species (positive and negative ions, atoms, neutrals, metastables and free radicals) are generated by ionization, fragmentation, and excitation. These species lead to chemical and physical interactions between plasma and the substrate surface, depending on plasma conditions such as gas, power, pressure, frequency, and exposure time. The depth of interaction and modification, however, is independent of gas type and is limited to 5 mm (Rakowski 1997).

4.2 MECHANISMS OF PLASMA TREATMENTS ON MATERIAL

As plasma is a gaseous mixture consisting of electrons, equally charged ions, molecules and atoms, many reactions occur simultaneously in a plasma system. There are two major processes with opposite effects, namely (a) polymer formation, leading to the deposition of materials and termed as *plasma polymerisation*, and (b) ablation, leading to the removal of materials. Besides the conditions of discharge, such as the energy density, the plasma gas mainly determines which of the two processes is dominant (Luo and van Ooij 2002). If the plasma gas has high proportions of carbon and hydrogen atoms in its composition, such as methane, ethylene and ethanol, the plasma will result in plasma polymerisation. The plasma polymer films are typically pinhole-free, highly cross-linked, and insoluble. It is easy to obtain very thin films (Rakowski 1997). Ablation of materials by plasma can occur by means of two principal processes; one is physical sputtering, and the other is chemical etching. The sputtering of materials by chemically non-reactive plasma, such as argon gas plasma is a typical example of physical sputtering. Chemical etching occurs in chemically reactive types of plasma. This type of plasma gas includes inorganic and organic molecular gases, such as O_2, N_2 and CF_4, which are chemically reactive but do not deposit polymers in their pure gas plasmas. Plasma ablation competes with polymer formation in almost all cases when plasma is used to treat the surfaces of solid materials (Luo and van Ooij 2002). A scheme of interaction between a solid phase and a plasma phase is summarised in Figure 4.1 (Luo and van Ooij 2002).

After plasma treatments, there are still a lot of free radicals remaining on the treated fibre surfaces. These free radicals play an important role in forming functional groups and bonds between fibre and matrix. They will also be extinguished when exposed to atmosphere, especially oxygen, by decreasing the extent of bonding between the fibre and matrix. Hence, the time lapse between plasma treatments and composite fabrication should be as short as possible.

4.2.1 POLYMERISATION

Plasma polymerisation is a unique technique for modifying polymer and other material surfaces by depositing a thin polymer film (d'Agostino et al. 1983; Wertheimer and Schreiber 1981; Moshonov and Avny 1980; Inagaki, Itami, and Katsuura 1982;

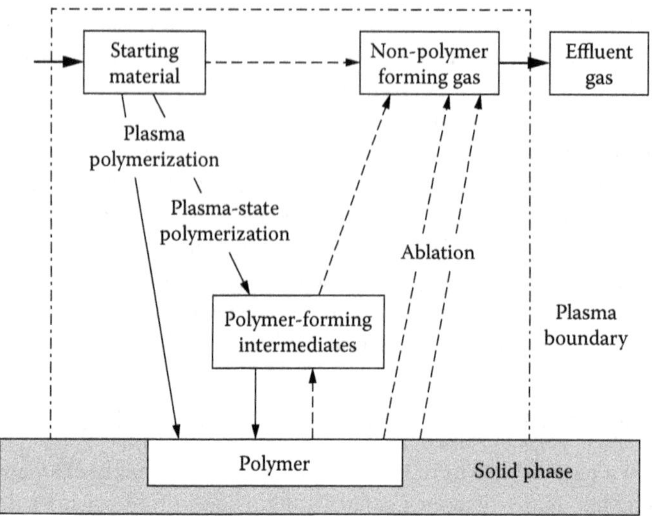

FIGURE 4.1 Polymerisation-ablation competition of plasma treatment. (From Luo and van Ooij 2002.)

Chen, Inagaki, and Katsuura 1982; Inagaki, Ohnishi, and Chen 1983; Cho and Yasuda 1988; Ho and Yasuda 1988, 1990; Urrutia, Schreiber, and Wertheimer 1988; Ertel, Ratner, and Horbett 1990; Yeh et al. 1988; Iriyama and Yasuda 1988). Table 4.1 lists some examples showing various applications of this technology in polymer surface modification. Plasma-deposited films have many special advantages:

1. A thin conformal film of thickness of a few hundred angstroms to one micrometer can be easily prepared.
2. Films can be prepared with unique physical and chemical properties. Such films, highly cross-linked and pinhole-free, can be used as very effective barriers.
3. Films can be formed on practically any kind of substrate, including polymers, metal, glass and ceramics. Generally speaking, good adhesion between the film and substrate can be easily achieved.

Plasma polymerisation is a very complex process that is not well understood. The structure of plasma-deposited films is highly complex and depends on many factors, including reactor design (Chan, Ko, and Hiraoka 1996), power level (Chan, Ko, and Hiraoka 1996), substrate temperature (Ohkubo and Inagaki 1990), frequency (Charlson et al. 1984), monomer structure and monomer pressure (Donohoe and Wydeven 1979), and monomer flow rate (H. Yasuda and Hirotsu 1977a, 1977b). Two types of polymerisation reactions can occur simultaneously, namely plasma-induced polymerisation and polymer-state polymerisation. In the former case, the plasma initiates polymerisation at the surface of liquid or solid monomers (Bradley and Czuha 1975; Epaillard, Broose, and Legeary 1989; Simionescu et al. 1981). For this to occur, monomers must contain polymerisable structures, such as double bonds, triple

TABLE 4.1
Examples of Polymer Surface Modification via Plasma Polymerisation

Application	Substrate	Monomer	Reference
Adhesion	Polyamide	Allyl amine, propane epoxy, hexamethyldisoloxane	Wertheimer and Schreiber (1981)
Adhesion	Polyethylene, poly(vinyl fluoride), polyterafluoroethylene, poly(vinyl chloride)	Acetylene	Moshonov and Avny (1980)
Adhesion	Polyethylene, polycarbonate, poly(methyl methacrylate), polytetrafluoroethylene, polypropylene, ABS rubber	Tetramethylsilane, tetramethyltin	Inagaki, Itami, and Katsuura (1982)
Adhesion	Polyethylene, polycarbonate, polytetrafluoroethylene	Tetramethylsilane + O_2, tetramethoxysilane	Chen, Inagaki, and Katsuura (1982)
Surface hardening	Polyethylene sheet	Tetramethylsilane	Inagaki, Ohnishi, and Chen (1983)
Tribology	Silicon rubber	Methane, perfluoro-1-methyldecalin	Cho and Yasuda (1988)
Water vapour barrier	Silicon rubber	Methane	Ho and Yasuda (1990)
Control permeability	Silicon rubber	Mixed vapour of hexamethyldisiloxane and methyl methacrylate	Urrutia, Schreiber, and Wertheimer (1988)
Contact lens coating	Silicon rubber	Methane	Ho and Yasuda (1988)
Blood compatibility	Poly(ethylene terephthalate)	Acetone, ethylene oxide, methanol, glutaraldehyde, formic acid, allyl alcohol	Ertel, Ratner, and Horbett (1990)
Blood compatibility	Silicon rubber	Tetrafluoroethylene, hexafluoroethane, hexafluoroethane + H_2, methane	Yeh et al. (1988)
Diffusion barrier	Poly(vinyl chloride)	Methane, acetylene	Iriyama and Yasuda (1988)

bonds, or cyclic structures. In the latter case, polymerisation occurs in plasma in which electrons and other reactive species have enough energy to break any bond. Any organic compound, even those without a polymerisable structure needed for conventional types of polymerisation, can be used in plasma-state polymerisation. The rates at which monomers polymerise are relatively similar regardless of the structure of the monomer (Chan, Ko, and Hiraoka 1996).

4.2.2 Plasma Ablation (Physical Sputtering and Chemical Etching)

In the case of glow discharge, plasmas with different ionisation extents can be produced. The active plasma species produced carry high kinetic energy (1 eV to several eV). This energy can initiate reactions of not only the saturated organic compounds, but also unsaturated ones. Although the kinetic energy is high, the temperature of the plasma is relatively low. The active species in plasma will lose their energy once they interact with the polymer material. As a result, the penetration of the plasma into the polymer material is rather shallow (beyond 1000 Å) (Yan and Guo 1989), and the interior of the material is only slightly affected. Hence, plasma treatment can be considered to be a surface treatment, such as (1) sputtering; (2) chemical etching; (3) ion-enhanced energetic etching and (4) ion-enhanced protective etching (Manos and Flamm 1989). The plasma species, carrying high kinetic energy, bombard the polymer, causing a sputtering or etching effect on the surface (Manos and Flamm 1989). This bombardment, therefore, alters the surface characteristics of the polymeric material.

4.2.2.1 Sputtering

In sputtering, ions accelerated across the sheath potential bombard a surface with high energy. The sudden energy impulse can immediately eject surface atoms outward or, by a billiard ball–like collision cascade, can even stimulate the ejection of subsurface species. If there is to be net material removal, however, molecules sputtered from the surface must not return. This requires a low gas pressure, or equivalently, a mean free path that is comparable to the vessel dimensions. If the mean free path is too short, collision in the gas phase will reflect and redeposit the sputtered species. Sputtering requires plasma conditions with high ion energies. These conditions exist in low pressure (<50 mTorr) where mean free paths are long as well. As a mechanical process, sputtering lacks selectivity. It is sensitive to the magnitude of bonding forces and structure of a surface rather than its chemical nature, and quite different materials can also sputter at similar rates. In a way, this is symptomatic of using ion bombardment with energy far higher than the surface binding energy (Manos and Flamm 1989; Lieberman and Lichtenberg 1994).

4.2.2.2 Chemical Etching

In chemical etching, gas-phase species merely react with a surface according to elementary chemistry. Fluorine atom etching of silicon is a good example of this mechanism. The key, and really the only, requirement for this kind of process is that a volatile reaction product can be formed. In silicon/F-atom etching, spontaneous reactions between F-atoms and the substrate form SF_4, a gas. The only purpose of the plasma in chemical etching is to make the reactive etchant species, e.g., F-atoms. The etchant species are formed through collisions between energetic free electrons and gas molecules, which stimulate dissociation and reaction of the feed gas, i.e., plasma feeds such as F_2, NF_3 and CF_4/O_2 all make F atoms. Chemical etching is the most selective kind of process because it is inherently sensitive to differences in bonds and the chemical consistency of a substrate. However, the process is usually isotropic or non-directional, which is sometimes a disadvantage. With isotropic etching, both vertical and horizontal material removal proceed at the same rate, making it impossible to form the fine lines (less than about 3 microns in the usual ≈1-μm-thick films) (Manos and Flamm 1989; Lieberman and Lichtenberg 1994).

4.2.2.3 Ion-Enhanced Energetic Etching

In ion-enhanced energetic etching, which is a directional etching mechanism, the impinging ions damage the surface and increase its reactivity. For example, an undoped single-crystal silicon surface is not etched by Cl_2 or Cl atoms at room temperature. When the surface is simultaneously exposed to a high-energy ion flux, the result is a rapid reaction that forms silicon chlorides and removes material much faster than the physical sputtering rate (Bogaerts et al. 2002). The word *damage* in this case can be referred to the partial dissociation of the surface compound. Whatever the microscopic details (which undoubtedly can vary greatly from one surface/etchant system to another), the generic mechanism is one in which ions impart energy to the surface, which serves to modify it and to render the impact zone as well as its environment more reactive (Rossnagel, Cuomo, and Westwood 1990).

4.2.2.4 Ion-Enhanced Protective Etching

This kind of etching mechanism can be classified as inhibitor ion-enhanced etching requiring two conceptually different species, i.e., etchants and inhibitors. The substrates and etchants in this mechanism will react spontaneously and etch isotropically if not for the presence of inhibitor species. The inhibitors form a very thin film on surfaces that cause little or no ion bombardment. The film acts as a barrier to etchant and prevents the attack of the feature sidewalls, thereby making the process anisotropic (Shul and Pearton 2000).

4.2.3 Advantages and Disadvantages of Plasma Treatment

The advantages of plasma treatment include the following (Chan, Ko, and Hiraoka 1996):

- Modification can be confined to the surface layer without modifying the bulk properties of the polymer. Typically, the depth of modification is restricted to few hundred Å only.
- Excited species in a plasma can modify the surfaces of all polymers, irrespective of their structures and chemical reactivity.
- By the selection of the feed gas to the plasma reactor, it is possible to achieve the desired type of chemical modification for the polymer surface.

- The use of plasma can avoid the problems encountered in wet chemical treatments such as residual chemical in the effluent and swelling of the substrate.
- Modification is fairly uniform over the whole surface.

The disadvantages of the plasma treatment are summarised as follows (Chan, Ko, and Hiraoka 1996):

- The plasma treatment is normally carried out in vacuum, although atmospheric type is being developed, thereby increasing the cost of operation.
- The processing parameters are highly system dependent, i.e., the optimal parameters developed and optimised for one system usually need to be modified for application to another system.
- The scale-up of an experimental setup to a large production reactor is not a simple process.
- The plasma process is so complex that it is difficult to achieve a good understanding of the interactions between the plasma and the surface necessary for a good control of the plasma parameters such as power level, gas flow rate, gas composition, gas pressure and sample temperature.
- It is very difficult to control precisely the amount of a specific functional group formed on the sample surface.

4.3 METHOD OF GENERATING PLASMA

Plasma is electrically neutral and generated by electrical discharge, high-frequency electromagnetic oscillation, high-energy radiation such as α and γ rays, etc. (Grill 1994). Electrical discharge is commonly used in industrial application. Plasma is usually excited and sustained electrically by the methods of (a) direct current (DC), (b) radio frequency (RF), or (c) microwave (MW) applied to the gas. Plasma density is controlled mainly by electron energy and gas temperature. Therefore, as long as identical energies and temperatures can be achieved, the type of discharge used to create the plasma is of little importance. The choice of a specific method and equipment to produce discharges is determined by the requirements of flexibility, process uniformity, cost and process rate. Various methods used for the generation of plasma are described in the following subsections.

4.3.1 DC Glow Discharge

A DC glow discharge is produced by applying a DC voltage between two conductive electrodes inserted into a gas at low pressure (Grill 1994). A high-impedance power supply is used to provide the electrical field (Rossnagel, Cuomo, and Westwood 1990).

A small quantity of free electrons is always present in the gas as a result of ionisation by naturally occurring radioactivity or cosmic rays. Free electrons can be produced by photoionisation or field emission. As the voltage applied to the gas in the discharge tube is gradually increased, the available free electrons as produced by radioactivity or cosmic rays are accelerated in the electric field, thereby gaining kinetic energy. The free electrons may lose energy upon inelastic collision with the atoms or molecules of the gas. These atoms or molecules will also be referred to as *collision targets.*

Initially, when the energy of the electrons is too low to excite or ionise a target, the collision will necessarily be elastic. The average function of electron energy lost in an elastic collision with a gas atom or molecule is $-2m_e/M$ eV (Grill 1994) where m_e and M are the mass of the electron and the target, respectively. Hence, only a very small fraction of the total kinetic energy of the electron, typically only 10^{-5}, is lost per elastic collision. Meanwhile, the electron continues to gain energy between collisions until it attains sufficient energy to cause ionisation of the targets through inelastic collisions. Large amounts of energy are transferred to the target in the inelastic collisions, making those collisions an efficient means of energy transfer. The new electrons produced in the ionisation process are in turn accelerated by the electric field to produce further ionisation.

When the number of electrons is sufficient to produce just enough ions to regenerate the number of lost electrons, a steady state is reached in which an equilibrium is established between the rate of formation of ions and the rate of their recombination with electrons. At this stage, the discharge is self-sustaining. Extensive breakdown occurs in the gas and the glow discharge is thus established.

4.3.2 Radio-Frequency (RF) Discharge

Although a DC discharge may be initiated, it will be quickly extinguished as the electrons accumulate on the insulator and recombine with the available ions. In some cases, it is preferable to have the electrodes located outside the reactor to avoid or minimise the contamination of the process caused by the material removed from the electrodes. Such problems can also be solved by alternating the polarity of the discharge (Rossnagel, Cuomo, and Westwood 1990).

When an alternating electric field of low frequency ($<$100 Hz) is applied between the two electrodes of the discharge tube, each electrode acts alternately as cathode or anode. Once the breakdown potential is surpassed on each half cycle, a temporary DC glow discharge is obtained. When the voltage drops during the cycle below the breakdown value, the discharge is extinguished and sufficiently low frequencies are reinitiated with inverse polarity (Grill 1994). Hence, high frequency is used to maintain the discharge process. The frequencies used in the high-frequency discharges are in the range of radio transmission, giving the high-frequency discharges the name of radio frequency, or RF, discharge.

The elastic collision frequency, ν, in gases at glow discharge conditions is normally between 10^9 to 10^{11} collisions/second (Manos and Flamm 1989). This makes the collision frequency much higher than the applied radio frequency, even for 13.56-MHz discharges, and so electrons will experience many collisions during each applied field cycle. They will be generated by impact ionisation in the body of plasma. Therefore, the loss of electrical carriers from the RF discharge is controlled by ambipolar diffusion and homogeneous recombination (recombination in the gas phase) but not by the electric field. Newly charged particles are produced mainly through electron impact ionisation of neutral gas and molecules. If an electron makes an elastic collision with an atom, reversing its motion while at the same time making the electric field change direction, it will continue to gain speed and energy (Manos and Flamm 1989). Electrons in a RF discharge could thus accumulate enough energy to cause ionization even at low electric field. As a result of this behaviour, the RF discharge is more efficient than the DC discharge in promoting ionisation and sustaining the discharge.

4.3.3 Microwave Plasma

Microwave plasma is sustained by the power supply operating at a frequency of 2.45 GHz. This frequency, which is commonly used for industrial or home heating applications, makes a suitable power supply readily available. The excitation of plasma by microwaves is similar to the excitation with RF, while their differences result from the range of frequencies. However, microwave discharge is more difficult to sustain at low pressures (<1 torr) than DC or RF discharge (Manos and Flamm 1989).

While the RF glow discharge can be made to extend virtually throughout the entire reactor, its dimensions are much smaller than the wavelength of the RF field (≈22 m at 13.56 MHz). The microwave plasma has its greatest glow intensity at the coupling microwave cavity and diminishes rapidly outside it because of the much smaller wavelength of the microwave (λ = 12.24 cm for frequency of 2.45 GHz). In a microwave plasma, the magnitude of the electric field can vary within the reactor, and the dimensions have the same order of magnitude as the wavelength. One can thus find active species from the discharge still persisting in a region free of the glow of the plasma, that is, in the afterglow.

4.4 VARIOUS PLASMA TREATMENT OPERATING SYSTEMS

4.4.1 Low-Pressure (Vacuum) Plasma Devices

Low-pressure plasmas are proven methods for surface modification. Vacuum devices provide a microscopically thorough, chemically mild, and mechanically non-destructive means for the removal of adsorbates such as dust, grease, and fatty acids or bacteria. The typical operating pressure range for vacuum plasma devices is between 10 mTorr and 10 Torr (Lieberman and Lichtenberg 1994).

Low-pressure plasmas may be generated using a DC power supply, thus providing a glow discharge, or a RF power supply producing a quasi-glow discharge or, more specifically, a quasi-neutral plasma bulk formed between two sheaths.

4.4.1.1 DC Glow Discharge for Low-Pressure Plasma

A typical DC glow discharge is used as a sputtering source for metallic materials, but is limited in use for many industrial applications as it has a narrow range of pressure applicability for sputtering. Most DC glow discharges for sputter deposition operate in what is known as the abnormal regime, which has a narrow range of pressure. The limitation of applicability is also due, in part, to the continuous need to conduct a net current to sustain the discharge (Chapman 1980). Another important limiting factor is the inability to use insulating materials over the electrodes, which will inhibit current conduction. By eliminating the contact between the electrode and plasma, both reliability and reproducibility are improved. The lifetime of plasma is also enhanced, and the chance of impurities is dramatically lowered (Roth 1995). For these reasons, more practical plasma processes are typically RF excited; however, DC glow discharges are commonly used for sputter deposition of metallic materials such as sputter deposition of gold on non-metallic surface for the purpose of scanning electron microscopy (SEM) to eliminate static charges and glaring under SEM.

4.4.1.2 RF (AC) Discharges for Low-Pressure Plasma

Radio-frequency discharges can be subdivided into inductive and capacitive discharges, differing in the way the RF field is induced in the discharge space. Inductive methods are based on electromagnetic induction so that the created electric field is a vortex field with closed lines. In capacitive methods, the voltage from the RF generator is applied across the electrodes, where the lines of force strike them, and the resultant field is essentially a potential field (Raizer, Shneider, and Yatsenko 1995). RF discharges can also take the form of a microwave discharge, known as *wave-heated discharge*, using higher frequency (≈2.45 GHz), but only the previous two will be discussed in this work.

4.4.1.2.1 Inductively Coupled RF Discharge

Inductively coupled plasma sources (ICP) have been researched for over a century (Lieberman and Lichtenberg 1994). The simplest form of an inductive discharge is a quartz tube placed inside a solenoid (the primary coil), through which a current is applied to generate plasma (Gudmundsson and Lieberman 1997). A circular electric field is induced in the coil, which can then initiate and maintain the discharge. The RF power, which is transferred by ohmic dissipation (joule heating) of induced RF currents, causes flow in plasma by high-frequency transformer action (Roth 1995). Most of the RF power is dissipated in the skin-depth layer, which is the layer between the plasma bulk and the containing chamber. This means that the interaction of the electromagnetic field with electrons is governed by electron thermal motion rather than electron–atom collisions.

ICPs have been studied in both high- and low-pressure regimes. At high gas pressure (atmospheric) a near-equilibrium plasma is generated, whereas at low-pressure (vacuum) regimes, a non-equilibrium plasma is created. Low-pressure ICPs have been used as ion sources for particle accelerators and ion thrusters for space propulsion. Due to the high density of the low-pressure plasma, more research has been conducted using this regime. The main interaction between an electromagnetic field and plasma, and thus the RF power dissipation, takes place in the skin layer near the plasma boundary. Depending on the plasma size, gas pressure, and driving frequency, various interactions between the electromagnetic field and plasma may occur. Under such conditions, ICPs manifest a variety of plasma physics effects typical for both low-temperature collisional gas discharge plasmas and for hot fusion and space plasmas (Godyak 2003).

4.4.1.2.2 Capacitively Coupled RF Discharge

Capacitively coupled devices are the most widely used plasma source for materials processing in which two parallel electrodes are placed in a vessel filled with gas at a certain pressure, and an RF voltage is applied across the electrodes. These electrodes may be identical (symmetric geometry) or may vary (asymmetric geometry) in diameter. They may also be insulated from the conducting discharge with a dielectric material, creating an "electrodeless" discharge (Raizer, Shneider, and Yatsenko 1995).

Typical devices carry an alternatively driven applied RF frequency across the electrodes in the range of 1–50 MHz, with an RF voltage of 100 to 1000 V. The RF power is transferred to the plasma by the randomization of kinetic energy imparted to the electron population by RF electric fields (Roth 1995).

Although capacitive sources work well for many processing applications, there are several drawbacks to note. Due to their configuration, capacitive devices are limited to plasma densities of 10^{16} m^{-3}. Higher densities may be obtained by increasing the RF power, but this will increase the corresponding voltage across the electrodes. Electrons are then accelerated by a higher sheath voltage, resulting in less efficient heating of the plasma bulk. This leads to overall inefficiency of the device. Capacitive sources also offer no independent control of the plasma density. Despite these drawbacks, however, the use of capacitive devices for commercial applications has continually increased (Shul and Pearton 2000).

4.4.2 Atmospheric-Pressure Plasma

Although most previous research has focused on vacuum systems, their high cost and production limitations have led to the development of higher-pressure (atmospheric) devices. To create a more economical, continuous, or high-speed process, a working pressure at or near 1 atm is a necessity. Atmospheric-pressure plasmas offer industry open-perimeter, on-line, continuous, large-area processing, unlike closed-perimeter vacuum systems (Herbert and Bourdin 1999).

Current atmospheric devices include corona/dielectric barrier discharges (DBD), atmospheric-pressure glow discharge (APGD), atmospheric-pressure non-equilibrium plasma (APNEP), and atmospheric-pressure plasma jets (APPJ). Like their vacuum counterparts, corona discharge/DBD systems induce surface modification through oxidation and radical formation. These processes have shown improvement in polymer adhesion, shrink resistance, dyeability, printability, and sterilization (Wakida et al. 1993; Ryu, Wakida, and Takagishi 1991; Pochner, Neff, and Lebert 1995; Carneiro et al. 2001). Although offering the same treatment at a lower cost and with faster production, these systems still suffer from some of the disadvantages previously mentioned for vacuum and corona systems: specifically, pinhole formation and non-uniformity of treatment.

Following the development of an atmospheric-pressure corona discharge, APGD was developed and reported by Kanazawa et al. (1988). This system employs a similar parallel-plate discharge design. Unlike the inhomogeneous plasma formed via the previous corona discharge/DBD atmospheric systems, APGD provides a homogeneous and streamer-free discharge (Herbert and Bourdin 1999). This discharge is maintained by controlling the following three conditions:

1. Inclusion of a helium seed/dilution gas.
2. A high-frequency source, kilohertz to megahertz.
3. An insulating plate/dielectric covering the electrodes.

The first and foremost in importance is the inclusion of helium as the seed gas. This system requires the presence of metastable helium, which will dissociate other atoms/molecules such as oxygen, resulting in ionization of the mixed molecules:

$$He^m + O_2 \rightarrow He + O^* + O^+ + e$$

$$He^+ + O_2 \rightarrow He^+ + O + O^*$$

$$He^+ + O_2 \rightarrow He + O^+ + O^*$$

A high-frequency source such as kilohertz (kHz) or radio frequency (MHz) and the dielectric plates are necessary to form a uniform discharge and prevent arc formation (Kanazawa et al. 1990; Placinta et al. 1997). However, it is common in such discharges to have streamers in the plasma, which may cause localised effects on the treated substrate.

Another independently developed APGD device was called PArallel pLate Atmospheric plasma Device for INdustry (PALADIN) (McCord et al. 2002), which is shown schematically in Figure 4.2. It is a capacitively coupled device designed with an audio frequency (AF) power supply. APGD devices have shown a wide range of surface-modification capabilities such as changes in wettability, adhesion, biocompatibility, and flame retardancy (Herbert and Bourdin 1999).

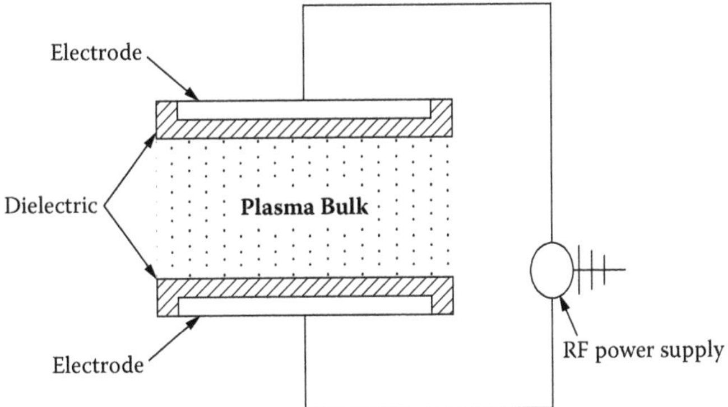

FIGURE 4.2 Schematic diagram of PALADIN. (From McCord et al. 2002.)

More recently, newer atmospheric devices have been developed to include APNEP and APPJ (Shenton and Stevens 2001; Park et al. 2001). Although neither system is available for industrial applications, they have already demonstrated great potential and a wide range of applications. According to Shenton et al. (2001), APNEP resulted in near-identical surface-chemical modification as seen by vacuum systems, and it was a viable system for surface cleaning, altering surface energy, surface oxidation and cross-linking. Comparisons between vacuum and atmospheric pressure revealed similar reactions despite the differences in pressure, gas throughput, and plasma thermal properties. The current vacuum plasma treatment of textile materials may be pertinent to atmospheric plasma treatments of the same materials. However, it is important to indicate that the physics of vacuum plasmas differ from those of atmospheric plasmas due to the high collisionality of the latter. Also, the fractional ionization of atmospheric plasmas is much lower than vacuum discharges, and most standard plasma-diagnostic techniques are not applicable to atmospheric discharges.

4.5 FACTORS AFFECTING PLASMA TREATMENT

In plasma treatment, the effectiveness of the treatment greatly depends on different factors (Kan and Yuen 2007). Several factors are commonly governing the effect of the plasma treatment, namely:

1. Nature of gas used
2. Flow rate
3. System pressure
4. Discharge power
5. Duration of treatment
6. Ageing of plasma-treated surface
7. Temperature change during plasma treatment
8. Sample distance

4.5.1 Nature of Gas Used

The result of plasma treatment depends strongly on the nature of the gas or the vapour used in glow discharge (H. Yasuda 1981; T. Yasuda, Gazicki, and Yasuda 1984; Kan, Chan, and Yuen 2004). Most organic, organosilicone, or organometallic vapours tend to form a thin film on the surfaces which are subjected to glow discharge, and the deposition of these films is the main factor modifying the polymer surface in such cases. On the other hand, glow discharges of non-polymerising gases, e.g., noble gas, nitrogen, oxygen, hydrogen, ammonia or water vapour, modify polymer surfaces through processes such as oxidation, ablation, cross-linking, and perhaps grafting. It should be recognised, however, that ablation supplies the gas phase with various chemical species, some of which may be able to form deposits, especially in mixtures with such non-reactive gases as hydrogen or ammonia. In short, the characteristics of plasma are varied by changing the gas that is used for electrical discharge;

for example, oxygen plasma is oxidising in nature, whereas hydrogen is reductive in nature. Furthermore, the surface composition and characteristics of the polymeric material also vary with gas feed of different natures.

4.5.1.1 Inert Gas

Helium, neon, and argon are the three inert gases commonly used in plasma technology. Due to the relatively lower cost, argon is by far the most common inert gas being used. The direct and radioactive energy-transfer (momentum transfer) processes created by inert gas plasma can cause physical modification of the surface. Inert gas plasma has been used for the pretreatment of substrates for cleaning purposes before reactive gases are applied. If a plasma reaction is to be carried out with a high system pressure but at a low reactive gas flow rate, an inert gas can serve as a diluent. Treatment of polymer surfaces by exposure to inert gas plasma has been utilised to improve the adhesive characteristics of polymers. Polymers have been subjected to low-power plasma of noble gases for certain periods, typically from 1 second to several minutes. This exposure is sufficient to abstract hydrogen and to form free radicals at or near the surface, which then interact to form cross-links and unsaturated groups with chain scission. The gas plasma also removes low-molecular-weight materials or converts them to a high molecular weight by cross-linking reactions. As a result, the weak boundary layer formed by the low-molecular-weight materials is removed. Consequently, greater adhesive joint strengths are observed. This treatment has been known as CASING (cross-linking by activated species of inert gases) (Schonhorn and Hansen 1967).

4.5.1.2 Oxygen-Containing Gas

Oxygen and oxygen-containing plasma are most commonly employed to modify polymer surfaces. It is well known that an oxygen plasma can react with a wide range of polymers to produce a variety of oxygen functional groups, including C–O, C=O, O–C–O and C–O–O at the surface. In oxygen plasma, two processes occur simultaneously, including (a) etching of the polymer surface through the reactions of atomic oxygen with the surface carbon atoms, giving volatile reaction products, and (b) formation of oxygen functional groups at the polymer surface through the reactions between the active species obtained from the plasma and surface atoms. The balance of these two processes depends on the operation parameters of a given experiment.

Oxygen-plasma treatment of PTFE illustrates the competitive nature of these two processes (Morra, Occhiello, and Garbassi 1990). The surface chemical composition of oxygen-plasma-treated PTFE as a function of treatment time is shown in Table 4.2. After a short treatment time of 0.5–2 min, the fluorine concentration decreased and the oxygen concentration increased; whereas after a long treatment time, the trend was reversed.

The interaction of a microwave plasma of carbon dioxide and polypropylene leads to two competitive reactions, namely (a) modification and (b) degradation (Chappel et al. 1991). Surface modification produces ketone, acid and ester on the polymer surface, whereas degradation generates volatile products and a layer of oxidised oligomers of polypropylene. The conditions favouring surface modification are low gas pressure, power and treatment time. Water plasma may be used to incorporate hydroxyl functionality into a material surface. Usually, oxidation reactions

TABLE 4.2
Surface Composition of Oxygen–Plasma-Treated PTFE as a Function of Treatment Time

Treatment Time	Chemical Composition (at%)		
(min)	C	F	O
0	39.8	60.4	0.8
0.5	44.6	48.9	6.4
1.0	42.7	51.1	7.1
2.0	42.6	50.9	6.5
5.0	40.9	57.0	2.1
10.0	38.3	60.5	1.2
15.0	38.3	61.4	0.3

Source: Morra, Occhiello, and Garbassi (1990).

rather than reduction reactions are obtained in H_2O plasma, for example, to create a hydrophilic surface on PMMA by the incorporation of hydroxyl and carbonyl functionalities (Vargo, Gardella, and Salvati 1989).

4.5.1.3 Nitrogen-Containing Gas

Nitrogen-containing plasma is widely used to improve wettability, printability, bondability and biocompatibility of polymer surfaces. For example, to improve the interfacial strength between polyethylene fibres and epoxy resins which are cured by amine cross-linking, amino groups are introduced on the fibre surface to promote covalent bonding (Chappel et al. 1991). The introduction of amino groups on the surface of polystyrene films with ammonia–plasma treatment is reported to improve cell affinity (Nakayama et al. 1988). Ammonia–plasma and nitrogen–plasma have been used to provide surface amino binding sites for immobilisation of heparin on a variety of polymer surfaces (Chan, Ko, and Hiraoka 1996). Ammonia–plasma treatment is reported to increase the peel strength between polytetrafluoroethylene and nitrile rubber when a phenol-type adhesive is used (Chan, Ko, and Hiraoka 1996).

Oxygen functionalities are always incorporated into nitrogen–plasma-treated polymer surfaces. It is a common phenomenon that oxygen is incorporated on polymer surfaces after and during non-oxygen–plasma treatments. Free radicals that are created on a polymer surface can react with oxygen during a plasma treatment. In addition, free radicals that remain on a polymer surface after a plasma treatment will react with oxygen when the surface is exposed to the atmosphere.

4.5.1.4 Fluorine-Containing Gas

In a fluorine-containing plasma, surface reactions, etching and plasma polymerisation can occur simultaneously. Which reactions predominate will depend on the gas feed, the operating parameters, and the chemical nature of the polymer substrate and electrode.

CF_x radicals play important roles as polymerisation promoters, etchants of silicon dioxide, and recombining species during anisotropic etching. Halogen atoms, especially fluorine and chlorine atoms, are the major etching species (i.e., non-polymerisable etching species) for a variety of materials. Ions and electrons can influence the plasma–substrate interaction process. Their bombardment of the surface can either alter the surface bonds of the lattice or promote the desorption of some chemisorbed species. The extent of bombardment can influence the etch rates, degree of anisotropy and polymerisation.

Tetrafluoromethane shows the highest relative etching characteristics for a material reactive with fluorine atoms. Its decomposition in the plasma is characterised by the highest concentration of fluorine atoms ([F]) and the lowest concentrations of CF and CF_2 radicals ([CF] and [CF_2], respectively). If the F/C atomic ratio of the feed-in monomer decreases (e.g., tetrafluoroethylene), [CF] and [CF_2] will be much higher at the expense of [F], and the fluorocarbon plasma becomes a polymerising plasma rather than an etching plasma (Chan, Ko, and Hiraoka 1996).

4.5.1.5 Hydrocarbon

Hydrocarbons such as methane, ethane, ethylene, acetylene and benzene have been widely used in the generation of plasma-polymerised hydrogenated carbon films. The outstanding physical properties of these films—such as micro hardness, optical refractive index and impermeability—provide them numerous potential applications such as anti-reflection and abrasion-resistant coatings (Kan and Yuen 2007).

4.5.1.6 Halocarbon

Plasma of fluorine-containing inorganic gases—such as fluorine, hydrogen fluoride, NF_3, bromine trifluoride, sulphur tetrafluoride and SF_6—and monomers are used to incorporate fluorine atoms into polymer surfaces to produce hydrophobic materials. The wide range of F/C ratios obtained by plasma-polymerising various fluorocarbon monomers provides tremendous potential for a variety of applications (Kan and Yuen 2007).

4.5.1.7 Organosilicon Plasma

Plasma polymers obtained from organosilicon monomers have demonstrated excellent thermal and chemical resistance as well as outstanding electrical, optical and biomedical properties. They may find uses in many branches of modern technology; for example, dielectric coatings or encapsulants in microelectronics, anti-reflection coatings in conventional optics, thin-film light guides in integrated optics, and biocompatible materials in medicine. Various organosilicon precursors frequently used include silanes, disilanes (SiSi), disiloxanes (SiOSi), disilazanes (SiNHSi), and disilthianes (SiSSi) (Kan and Yuen 2007).

4.5.2 Flow Rate

Figure 4.3 shows the effect of gas flow rate on the weight loss of nylon 6 treated by means of air plasma for 5 min with different degrees of discharge power (T. Yasuda, Gazicki, and Yasuda 1984). Note that the weight loss increases with the flow rate

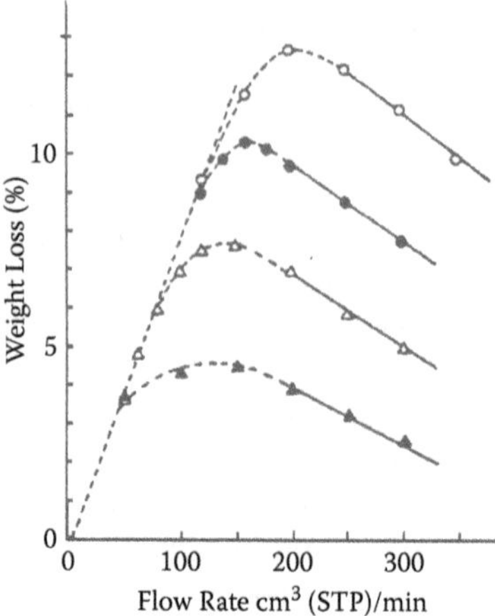

FIGURE 4.3 Effect of gas flow rate on weight loss of nylon 6 treated by air plasma for 5 min: (o) plasma discharge at 100 W, (•) 70 W, (△) 50 W, (▲) 30 W. (From T. Yasuda, Gazicki, and Yasuda 1984.)

at every discharge wattage in the low-flow-rate region. As the flow rate is further increased, the weight loss deviates from linearity and starts to decrease. The initial increase with the flow rate can be explained by the increase of the number of reactive species, particularly of O and O_3 in air. The deviation from linearity and the further decrease occur at lower flow rates when a lower wattage is applied. It is clear from the plot that, at the higher flow rate, the concentration of the active species decreases despite the increase in gas flow rate.

4.5.3 System Pressure

The system pressure is perhaps the least understood parameter of plasma treatment. This misunderstanding stems largely from the lack of distinction between non-polymer-forming plasma and polymer-forming plasma. The polymerisation itself can change the system pressure. Another factor contributing to the misunderstanding is the failure to recognise the effect of the gas. In many cases, the system pressure observed before plasma treatment, P_0, is cited as the system pressure throughout the treatment, P_g. However, it has been claimed that P_g is adjusted to P_0 by controlling the pumping rate. Since P_g is dependent on the production rate of the plasma, such an operation is not always possible. Furthermore, in view of the etching effect of gas, such an operation does not seem to have any advantage or significance in controlling the process. In summary, the system pressure used may affect the energy of the plasma species. If the pressure is high, the probability of collision between plasma species will be increased, leading to the loss of energy of the species before interacting with the material. At a pressure below approximately 0.02 torr, the transport of species becomes collisionless (T. Yasuda, Gaziciki, and Yasuda 1984; Manos and Flamm 1989).

4.5.4 Discharge Power

The intensity of plasma is a combined factor of pressure and discharge power (T. Yasuda, Gazicki, and Yasuda 1984; Jin, Lu, and Dai 2002). The breakdown energy necessary to produce plasma varies from one gas to another. Hence, the initiating energy is not a constant, but is primarily dependent on the nature of the gas fed. Normally, the higher the discharge power applied, the more kinetic energy the plasma species will carry, resulting in strong intensity of plasma action.

In general, there will be a change in the total amount of the excited particles inside the plasma, and their energy level accordingly, when the input power increases under a constant pressure (Jin, Lu, and Dai 2002).

4.5.5 Duration of Treatment

The duration of treatment plays an important role in plasma treatment (Molina et al. 2003; Binias, Wlochowicz, and Binias 2004). Generally speaking, the longer the duration of treatment, the more severe the modification of the material surface, e.g., sputtering or etching, will be. A longer duration will not only affect the material surface, but also provide an opportunity for the plasma species to penetrate into the interior region of the material. This may alter the morphology of the polymeric material. However, when the treatment duration is too long, this will adversely affect the material, and therefore careful control of treatment duration is required.

Jin, Lu, and Dai (2002) studied the relationship between duration of plasma treatment and dyeing behaviour of reactive dyeing on wool which illustrates the percentage of exhaustion against the rate of dyeing of plasma-treated wool. At 10 to 15 min of duration of treatment, there is an obvious increase of dyeing behaviour. According to their study, the increased duration of treatment results in a decease in dyeing behaviour. This result confirms that long duration of treatment would not give better effects and that there is an optimum treatment time. Such phenomena led to the supposition that the substrate produces a plasma-like region where all bonds are broken during the treatment. Most of these bonds tend to recombine within the same molecule rather than with the neighbouring molecules. This recombination can eventually lead to carbon cluster formation (Kan and Yuen 2007). Hence, the overly long treatment time might barrage and damage the dyeing path, resulting in a decrease in the dyeing behaviour.

4.5.6 Ageing of Plasma-Treated Surface

In general, the concentration of functional groups introduced to a polymer surface by plasma treatment may change as a function of time, depending on the environment and temperature. This is due to the fact that polymer chains have much greater mobility at the surface than in the bulk, allowing the surface to reorient in response to different environments. Surface orientation can be accomplished by the diffusion of low-molecular-weight oxidised materials into the bulk and the migration of polar function groups away from the surface. Ageing of plasma-treated polymer surfaces can be minimised in a number of ways. An increase in the crystallinity and orientation of a polymer surface enhances the degree of order and thus reduces the mobility of polymer chains, resulting in slower ageing. A highly cross-linked surface also restricts the mobility of polymer chains and helps reduce the rate of ageing (Brennan et al. 1991; Tsoi, Kan, and Yuen 2011).

4.5.6.1 Effects of Environment

When a polymer is exposed to an oxygen-containing plasma, the surface changes to a high-energy state, i.e., an increase in surface tension as a result of the formation of polar groups. Various surface studies indicate that when the treated surface is placed in a low-energy medium such as air or a vacuum, the decrease in surface energy is caused by the rotation of the polar groups in the bulk or the migration of low-molecular-weight fragments to the surface to reduce the interfacial energy (Morra, Occhiello, and Garbassi 1989; Munro and McBriar 1988; Morra et al. 1990; Rashidi et al. 2004). When a low-energy surface formed by treating a polymer in fluorine-containing plasma is placed in a high-energy medium such as water, the apolar groups tend to minimise the interfacial energy by moving away from the surface into the bulk. This phenomenon is usually described as the ageing of a treated surface. The ageing of plasma-treated polymer surfaces is a very complex phenomenon that is strongly affected by the treatment parameters, the nature of the polymer and the storage conditions. Contact-angle measurement, which is a very surface-sensitive technique, has been successfully used to study the dynamic characteristics of polymer surfaces in various environments.

For example, when an oxygen-plasma-treated poly(dimethyl siloxane) surface is aged in air, the surface returns to a low-energy state (Morra et al. 1990). In vacuum or air, the surface orients its apolar groups towards the interface, minimising the interfacial free energy. When ageing is performed in water or an aqueous phase, the plasma-treated surface, which has a high concentration of polar groups, maintains its polar groups at the surface, thereby minimising the interfacial free energy.

The ageing of the nitrogen functional groups on polyethylene has been studied (Foerch et al. 1990). The surface chemical composition of a nitrogen–plasma-treated polyethylene surface was monitored as a function of storage time in air. A rapid loss of nitrogen and a significant increase in oxygen was observed during the first few days. Subsequent experiments revealed no further loss of nitrogen, but a gradual increase in oxygen. The initial loss of nitrogen and increase of oxygen can be explained by the hydrolysis of imines by means of atmospheric water:

$$\mathrm{R{-}C({=}NH){-}R'} \xrightarrow{H_2O} \mathrm{R{-}C({=}O){-}R'} + NH_3$$

and the slower increase in oxygen is due to the following reaction:

$$\mathrm{R{-}CH{=}N{-}R'} \xrightarrow{H_2O} \mathrm{R{-}CH{=}O} + \mathrm{H_2N{-}R'}$$

4.5.6.2 Effect of Temperature

Another important factor that affects the ageing characteristics of a plasma-treated polymer surface is temperature. A lower storage temperature reduces the rate of ageing. Figure 4.4 shows the change of water advancing the contact angle on

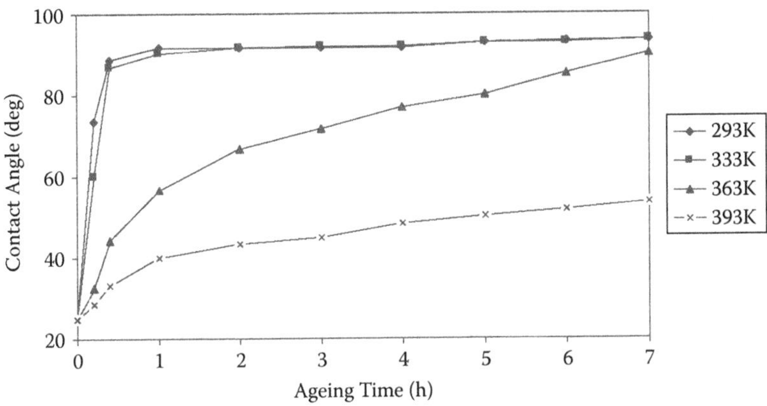

FIGURE 4.4 Contact angle of water on oxygen–plasma-treated polypropylene as a function of ageing time at different temperatures. (From Morra, Occhiello, and Garbassi 1989.)

an oxygen–plasma-treated polypropylene surface as a function of ageing time at different temperatures (Morra, Occhiello, and Garbassi 1989). The rapid change of the contact angle at high temperatures supports the idea that the changes in the surface structure are caused by polymer chain motion, resulting in reorienting the polar groups into the bulk.

4.5.7 Temperature Change during Plasma Treatment

The decomposition of fibres under plasma condition is a major concern during the plasma treatment. The temperature increase inside the plasma reactor is one possible way of explaining the thermal decomposition of fibre as a side effect (T. Yasuda, Gazicki, and Yasuda 1984). In most cases, the temperature inside the reactor reaches a saturation level, usually lower than 130°C, and remains relatively constant during the whole treatment. Air–plasma treatment is accompanied with a slightly higher temperature increase, which might be attributed to the evolution of heat of oxidation. Nevertheless, the temperatures of plasma treatment are always far below the thermal decomposition regions for the respective polymers. Hence, the possibility of the pyrolysis of fibres accompanying their plasma degradation should be ruled out (T. Yasuda, Gazicki, and Yasuda 1984). One should remember, however, that in certain cases the evolution of the low-molecular-weight species from the bulk of the fibre may be expected at elevated temperatures. Although this phenomenon is not significant, it may contribute somewhat to the total weight loss (Wong et al. 1999).

4.5.8 Sample Distance

The sample distance refers to the distance between the sample and the plasma source. If a plasma jet is used, sample distance is the defined as the jet-to-substrate distance (Wang and Qiu 2007; Kan et al. 2010). Figure 4.5 shows the influence of jet-to-substrate distance on water-absorption time of a treated wool fabric (Wang and

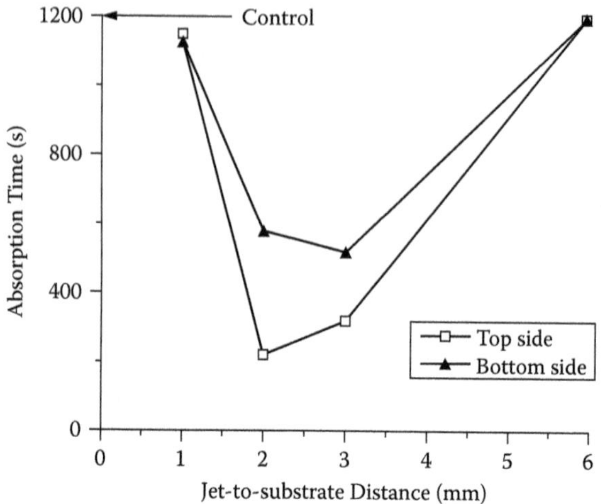

FIGURE 4.5 Influence of jet-to-substrate distance on water-absorption time on two sides of the treated fabric. (From Wang and Qiu 2007.)

Qiu 2007). Water-absorption time of the two sides of the treated wool fabric and the difference between them decreased first and then increased as the jet-to-substrate distance increased. When the distance was smaller than 1 mm or larger than 6 mm, the water-absorption time was hardly changed by the plasma treatment. When the distance was 2–3 mm, the water-absorption time of the two sides of the treated wool fabric was much shorter than that of the control, and there was no significant difference between the two sides. When the distance between the plasma jet nozzle and the fabric surface was too small, the flow of the gas from the nozzle was almost blocked by the fabric, and the gas could only be bounced off the surface and flew out in a direction more parallel to the fabric surface, which greatly reduced the effectiveness of the treatment. On the other hand, when the distance reached 6 mm, the velocity and the activity of the active species in the plasma jet greatly decreased when reaching the top side of the fabric and thus was not effective either (Rakowski 1997).

4.6 CONCLUSION

This chapter reviewed the development and types of plasma available for use in processes to treat materials. In addition, the factors affecting plasma treatment in textile materials were also discussed. The information in this chapter provides some idea about the possibility and flexibility of using plasma for treating textile materials so that the plasma treatment could be a replacement for conventional textile wet processing.

REFERENCES

Binias, D., A. Wlochowicz, and W. Binias. 2004. Selected properties of wool treated by low-temperature plasma. *Fibres and Textiles in Eastern Europe* 12(2): 58–62.

Bogaerts, A., E. Neyts, R. Gijbels, and J. van der Mullen. 2002. Gas discharge plasmas and their applications. *Spectrochimica Acta Part B* 57: 609–58.

Bradley, A., and M. Czuha Jr. 1975. Analytical methods for surface grafts. *Analytical Chemistry* 47: 1838–40.

Brennan, W.J., W.J. Feast, H.S. Munro, and S.A. Walker. 1991. Investigation of the ageing of plasma-oxidised PEEK, *Polymer*, 32: 1527–1530.

Carneiro, N., A. P. Souto, E. Silva, A. Marimba, B. Tena, H. Ferreira, and V. Magalhaes. 2001. Dyeability of corona-treated fabrics. *Coloration Technology* 117(5): 298–305.

Chan, C. M. 1994. *Polymer surface modification and characterization*. New York: Hanser Publisher.

Chan, C. M., T. M. Ko, and H. Hiraoka. 1996. Polymer surface modification by plasmas and photons. *Surface Science Reports* 24: 1–54.

Chapman, B. N. 1980. *Glow discharge processes: Sputtering and plasma etching*. New York: John Wiley and Sons.

Chappel, P. J. C., J. R. Brown, G. A. George, and H. A. Willis. 1991. Surface modification of extended chain polyethylene fibres to improve adhesion to epoxy and unsaturated polyester resins. *Surface and Interface Analysis* 17: 143–50.

Charlson, E. J., E. M. Charlson, A. K. Sharma, and H. K. Yasuda. 1984. Electrical properties of glow-discharge polymers, parylenes, and composite films. *Journal of Applied Polymer Science, Applied Polymer Symposium* 38: 137–48.

Chen, K. S., N. Inagaki, and K. Katsuura. 1982. Preliminary experiment of surface hardening of polymers by glow discharge polymerization. *Journal of Applied Polymer Science* 27: 4655–60.

Cho, D. L., and H. Yasuda. 1988. Tribological application of plasma polymers. *Journal of Applied Polymer Science, Applied Polymer Symposium* 42: 139–56.

Clark, D. T., A. Dilks, and D. Shuttleworth. 1978. *Polymer surface*. New York: John Wiley and Sons.

d'Agostino, R., F. Cramarossa, V. Colaprico, and R. d'Ettole. 1983. Mechanisms of etching and polymerisation in radiofrequency discharges of CF_4-H_2, CF_4-C_2F_4, C_2F_6 and C_3F_8-H_2. *Journal of Applied Physics* 54: 1284–88.

Donohoe, K. G., and T. Wydeven. 1979. Plasma polymerization of ethylene in an atmospheric pressure-pulsed discharge. *Journal of Applied Polymer Science* 23: 2591–2601.

Epaillard, F., J. C. Broose, and G. Legeary. 1989. Plasma-induced polymerisation. *Journal of Applied Polymer Science* 38: 887–98.

Ertel, S. I., B. D. Ratner, and T. A. Horbett. 1990. Radiofrequency plasma deposition of oxygen-containing films on polystyrene and poly(ethylene terephthalate) substrates improves endothelial cell growth. *Journal of Biomedical Materials Research* 24: 1637–59.

Foerch, R., N. S. McIntyre, R. N. S. Sodhi, and D. H. Hunter. 1990. Nitrogen plasma treatment of polyethylene and polystyrene in a remote plasma reactor. *Journal of Applied Polymer Science* 40: 1903–15.

Godyak, V. 2003. Plasma phenomena in inductive discharges. *Plasma Physics and Controlled Fusion* 45(12A): A399–A424.

Grill, A. 1994. *Cold plasma in materials fabrication: From fundamentals to applications*. New York: IEEE Press.

Gudmundsson, J. T., and M. A. Lieberman. 1997. Magnetic induction and plasma impedance in a cylindrical inductive discharge. *Plasma Sources Science and Technology* 6(4): 540–50.

Herbert, P. A. F., and E. Bourdin. 1999. New generation atmospheric pressure plasma technology for industrial on-line processing. *Journal of Coated Fabrics* 28: 170–82.

Ho, C. P., and H. Yasuda. 1988. Ultrathin coating of plasma polymer of methane applied on the surface of silicone contact lenses. *Journal of Biomedical Materials Research* 22: 919–37.

Ho, C. P., and H. Yasuda. 1990. Coatings and surface modification by methane plasma polymerization. *Journal of Applied Polymer Science* 39: 1541–52.

Inagaki, N., M. ltami, and K. Katsuura. 1982. Adhesion between polymer substrates and plasma films from tetramethylsilane and tetramethyltin by glow discharge polymerisation. *International Journal of Adhesion and Adhesives* 2(3): 169–74.

Inagaki, N., Y. Ohnishi, and K. S. Chen. 1983. Glow discharge polymerisation of tetramethylsilane by capacitive coupling of 20 kHz frequency and surface hardening of polyethylene sheet. *Journal of Applied Polymer Science* 28: 3629–40.

Iriyama, Y., and H. Yasuda. 1988. Plasma treatment and plasma polymerisation for surface modification of flexible poly(vinyl chloride). *Journal of Applied Polymer Science, Applied Polymer Symposium* 42: 97–124.

Jin, J. C., W. Lu, and J. J. Dai. 2002. Study on the dyeing behaviour of low temperature glow discharge treated wool. *Journal of Dong Hua University* (Eng. Ed.) 19(4): 12–15.

Kan, C. W., K. Chan, and C. W. M. Yuen. 2000. Application of low temperature plasma on wool, Part I: Review. *The Nucleus* 37(1–2): 9–21.

Kan, C. W., K. Chan, and C. W. M. Yuen. 2004. Surface characterisation of low temperature plasma treated wool fibre: The effect of the nature of gas. *Fibers and Polymers* 5: 52–58.

Kan, C. W., and C. W. M. Yuen. 2007. Plasma technology in wool. *Textile Progress* 39(3):121–87.

Kan, C. W., C. W. M. Yuen, W. Y. I. Tsoi, and T. B. Tang. 2010. Plasma pretreatment for polymer deposition: Improving anti-felting properties of wool. *IEEE Transactions on Plasma Science* 38(6): 1505–11.

Kanazawa, S., M. Kogoma, T. Moriwaki, and S. Okazaki. 1988. Stable glow plasma at atmospheric pressure. *Journal of Physics D: Applied Physics* 21(5): 838–40.

Kanazawa, S., M. Kogoma, T. Moriwaki, and S. Okazaki. 1990. The mechanism of the stabilization of glow plasma at atmospheric pressure. *Journal of Physics D: Applied Physics* 23(8): 1125–28.

Lieberman, M. A., and A. J. Lichtenberg. 1994, *Principles of plasma discharges and materials processing*. New York: John Wiley and Sons.

Luo, S., and W. J. van Ooij. 2002. Surface modification of textile fibres for improvement of adhesion to polymeric matrices: A review. *Journal of Adhesion Science and Technology* 16(13): 1715–35.

Manos, D. M., and D. L. Flamm. 1989. *Plasma etching: An introduction.* London: Academic Press.

McCord, M. G., Y. J. Hwang, P. J. Hauser, Y. Qiu, J. J. Cuomo, O. E. Hankins, M. A. Bourham, and L. K. Canup. 2002. Modification of nylon and polypropylene fabrics with atmospheric pressure plasmas. *Textile Research Journal* 72(6): 491–98.

Molina, R., P. Jovancic, D. Jocic, E. Bertan, and P. Erra. 2003. Surface characterization of keratin fibres treated by water vapour plasma. *Surface and Interface Analysis* 35: 128–35.

Morra, M., E. Occhiello, and F. Garbassi. 1989. Contact angle hysteresis on oxygen plasma treated polypropylene surfaces. *Journal of Colloid and Interface Science* 132: 504–8.

Morra, M., E. Occhiello, and F. Garbassi. 1990. Surface characterization of plasma-treated PTFE. *Surface and Interface Analysis* 16: 2412–17.

Morra, M., E. Occhiello, R. Marola, F. Garbassi, P. Humphrey, and D. Johnson. 1990. On the ageing of oxygen plasma-treated polydimethylsiloxane surfaces. *Journal of Colloid and Interface Science* 137: 11–24.

Moshonov, A., and Y. Avny. 1980. The use of acetylene glow discharge for improving adhesive bonding of polymeric films. *Journal of Applied Polymer Science* 25: 771–81.

Munro, H. S., and D. I. McBriar. 1988. Influence of post-treatment storage on the surface chemistry of plasma oxidised polymers. *Journal of Coating Technology* 60: 41–46.

Nakayama, Y., T. Takahagi, F. Soeda, K. Hatada, S. Nagaoka, J. Suzuki, and A. Ishitani. 1988. XPS analysis of NH_3 plasma-treated polystyrene films utilising gas phase chemical modification. *Journal of Polymer Science Part A: Polymer Chemistry* 26: 559–72.

Ohkubo, J., and N. Inagaki. 1990. Influences of the system pressure and the substrate temperature on plasma polymers. *Journal of Applied Polymer Science* 41: 349–59.

Park, J., I. Henins, H. W. Herrmann, G. S. Selwyn, and R. F. Hicks. 2001. Discharge phenomena of an atmospheric pressure radio-frequency capacitive plasma source. *Journal of Applied Physics* 89(1): 20–25.

Placinta, G., F. Arefi-Khonsari, M. Gheorghui, J. Amouroux, and G. Popa. 1997. Surface properties and the stability of PET films treated in plasmas of helium-oxygen mixtures. *Journal of Applied Polymer Science* 66(7): 1367–75.

Pochner, K., W. Neff, and R. Lebert. 1995. Atmospheric pressure gas discharges for surface treatment. *Surface and Coatings Technology* 74–75: 394–98.

Raizer, Y. P., M. N. Shneider, and A. Y. Yatsenko. 1995. *Radio-frequency capacitive discharges*. Boca Raton, FL: CRC Press.

Rakowski, W. 1997. Plasma treatment of wool today, Part I: Fibre properties, spinning and shrinkproofing. *Journal of Society of Dyers and Colourists* 113: 250–55.

Rashidi, A., H. Moussavipourgharbi, M. Mirjalili, and M. Ghoranneviss. 2004. Effect of low-temperature plasma treatment on surface modification of cotton and polyester fabrics. *Indian Journal of Fibre and Textile Research* 29: 74–78.

Rossnagel, S. M., J. J. Cuomo, and W. D. Westwood. 1990. *Handbook of plasma processing technology*. New York: Noyes Publishing.

Roth, J. R. 1995. *Industrial plasma engineering*. Vol. 1, *Principles*. Bristol, UK: IOP Publishing.

Ryu, J., T. Wakida, and T. Takagishi. 1991. Effect of corona discharge on the surface of wool and its application to printing. *Textile Research Journal* 61(10): 595–601.

Schonhorn, H., and R. H. Hansen. 1967. Surface treatment of polymers for adhesion bonding. *Journal of Applied Polymer Science* 11: 1461–74.

Shenton, M. J., and G. C. Stevens. 2001. Surface modification of polymer surfaces: Atmospheric plasma versus vacuum plasma treatments. *Journal of Physics D: Applied Physics* 34: 2761–65.

Shenton, M. J., G. C. Stevens, N. P. Wright, and X. Duan. 2001. Chemical-surface modification of polymers using atmospheric pressure nonequilibrium plasma and comparisons with vacuum plasmas. *Journal of Polymer Science: Part A: Chemistry* 40: 95–98.

Shul, R. J., and S. J. Pearton. 2000. *Handbook of advanced plasma processing techniques*. Berlin/Heidelberg, Germany: Springer-Verlag.

Simionescu, B. C., M. Leanca, S. Loan, and C. I. Simionescu. 1981. Plasma-induced polymerisation, 4: Low conversion bulk polymerisation of styrene. *Polymer Bulletin* 4: 415–19.

Tonks, L., and L. Langmuir. 1929a. Oscillations in ionised gases. *Physics Review* 33:195–210.

Tonks, L., and L. Langmuir. 1929b. The interaction of electron and positive ion space charges in cathode sheaths. *Physics Review* 33: 954–89.

Tonks, L., and L. Langmuir. 1929c. A general theory of the plasma of an arc. *Physics Review* 34: 876–922.

Tsoi, W. Y. I., C. W. Kan, and C. W. M. Yuen. 2011. Using ageing effect for hydrophobic modification of cotton fabric with atmospheric pressure plasma. *BioResources* 6(3): 3424–39.

Urrutia, M. S., H. P. Schreiber, and M. R. Wertheimer. 1988. Plasma deposition of copolymer and their permeation characteristics. *Journal of Applied Polymer Science, Applied Polymer Symposium* 42: 305–25.

Vargo, T. G., J. A. Gardella Jr., and L. Salvati Jr. 1989. Multitechnique surface spectroscopic studies of plasma modified polymers, III: H_2O and O_2/H_2O plasma modified poly(methyl methacrylate)s. *Journal of Polymer Science Part A: Polymer Chemistry* 27: 1267–86.

Wakida, T., S. Tokino, S. Niu, and H. Kawamura. 1993. Surface characteristics of wool and PET fabrics and film treated with low-temperature plasma under atmospheric pressure. *Textile Research Journal* 63(8): 433–38.

Wang C. X., and Qiu, Y. P. 2007. Two sided modification of wool fabrics by atmospheric pressure plasma jet: Influence of processing parameters on plasma penetration. *Surface and Coatings Technology* 201: 6273–77.

Wertheimer, M. R., and H. P. Schreiber. 1981. Surface property modification of aromatic polyamides by microwave plasmas. *Journal of Applied Polymer Science* 26: 2087–96.

Wong, K. K., X. M. Tao, C. W. M. Yuen, and K. W. Yeung. 1999. Low temperature plasma treatment of linen. *Textile Research Journal* 69: 846–55.

Yan, H. J., and W. Y. Guo. 1989. A study on change of fibre structure caused by plasma action. In *Proceedings of the Fourth Annual International Conference of Plasma Chemistry and Technology*, ed. H. V. Boenig, 181–88. Lancaster, PA: Technomic.

Yasuda, H. 1981. Glow discharges polymerisation. *Macromolecule Review* 16: 199–293.

Yasuda, H., and T. Hirotsu. 1977a. Polymerisation of organic compounds in an electrodeless glow discharge, VIII: Dependence of plasma polymerisation of acrylonitrile on glow characteristic. *Journal of Applied Polymer Science* 21: 3139–45.

Yasuda, H., and T. Hirotsu. 1977b. Polymerisation of organic compounds in an electrodeless glow discharge, IX: Flow-rate dependence of properties of plasma polymers of acetylene and acrylonitrile. *Journal of Applied Polymer Science* 21: 3167–77.

Yasuda, T., M. Gazicki, and H. Yasuda. 1984. Effect of glow discharge on fibres and fabrics. *Journal of Applied Polymer Science, Appl. Polym. Symp.* 38: 201–14.

Yeh, Y. S., Y. Iriyama, Y. Matsuzawa, S. R. Hanson, and H. Yasuda. 1988. Blood compatibility of surfaces modified by plasma polymerisation. *Journal of Biomedical Materials Research* 22: 795–818.

5 Application of Plasma in the Pretreatment of Textiles

Pretreatment of textiles is the first process involved in textile wet processing, and its aim is to remove natural impurities and other impurities, such as sizing agents and oil residues that are added or stained during the fabric manufacturing process. Removal of these impurities enhances water penetration and thus the fabric's wettability, thereby improving the subsequent colouration (dyeing or printing) and finishing processes that are carried out to achieve the desired properties and end uses of the textile fabric (Leung, Lo, and Yeung 1996).

Different types of fibre contain different impurities, and thus the conditions and requirements of the pretreatment processes vary greatly with the type of fibre. For cotton fibre, conventional pretreatment processes include grey inspection, singeing, desizing, scouring, bleaching and mercerization. These pretreatments remove impurities such as pectin, wax and oil, cotton seed residues, sizing materials, etc., thereby improving the aesthetic quality of the cotton materials. For bast fibre, pretreatment processes include retting, scouring and bleaching to remove natural impurities such as pectin. Pretreatment processes for wool fibre include scouring and carbonising to remove wool suint, wool grease, dirt and cellulosic impurities. Pretreatment processes for silk include a degumming process to remove most of the sericin and natural colouring impurities. For manufactured fibres, which contain no natural impurities but do contain added sizing materials and oil stains, the pretreatment process is comparatively simple (Roy Choudhury 2006).

Generally speaking, pretreatment processes require a long treatment time, huge amounts of chemicals and high treatment temperature, resulting in low production efficiency, high energy consumption and heavy loading in effluent. Plasma treatment has showed a very significant effect in the textile industry, and its etching effect could be used for surface cleaning. As a result, it is possible to use plasma treatment as an alternative as a single process to the conventional chemical-based pretreatment process. On the other hand, we can also use plasma treatment to pretreat the textile material first, followed by mild chemical pretreatment. This has the effect of reducing treatment time, shortening the processing time, reducing the amount of chemical used and lowering the treatment temperature. As a result, a higher production rate, a lower effluent load and an energy-saving pretreatment process can be achieved. Therefore, it is possible to use plasma treatment as a green technology in pretreatment of textile materials (Shishoo 2007).

5.1 PLASMA PRETREATMENT FOR CELLULOSIC FIBRES

In this section, we discuss the application of plasma for treating cellulosic fibres such as cotton and bast.

5.1.1 Application of Plasma Pretreatment for Cotton Fibre

In the cotton grey fabric, we can find sizing material, pectin, wax and oil impurities, etc., adhered on the fabric surface, which contribute to the poor wettability of the grey fabric. In such cases, the cotton grey fabric cannot be used for dyeing and printing directly unless we conduct the desizing, scouring and bleaching processes to remove these impurities. The use of oxygen and air in plasma treatment can produce high-energy plasma species to bombard the surface of the cotton fibre, where the plasma species can oxidise and subsequently break down the impurities. The active species of oxygen in the plasma can also partially break down polyvinyl alcohol (PVA) sizing agents in the warp yarn of cotton fabric. The molecular chain of the PVA will be partially broken down by oxygen species in plasma and change to carbon dioxide and water and finally be removed. On the other hand, part of the cleaved molecular chain of PVA will be converted to hydroxyl-based and carboxyl-based water-soluble fragments which can be easily be removed in the subsequent textile wet processing (Peng and Qiu 2009; Cai et al. 2002; Matthews, McCord, and Bourham 2008).

At the same time, the active species in plasma provides an etching effect on the cotton fibre surface. The wax and oil impurities, which are present as a continuous covering on the cotton fibre surface, are also broken down and hence their solubilities will be increased (Matthews, McCord, and Bourham 2008). This plasma reaction can provide an alternative to the conventional scouring process.

From the results by J. Chen (2005), it is clear that plasma treatment can effectively remove the sizing material and wax impurities from the grey fabric, comparable with the performance of the conventional process. With the help of an iodine spot test, sizing material can be effectively removed by the plasma treatment. In addition, if an oxygen mixture is used under lower discharge power, the desizing process can still be conducted, and the sizing material can also be removed in a subsequent bleaching process. However, it should be noted that the effectiveness of plasma treatment depends much on the nature of plasma gas used (Bhat et al. 2012). Under the same plasma treatment condition, oxygen gas will achieve a better etching effect than air (J. Chen 2005).

If plasma treatment is conducted with different gases, the improvement of wettability is not the same, even if the same treatment conditions are applied. The improved wettability of cotton after plasma treatment can be explained by a study using electron spectroscopy for chemical analysis (ESCA) (J. Chen 2005). Chen noted that after oxygen plasma treatment, the oxygen content of the grey fabric increased but the carbon content was reduced significantly compared with conventional alkaline-based scouring. The increased in the oxygen content of the plasma-treated grey fabric was due to the increase in the amount of the carbonyl (–CO–) and carboxyl (–COO–) functional groups. When oxygen was used as the plasma gas, more carbonyl groups had been introduced to the cotton grey fabric, hence improving the wettability. These results indicate that oxygen plasma can achieve a better wettability than a conventional alkaline-based scouring process (J. Chen 2005).

Beyond the improved wettability, plasma treatment also produces a lower degree of fibre damage, and the use of oxygen plasma lowers the degree of polymerisation (DP) of cotton fibre (Ward and Benerito 1982). After oxygen plasma treatment, the DP of the cotton fibre was reduced slightly because of the breakdown and cleavage of 1,4-glycosidic bonds connecting the β-D-glucopyranose molecular chain due to the bombardment of active species in the plasma. However, the reduction in DP is not as high as expected and, therefore, it could be concluded that the fibre damage is not severe because the effect of the plasma treatment extends only to a depth of 10 μm in the fibre surface. On the other hand, if the conditions of the plasma treatment are not controlled properly, fading of colour will occur in the dyed cotton material, accompanied by severe weight loss and a reduction in tensile strength (Ward and Benerito 1982). A study by J. Chen (2005) showed the effect of oxygen plasma treatment on the tensile strength and elongation at break of a cotton grey fabric. The results show that the breaking of plasma-treated fabric is higher than that for conventional alkaline-scoured fabric, but the level of damage depends on the fabric construction.

Industrial application of plasma pretreatment for cotton grey fabric results in poorer hand feel and softness, but these adverse effects can be compensated by the subsequent finishing process. Plasma treatment can reduce the effluent load as well as the use (and cost) of energy when compared with conventionally pretreated cotton grey fabric. In term of energy use, plasma treatment consumes about 9.8 mL of gasoline per metre, while the conventional pretreatment method consumes 62.5 mL of gasoline. In addition, the residues remaining in the surface of the cotton grey fabric after plasma treatment can be easily removed by the subsequent washing process. The duration of plasma treatment is less than that required for conventional pretreatment of cotton grey fabric with chemical, but similar results can be achieved (J. Chen 2005).

5.1.2 Application of Plasma Pretreatment for Bast Fibre

Bast fibre (e.g., ramie and flax) is a kind of natural cellulosic fibre, and its textile products have very good air permeability, comfort and moisture-absorption properties. However, the impurity content of bast fibre is very high, and it also has very high crystallinity and orientation in the molecular structure. Therefore, if the pretreatment process is not conducted properly, a lower depth of shade will be obtained in the latter dyeing process. As a result, the requirements for pretreatment of bast fibre are very demanding, such that energy consumption and pollution are serious problems. There is a clear need to improve the pretreatment processes for bast fibre (J. Chen 2005).

The effect of plasma treatment as a pretreatment process on ramie fibre has been studied, and the results show that oxygen plasma could alter the ramie fibre properties. With a longer treatment time and higher discharge power, a high weight loss was achieved. This is because when a higher discharge power is used, the plasma species have more energy with which to bombard the material surface. In addition, longer treatment time allows more time for the plasma species to react with the material surface, resulting in a greater surface-etching effect. This etching effect can be used to remove the impurities in the ramie fibre surface such as sizing material, wax and oil impurities, hemicellulose, pectin, etc. In addition, after oxygen plasma treatment, free radicals are formed in the ramie fibre surface, and these free radicals undergo oxidation upon contact with the air. The introduction of carbonyl and carboxyl functional groups to the ramie fibre surface results in improved wettability. The remaining residues can then be removed by a subsequent washing process. Oxygen plasma-treated ramie fabric also has a better dyeing behaviour (J. Chen 2005).

J. Chen (2005) shows that the improved wettability of the ramie grey fabric was greater for the oxygen plasma treatment than for the conventional scouring method. However, the effect on whiteness and dyeability are not improved as much because of the incomplete removal of impurities from the fibre surface. However, when oxygen plasma treatment was applied first followed by a conventional scouring process with a short treatment time, the wettability, whiteness and dyeability were increased significantly because of the removal of the impurities. Furthermore, when oxygen plasma-treated ramie grey fabric was further treated with a one-bath scouring and bleaching process, all the properties were improved to levels that were much better than those obtained with a conventional one-bath scouring and bleaching process.

The nature of the plasma gas used in treatment affects the results that are achieved. When nitrogen gas was used for treating a ramie fabric, similar to oxygen plasma treatment, a longer treatment time resulted in higher weight loss and increased the rate of desizing (Xa 1997). In addition, nitrogen plasma imparts an etching effect on the ramie fibre that removes the impurities in the fibre surface such as sizing material, wax and oil impurities, hemicellulose, pectin, etc. Moreover, grooves and cracks were found in the nitrogen plasma-treated ramie fibre surface, which increased the contact between water molecules and the fibre surface. The introduction of hydrophilic groups to the fibre surface facilitated the entry of water molecules to the inner part of the fibre, resulting in increased wettability (J. Chen 2005).

The use of plasma treatment on ramie fibre will reduce the use of chemicals, energy consumption in heating and also the desizing and scouring time. This, in turn, will reduce the pollution problem while saving energy and lowering production costs.

Plasma treatment can also be used for treating flax fibre (K. Wong et al. 1999). Oxygen and argon plasmas are known to be effective in modifying the flax fibre surface and improving the hydrophilic properties by inducing the formation of polar functional group such as –CO–, –C=O and –COOH. ESCA was applied to provide more information concerning the changes in chemical composition (via elemental analysis) and chemical state (via bonding and oxidation) of atom types on the fabric surface (typical penetration depth of sampling is 3–5 nm) both qualitatively and quantitatively.

Table 5.1 shows the relative intensities of C1s and O1s representing the chemical composition percentages on the fibre surface (K. Wong et al. 1999). These results show that all the plasma treatments, with the exception of argon plasma at 15 Pa and 200 W for 2.5 min, can lead to lower C1s and higher O1s intensities.

TABLE 5.1
Effect of Plasma, Using Argon and Oxygen Gases, on the Surface Chemical Composition of Flax Fibre Analysed by ESCA Method

Treatment Condition		Surface Chemical Composition		
		C1s	O1s	O1s/C1s
Untreated flax		67.65	32.35	0.48
Oxygen plasma	15 Pa, 100 W, 2.5 min	62.50	37.50	0.60
	15 Pa, 100 W, 60 min	52.12	47.88	0.92
	15 Pa, 200 W, 2.5 min	67.32	32.68	0.49
	15 Pa, 200 W, 60 min	47.38	52.62	1.11
Argon plasma	15 Pa, 100 W, 2.5 min	67.58	32.42	0.48
	15 Pa, 100 W, 60 min	61.02	38.98	0.64
	15 Pa, 200 W, 2.5 min	68.85	31.15	0.45
	15 Pa, 200 W, 60 min	44.03	55.97	1.27

Source: K. Wong et al. (1999).

As a result, the O1s/C1s ratio (the O/C ratio) increases considerably for most of the cases. For the samples treated with a shorter exposure time (2.5 min), the change in chemical composition is not very significant with respect to all gases and discharge power levels. After a prolonged exposure time for 60 min, the O/C ratio increases significantly in the ascending order of 100-W argon plasma < 100-W oxygen plasma < 200-W oxygen plasma < 200-W argon plasma. From this result, it is possible to discriminate the effect into two different aspects according to the discharge power used. After 60 min of exposure at the lower discharge power (100 W), the oxygen plasma is more effective than the argon plasma in term of increasing the O/C ratio because of the higher incorporation of oxygen components into the fibre structure under the Zoxygen plasma. It was initially assumed that the oxygen plasma at a higher discharge power would further lead to stronger incorporation of oxygen atoms on the fibre surface. However, the results showed that the argon plasma was more effective than the oxygen plasma (200 W, 60 min) at the higher discharge power. This phenomenon might be related to the slow ablation rate of the physical sputtering caused by the argon plasma. Under the same condition of 200-W discharge power for 60 min, the weight loss for the oxygen plasma-treated sample was 10.3%, while the weight loss for the argon plasma was only 1.7%. Apparently, the oxidised components that could be removed effectively by the argon plasma etching tended to accumulate on the fibre surface (K. Wong et al. 1999).

In the case of oxygen plasma, the weight loss increased with both the discharge power and the treatment time. When doubling the power from 100 to 200 W, nearly twice the weight loss is obtained, which is also linearly correlated with the exposure time. The highest weight loss occurred with 200-W oxygen plasma, followed by the 100-W oxygen plasma and the 200-W and 100-W argon plasma samples for

60 min of exposure. In the case of argon plasma, it seems that there is no significant difference between the weight-loss rates at 100 and 200 W. The oxygen plasma treatment brings about a much higher loss (10.3%) than the argon plasma (1.7%) under the same conditions (200 W, 60 min) (K. Wong et al. 1999).

The chemical states of atoms represented by the relative peak area can be obtained by wave separation of the C1s spectrum. The carbon component was further divided into four sub-components, assuming that the peaks at 285, 286.5, 287.9 and 289.1 eV corresponded to –CH, –CO–, –C=O and –COOH, respectively. Table 5.2 shows the relative peak areas of chemical component percentages (K. Wong et al. 1999).

The –COOH component increased dramatically on the fibre surface after being treated with either the oxygen or argon plasma. The –CH component also increased after most of the conditions, except for the oxygen plasma at 100 W for 60 min. The results agree with the previous finding on low-temperature plasma treatment of cotton fibre. The increment of oxidised and –CH components was attributed, respectively, to the increase of free-radical intensity and the breakage of glucoside bonds to form the activated carbonyls on the fibre surface. Different discharge power levels and gases seem to provide various degrees of polar-group incorporation. Higher percentages of –C=O and –COOH incorporation were found for the oxygen plasma than for the argon plasma when using lower discharge power for both 2.5-min and 60-min exposure times. These results are in agreement with the previous results where, at low discharge power, the oxygen plasma was more effective than the argon plasma in terms of increasing the O/C ratio. When using higher discharge power (200 W) with shorter exposure time, as shown in Table 5.2, the increment of oxidised component was smaller than that with the lower discharge power (100 W). The amount of increment for oxygen plasma was higher than those of the argon plasma at the same power and time of exposure. The reverse is observed when the exposure time reaches 60 min in the case of –COOH. In the case of argon plasma, the component

TABLE 5.2
Relative Peak Area of Chemical Component Percentages of Flax Fibre Treated with Oxygen and Argon Plasma

Treatment Condition		Surface Chemical Composition			
		–CH	–CO–	–C=O	–COOH
Untreated flax		45.6	40.9	12.4	1.1
Oxygen plasma	15 Pa, 100 W, 2.5 min	48.4	26.9	13.9	10.8
	15 Pa, 100 W, 60 min	27.2	38.6	16.3	17.9
	15 Pa, 200 W, 2.5 min	63.4	17.2	10.6	8.8
	15 Pa, 200 W, 60 min	53.6	24.1	7.9	14.4
Argon plasma	15 Pa, 100 W, 2.5 min	52.7	23.9	13.4	10.0
	15 Pa, 100 W, 60 min	65.3	15.6	9.0	10.1
	15 Pa, 200 W, 2.5 min	67.1	14.2	9.9	8.8
	15 Pa, 200 W, 60 min	73.8	9.6	5.5	11.1

Source: K. Wong et al. (1999).

of –CH significantly increased, especially when a higher discharge power was used. This may be explained by the accumulation of –CH due to the slower ablation rate and lower weight-loss value. Perhaps it may be assumed that the argon plasma mainly provides physical sputtering, while the oxygen plasma provides chemical etching. This explanation is consistent with the previously observed phenomenon that the –CH component increased sharply after argon plasma exposure and the oxidised component increased tremendously with the oxygen plasma treatment (K. Wong et al. 1999).

In the case of using higher discharge power (200 W), the oxygen plasma induced a 10.3% weight loss with an O/C ratio of 1.11 and 53.6% of –CH component, while the argon plasma induced only a 1.7% weight loss with an O/C ratio of 1.27% and 73.8% –CH component. This demonstrates that the strong incorporation of oxygen component by oxygen plasma under this condition is offset by the high weight loss. Thus the hypothesis of an accumulation of oxidised (yellowed) components can be adopted. The higher value of –CH for the argon-treated sample implies that physical sputtering is the main effect provided by the argon plasma (K. Wong et al. 1999).

In the case of using the lower discharge power (100 W), the oxygen plasma induced a 6.7% weight loss with an O/C ratio of 0.92% and 27.2% of –CH component, while the argon plasma induced a 1.5% weight loss with an O/C ratio of 0.64% and 65.3% of –CH component. Since the weight loss was not severe, it follows that a strong incorporation of oxygen component by the oxygen plasma under this condition can be retained. The principal effect of the argon plasma is the breakage of cellulose bonding with less oxidizing effect when a lower discharge power is used. Therefore, the hypothesis of oxidised component accumulation cannot be applied when a lower discharge power (100 W) is used (K. Wong et al. 1999).

Scanning electron micrographs of flax treated with oxygen plasma illustrated a progressive change in the fibre surface morphology with the treatment time. These micrographs revealed the formation of voids and cracks on the fibre surface caused by the plasma ablation. Progressive pitting and surface damage appeared on the fibre surface, and the surface area was significantly increased when using a higher discharge power. It seems that certain spots on the fibre surface are more susceptible to etching, resulting in the formation of cracks with a pronounced empty corncob structure. The typical dimension of the surface structure is in the micrometer range. The effect becomes more significant when the sample is treated with high-discharge power. The changes in fibre surface morphology observed after the plasma treatment could be explained by the localised ablation of the surface layer. The presence of micropores indicates the predominant effect of the oxygen plasma (chemical etching) on the fibre surface. Differential etching of crystalline and amorphous regions might be the origin of the roughness. This process leads to an almost complete breakdown of a relatively small number of molecules on the surface into very low-molecular-weight components which eventually vaporise in the low-pressure system when the exposure time is prolonged. If fabric strength is taken into consideration against exposure time, the strength increases at the beginning and then decreases with the time of exposure. This initial increase of strength may be due to the effects of plasma etching on the fibre surface, such as roughening of fibre, thereby increasing the inter-fibre friction. After prolonged exposure, a significant reduction in fabric strength was found for the oxygen plasma treatment (K. Wong et al. 1999).

Apart from the initial increase in fabric strength after a short period of exposure time, fabric weight loss is linearly correlated with the strength loss. Excessive strength loss occurs when the exposure time to an oxygen plasma goes beyond 20 min at 200-W discharge power. Under this condition, fibre damage (with possible formation of oxycellulose) may occur in the plasma-treated cellulose at prolonged exposure times and elevated temperature (K. Wong et al. 1999).

K. Wong et al. (1999) also examined the increase in flax fabric yellowness with higher discharge power and longer exposure time. The increase of yellowness was less severe with the oxygen plasma at 100 W, even when the exposure time was prolonged to 60 min, but severe yellowing occurred when using 200-W discharge power. The argon plasma made the fabric yellower than the oxygen plasma for all cases under investigation (K. Wong et al. 1999).

On the one hand, it was expected that the oxidative effect brought about by the oxygen plasma would be more prominent than that by the argon plasma in terms of fabric yellowing. On the other hand, the higher ablation rate caused by the oxygen plasma may lead to the subsequent and fast removal of the yellow component. The differences in yellowness brought about by the oxygen and argon plasma treatments might be due to the slower ablation rate of the inert gas (argon gas), resulting in the accumulation of yellow oxidised components on the fibre surface. However, this observation seems to be partially related to the O/C ratio measured by ESCA. The explanation is that the accumulation of oxidised components occurs under higher discharge power. Apart from the O/C ratio, the percentage of –CH component seems to match the sequence of fabric yellowness. The results showed that the sample treated with 100-W argon plasma was yellower than that obtained with 100-W oxygen plasma after 60 min of exposure, even though the latter had a higher O/C ratio. Moreover, the –CH percentage of the samples treated with 100-W argon plasma was much higher than that with 100-W oxygen plasma. Therefore, in addition to the lower weight loss (slow ablation) at 100 W, argon plasma could make a fabric yellower than the 100-W oxygen plasma. Irradiation of fibre in the plasma may cause the introduction of polar groups onto the fibre surfaces, thereby increasing the rate of water uptake and making the fibre more water absorbent (K. Wong et al. 1999).

In order to study the effect of storage time (under standard conditions) after the plasma treatment, a study was carried out by using two fabric samples treated by the oxygen and argon plasmas at 100 W for 5 min (K. Wong et al. 1999). The experimental results revealed that the effect of storage time on the fabric water uptake was small within the range from 0 to 1000 h. The highest fabric water uptake of the sample was found at zero storage time, which can probably be explained by the drying effect of the treatment. The fabric was dry initially when it was removed from the plasma chamber because the moisture content of the fibre was removed during the process of plasma treatment. Thus a maximum fabric water uptake would be expected from the immediate measurement. However, as the moisture content of the sample reaches its equilibrium, the stable water uptake values for the fabric can be obtained.

The demand for flax materials in apparel applications has increased recently because of the fabric qualities of comfort, hygiene and elegance. However, the flax fibre has a high degree of crystallinity, which translates to higher bending stiffness and greater processing difficulty. Therefore, there is a clear need for the development of fibre-modification and -finishing treatments by environmentally friendly processes. Plasma and enzyme treatments appear to be promising options. Enzymes have been used extensively on cotton textiles to remove small fibre ends from fabric surfaces, in order to create a smooth fabric appearance and introduce a degree of softness without using traditional chemical treatments (K. Wong et al. 1999). The mechanism of enzyme treatment was identified as a specifically catalytic action. In the course of enzymatic hydrolysis of cellulosic materials, key structural features, such as crystallinity and accessible surface area, are known to determine the susceptibility of cellulosic materials to enzymatic degradation. Exposure of materials to suitable plasma treatments can cause both chemical and physical changes on the surface layers, thereby providing a more reactive surface without interfering with the bulk properties simply because of the treatment's shallow depth of penetration (K. Wong et al. 1999).

The experimental results of the combined treatment with plasma and enzyme are shown in Table 5.3 (K. Wong et al. 2000), and there was no significant change of the cuprammonium fluidity and X-ray crystallinity for the plasma-treated samples. The moisture regain of the pretreated samples showed a slight reduction after the plasma treatment. This suggests that the plasma etching attacked only the ends of accessible chains on the crystallite surface and caused no significant change in the fibre bulk crystallinity.

The fabric strength increased slightly after the plasma pretreatments, as shown in Table 5.3. This increase of strength might be due to the effects caused by the plasma etching on the fibre surface such as roughening, thereby increasing the inter-fibre

TABLE 5.3
Cuprammonium Fluidity, X-Ray Crystallinity and Moisture Regain of Samples after Plasma and Enzymatic Treatments

Sample	Cuprammonium Fluidity	X-Ray Crystallinity	Moisture Regain (%)	Fabric Strength (%)[a]	Weight Loss (%)
Untreated	17.7	0.81	7.7	0	0
After enzyme treatment	18.2	0.82	7.5	−12.7	2.9
Plasma pretreated	16.7	0.80	7.4	+6.5	0.4
After enzyme treatment	16.9	0.81	7.3	−8.0	4.7

Source: K. Wong et al. (2000).

[a] Compared to the untreated control sample: "+" is percentage of fabric strength increment; "−" is percentage of fabric strength reduction.

friction. On the other hand, the plasma etching brought about small weight losses, i.e., 0.4%, as shown in Table 5.3 (K. Wong et al. 1999).

After the plasma pretreatment and subsequent enzymatic treatments, the cuprammonium fluidity and X-ray crystallinity did not show a significant change. As very little change occurred in the bulk structure of the fibres, the plasma and enzymatic reactions caused only a slight reduction in moisture regain, which might reflect the degree of removal of amorphous regions on the fibre surface (K. Wong et al. 1999).

The changes in fabric strength for plasma-pretreated and enzyme-treated samples are shown in Table 5.3. With the plasma pretreatment, the fabric strength loss for the enzyme-treated sample was 8.0%. When the plasma pretreatment was employed prior to the enzyme treatment, the accessibility of the crystalline region would be increased owing to the introduction of cracks and holes as well as the presence of more-reactive ends of molecular chains on the fibre surface. Thus, penetration and reaction of enzyme molecules with the fibre were increased, as reflected by the higher weight-loss percentage. A higher total weight-loss value of 4.7% was obtained for the oxygen plasma-pretreated sample with the subsequent enzymatic treatment. This implies that a suitable exposure of sample to plasma can effectively achieve a higher weight loss (faster enzymatic rate) and acceptable resultant strength loss (K. Wong et al. 1999).

5.2 PLASMA PRETREATMENT FOR PROTEIN FIBRES

In this section, we discuss the application of plasma for treating protein fibres such as wool and silk.

5.2.1 Application of Plasma Pretreatment for Wool Fibre

Research of plasma treatment on wool fibre as a pretreatment process dates back to 1956 (Rakowski 1997). Plasma treatment of wool fibre has been shown to improve antifelting properties, dyeability and surface wettability. Plasma treatment can alter the surface morphology and chemical composition of wool fibres, but the effect depends greatly on the plasma gas used, system pressure, discharge power and the treatment time. A major advantage of plasma treatment on wool fibre is that it is a dry process in which fibre alteration is concentrated at the fibre surface, with less damage to the bulk fibre.

After plasma treatment, the scanning electron microscope picture shows that the fibre surfaces are roughened, which will improve the yarn strength, the adhesion force between wool fibres and the fibre spinnability (Kan 2007). In addition, the directional frictional effect was changed, which will further improve the antifelting and antishrinkage properties of wool fibre. From a chemical point of view, after plasma treatment, the oxidation effect occurs on the fibre surface such that the cystine content in the cuticle layer of the fibre will be oxidised, which will reduce the degree of cross-linkage in the cystine. This will result in an increment of the polar function groups in the wool fibre surface, leading to improved wettability, dyeability and antishrinkage properties (Kan and Yuen 2007).

Figure 5.1 shows scanning electron microscope (SEM) images of wool surfaces treated under different conditions with the use of a plasma jet system

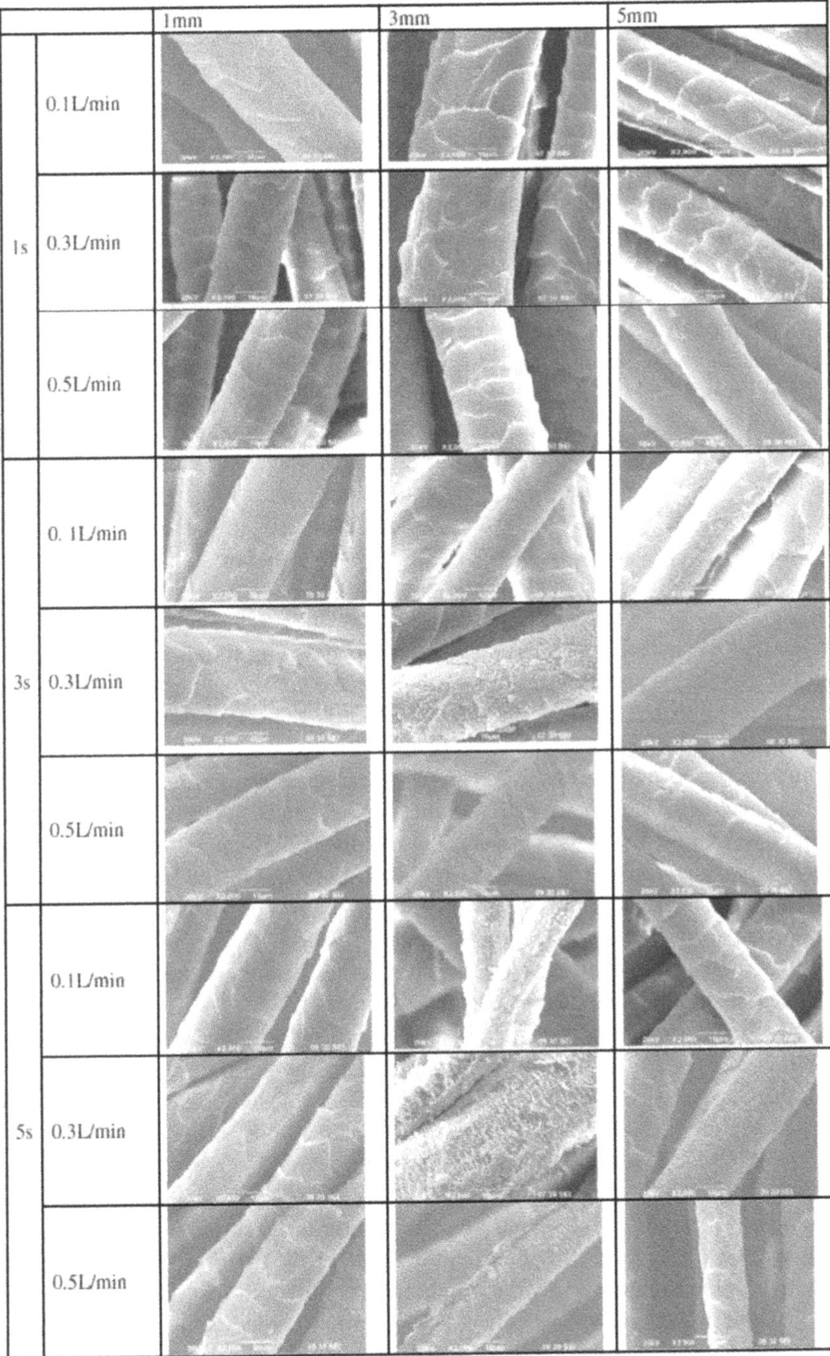

FIGURE 5.1 SEM images of wool fibre surface under different treatment conditions (From Kan, Yuen, and Hung 2013).

(Kan, Yuen, and Hung 2013). Some micropores appeared on the surface of wool fibres after plasma treatment and the number of micropores increased with oxygen flow rate in the order 0.3 L/min > 0.5 L/min > 0.1 L/min. The oxygen flow rate can be considered as the concentration of oxygen used for plasma treatment. It was expected that 0.5 L/min would provide the most significant plasma etching effect and produces more micropores, but the experimental results revealed that 0.3 L/min imparted the most significant plasma etching effect of the different flow rates used. This phenomenon can be attributed to an increase in oxygen flow rate, increasing the supply of active plasma species for reaction on the wool fabric. However, if a large amount of oxygen is supplied continuously, the active plasma species might react with the oxygen instead of the wool fabric surface, resulting in the amount of active plasma species getting reduced and lowering the surface reaction. As a result, 0.3 L/min was found to impart the most significant etching effect to wool fabric surfaces in plasma treatment. Density of micropores was the highest when the distance between plasma jet and fabric surface was 3 mm. The sequence of distance from plasma jet to substrate from the highest to the lowest density of micropores is 3 mm > 5 mm > 1 mm. When the distance was smaller than 1 mm or larger than 5 mm, the effect of plasma in terms of micropore density was not significant. When the distance between the plasma jet nozzle and fabric surface was too small, the effect of plasma treatment was greatly reduced because the gases bounced off the fabric surface and dispersed. When the distance reached 5 mm, the distance between jet nozzle and substrate surface was too large and the active plasma species required more time to reach the fabric surface. As a result, the effect of plasma treatment was greatly reduced because the velocity and activity of active plasma species were greatly decreased by the travelling time. The number of micropores found on the fibre surface increased when treatment time was increased. The order of the number of micropores was 3 s > 5 s > 1 s. As the treatment time increased, concentration of active plasma species from the plasma jet increased. Once the concentration of active plasma species increased to a critical level, the reaction between active plasma species and wool fabric surface was saturated. Excess active plasma species were then neutralised by the surrounding air species because the plasma treatment was conducted in an atmospheric pressure environment.

A number of studies (Kan, Chan, et al. 1998a, 1998b, 1998c; Yan and Guo 1989; Tokino et al. 1993) have shown that plasma treatment can improve the laundering properties of the wool fabric. This improvement is attributed to the reduction of the directional friction coefficient (DFE) of the fabric, resulting in a decrease in the felting tendency of the wool (Kan, Chan et al. 1999).

Table 5.4 shows that the plasma-treated wool can have a greater reduction in DFE than the other chemical treatments. This reduction in DFE implies that the felting tendency of wool is lower. In addition, the value of the density (D) of the felt ball is an inverse measure of the degree of felting. It is observed that the untreated wool fibre has the greatest D values, whereas the plasma-treated wool shows the greatest reduction in D among the oxidised, reduced and polymer-deposited wool (Kan, Chan et al. 1999).

From Table 5.5, it may be seen that the area shrinkage significantly decreased after the subsequent plasma treatments. Clearly, the area shrinkage increased as the

TABLE 5.4
DFE Value and Felt-Ball Density of Wool Fibre under Different Treatments

Sample	DFE (%)[a]	Felt-Ball Density, D (g/cm^3)[b]
Untreated	41.8	0.064
Potassium permanganate/salt	29.4	0.042 (↓34.4%)
Sodium metabisulphite treatment	35.1	0.054 (↓15.6%)
Plasma treatment (oxygen plasma)	23.7	0.022 (↓65.6%)
Basolan DC (chlorination) + Basolan MW Micro (polymer deposition) combined treatment	29.3	0.039 (↓39.1%)

Source: Kan, Chan, et al. (1998a, 1998b, 1998c).

[a] DFE was calculated by Mercer's equation: $\text{DFE} = \dfrac{\mu_a - \mu_w}{\mu_a + \mu_w} \times 100\%$.

[b] Felt-ball density was calculated by: $D = \dfrac{g}{V} = 0.524d^3$.

where D = density of felted ball (g/cm^3); g = weight of wool sample (g), i.e., 2 g; V = volume of felting ball (cm^3); d = average diameter of felted ball (cm).

TABLE 5.5
Results of Area Shrinkage of Different Plasma-Treated Samples

Sample[a]	Relaxation Dimensional Change in Area Shrinkage (%)	Consolidation Dimensional Change in Area Shrinkage (%)	Felting Dimensional Change in Area Shrinkage (%)
Untreated	6.90	9.22	12.28
PO	0.80	1.07	1.46
PN	0.73	1.93	2.06
PM	0.73	1.40	1.86

Source: Kan, Chan, et al. (1998c).
[a] PO = oxygen plasma; PN = nitrogen plasma; PM = gas mixture (25% hydrogen and 75% nitrogen) plasma.

processing changed from relaxation shrinkage to felting shrinkage. In the relaxation shrinkage, all of the plasma-treated fabrics showed a similar shrinkage effect. Of the three plasma treatments, the PO-treated fabric had the greatest relaxation shrinkage compared to the PN- and PM-treated fabrics, which had the same values for relaxation dimensional change. A different pattern was seen in the consolidation dimensional change, where the PN treatment showed the strongest influence. In this case, the PN-treated fabric showed the greatest shrinkage followed by the PM- and PO-treated fabric. Finally, the felting dimensional change, the result of area shrinkage of the oxygen plasma-treated (PO) fabric, was the least among the other plasma treatments. The same sequence was observed in the felting dimensional change as that shown in the consolidation dimensional change, i.e., PN > PM > PO in area shrinkage. From such results, it was found that a very different sequence was obtained in the fabric state when compared with the fibre. Therefore, the effect of the nature of the plasma gas in the fabric state differed from that in the fibre state in the case of felting and shrinkage properties measurement. The oxygen plasma showed the best improvement in the fabric shrinkage as comparing with the nitrogen plasma and the gas-mixture plasma.

The amino acid analysis methods can cause the breakdown of intermediate cystine products during the acid hydrolysis stage, but Attenuated Total Reflectance–Fourier Transform Infrared (ATR-FTIR) spectroscopy with second-order derivative spectroscopic analysis can offer the benefits of non-destructive testing. Comparison of zero- and second-order spectra of ATR-FTIR reveals that the intensities obtained in the zero-order derivative spectra were inverted in the second-order derivative spectra. The ATR-FTIR technique can analyse to a depth of 500 nm, which is good enough to detect the surface chemical components of the wool fibre. Therefore, the ATR-FTIR technique offers both qualitative and quantitative methods for measuring the composition of the wool surface. The absorbance of the selected band frequencies, i.e., 1600, 1121, 1071, 1040, 1022 and 1000 cm^{-1}, were divided by absorbance of the peptide frequency (Amide III, 1232 cm^{-1} which was used as an internal reference), and the absorbance ratio was related to the concentration of the surface component (Kan, Chan et al. 1999).

In the past, the determination of Bunte salt was based on a colourimetric technique which did not allow direct determination. However, the Bunte salt concentration can now be determined non-destructively by means of an ATR-FTIR technique (Kan, Chan, et al. 1999), which established that the amount of Bunte salt increased with increasing plasma treatment time. Of the three differently plasma-treated wool fabrics studied (oxygen plasma treated [PO], nitrogen plasma treated [PN] and gas mixture [25% hydrogen and 75% nitrogen] plasma [PM]), the highest absorbance ratio obtained was the case of oxygen plasma-treated fabric followed by nitrogen plasma and gas-mixture plasma. For the oxygen and nitrogen plasma treatments, the initial rate (1–5 min) of the formation of Bunte salt was much faster than that of the gas-mixture plasma treatment. However, in the case of gas-mixture plasma treatment, the formation rate of Bunte salt increased gradually throughout the duration of treatment. On the whole, this formation of Bunte salt probably relates to the improved shrink-resistance properties of wool (Gomez et al. 1994).

Besides Bunte salt formation, cysteic acid was also formed as a result of the cleavage of disulphide linkages. The presence of the cysteic acid on the polypeptide chain, together with the Bunte salt, provides a polar surface for the wool fabric, which in turn helps to improve the wettability of the wool fabric (Wakida et al. 1993). Furthermore, the cleavage of the disulphide bonds helps to remove the surface barrier of the wool fibre. Kan, Chan, et al. (1999) revealed that the cysteic acid content increased considerably after the plasma treatment. The three plasma treatments demonstrate similar graph patterns, i.e., after a rapid initial increase, the cysteic acid content continues to increase gradually throughout the treatment time. However, oxygen plasma gave the largest absorbance ratio, and the sequence was found to be the same as that of Bunte salt formation, i.e., PO > PN > PM.

Apart from the Bunte salt and cysteic acid, the other interesting cystine residues to be studied were cystine monoxide and cystine dioxide. Both cystine residues were believed to be intermediate cystine oxidation products (disulphide → monoxide → dioxide → sulphonic acid) (Douthwaite, Lewis, and Schumacher-Hamedat 1993). Cystine monoxide and cystine dioxide are interesting because they represent a more reactive form than the parent disulphide. The formation of cystine monoxide and cystine dioxide in wool thus generates a more reactive substrate, which provides a suitable site for introducing agents such as dyes and softeners carrying nucleophilic reactive groups (MacLaren and Kirkpatrick 1968).

A previous study (Yeung et al. 1997) showed that amino groups ($-NH_2$) were introduced to polyester after nitrogen or gas-mixture plasma treatment. Although the wool fabric itself contained amino groups ($-NH_2$), further introduction of amino groups may have enhanced the absorption of anionic dye during the dyeing process (Kan, Chan, et al. 1999). These results showed the NH content of the plasma-treated wool fabrics to be a function of treatment time. Obviously, the NH content increased in all of the cases studied. For oxygen plasma, the NH content increased only moderately, while the increase was very pronounced in the case of nitrogen and gas-mixture plasma treatments. The NH contents of nitrogen and gas-mixture plasma-treated wool fabric samples were comparable, but those treated with the nitrogen plasma showed more NH content than those treated with gas-mixture plasma. This increase of NH content may provide an explanation for the previous dyeing results (Kan, Chan, et al. 1998a, 1998b, 1998c; Kan, Chan, and Yuen 2000a, 2000b), i.e., the nitrogen plasma-treated and gas-mixture plasma-treated fibres had a higher percentage of exhaustion at equilibrium (%E at Em) value than the untreated fibre. This may be due to the increase of plasma-induced NH groups on the fibre. The induced NH groups on the wool fibre surface introduced new dye sites on the fibre, thereby enhancing the dye absorption ability of the wool fibre.

Hydrogen in gas-mixture plasma (25% hydrogen and 75% nitrogen) may be changed to reactive hydrogen such as H^+ and H_2^+ under conditions of electrical initiation. When these highly reactive hydrogen species bombard the fibre surface, a free radical may be formed by eliminating an atom from a saturated compound, as shown in Scheme 1(a). However, it was postulated that the free radicals on the polymer chain may instead combine, as shown in Scheme 1(b).

Scheme 1(a)

$$-C-C-H + H^+ \rightarrow -C-C\cdot + H_2^+$$

Scheme 1(b)

$$2-C-C\cdot \rightarrow -C-C-C-C-$$

These schemes demonstrate that, as a consequence of treatment, a cross-linking reaction may occur, and experimental results showed an increase in the carbon-carbon single-bond content on the fibre surface after plasma treatment. However, in the case of gas-mixture plasma, the carbon-carbon single-bond content increased considerably more as compared to oxygen and nitrogen plasma. Thus it may be proposed that the gas-mixture plasma treatment enhances the carbon-carbon single-bond formation on the wool fibre surface, as depicted in Scheme 1(b). These cross-linkages on the fibre surface may impart hydrophobicity and hinder diffusion through the surface. On the other hand, the gas-mixture plasma can improve the wettability and hence the dyeability by introducing amino groups to the fibre surface, causing the wool fibre to become more hydrophilic. The dyeing absorption behaviour of the gas-mixture plasma-treated wool fibre may be the compromise of these two opposing factors (Kan, Chan et al. 1999).

The microanalytical data of the surface elemental composition of low-temperature plasma (LTP)-treated wool samples were collected, and these are summarised in Table 5.6. The results show that the carbon content was significantly reduced after plasma treatment. This reduction is probably due to the etching effect of plasma treatment on the wool fibre, resulting in the removal of fibre surface material. After the etching process, the inner surface of the wool fibre was exposed along with the chemical effect due to the plasma species introduced by the new functional group. Both factors contributed together, causing a change of the surface composition. Scanning electron microscope (SEM) pictures in previous research (Kan, Chan, et al. 1998a, 1998b, 1998c; Kan, Chan, and Yuen 2000a, 2000b) have clearly shown that the oxygen plasma imparted the most significant surface etching effect by introducing grooves along the fibre axis. This grooving effect induced by oxygen plasma was most pronounced followed by nitrogen and then the gas mixture. This sequence agrees with the order of reduction in carbon content as obtained by X-ray photoelectron spectroscopy (XPS) analysis (Kan, Chan et al. 1999).

As seen in Table 5.6, the nitrogen content of the wool fibre increased to different extents after different LTP treatments. The nitrogen plasma induced the highest amount of nitrogen component into the wool fibre, followed by the gas mixture and then the oxygen plasma. This enhancement of nitrogen content on the wool fibre reflects an increase in the NH content of the wool fibre.

The oxygen content of the plasma-treated wool fibre was also found to increase. It may therefore be deduced that oxidation has occurred during the plasma

TABLE 5.6
Surface Elemental Analysis and Atomic Ratio of Wool Treated with Different Low-Temperature Plasma Gases

	Elemental Concentration (wt%)				Atomic Ratio	
Sample[a]	C1s	N1s	O1s	S2p	C/N	O/C
Untreated	74.72	8.78	13.55	2.58	8.51	0.18
PO	65.61	8.88	20.16	2.26	7.39	0.31
PN	68.31	10.19	18.86	2.23	6.70	0.27
PM	68.67	9.34	18.82	2.14	7.35	0.27

Source: Kan, Chan, et al. (1999).

[a] PO = oxygen plasma; PN = nitrogen plasma; PM = gas mixture (25% hydrogen and 75% nitrogen) plasma.

treatment and that the oxygen plasma showed the strongest effect, followed by nitrogen gas and then the gas mixture. However, nitrogen plasma and gas-mixture plasma produced a quite similar oxidation effect on the wool fibre. Since the gas mixture was composed of 75% nitrogen and 25% hydrogen, its oxidising effect would obviously be expected to be minimal, which may explain the sequence of results shown in Table 5.6. The increased amount of oxygen may enhance the hydrophilicity of the wool fibre, which increases the wettability of the wool. As a result, the dye uptake and polymer adhesion during finishing could be also enhanced (Kan, Chan et al. 1999).

As seen in Table 5.6, the content of sulphur after plasma treatment decreased slightly. One possible explanation is that the plasma treatment may have etched away the cuticle which contains a large number of disulphide bonds (–S–S–). Of the three plasma gases used, the gas mixture showed the largest reduction in sulphur content, followed by nitrogen plasma and then oxygen plasma. The ESCA spectrum showed two broad S2p peaks at binding-energy values of 163 and 168 eV. In the untreated wool fibre, the 163-eV peak intensity was stronger than the intensity at the 168-eV peak. After plasma treatment, the 168-eV peak intensity was stronger than that of the 163-eV peak. This shift of the S2p peak to higher binding energy is an indicator of the increase in the oxidation state of the sulphur atoms at the fibre surface (Lindberg et al. 1970), which suggests conversion of cystine residues to cysteic acid residues (Carr, Leaver, and Hughes 1986) according to the following equation:

$$\text{W-S(II)-S-W} \rightarrow \text{W-S(VI)O}_3\text{H} \tag{5.1}$$

where W = wool. Since the 168-eV peak is rather broad, it is possible that intermediate oxidation products of cystine may also have been present.

The ESCA analysis could be used to monitor the superficial chemical changes (depth about 10 nm) after the plasma treatment (Kan, Chan, and Yuen 1998). The plasma treatment has shown a decrease in the relative atomic concentration of carbon and an increase in the relative atomic concentration of oxygen, suggesting the oxidation of the fatty layer present on the outermost part of the epicuticle (Kan, Chan, and Yuen 1998).

As seen in Table 5.6, the C/N atomic ratio for untreated wool decreased from 8.51 to 7.39 for oxygen plasma-treated wool, to 6.70 for nitrogen plasma-treated wool, and to 7.35 for gas-mixture plasma-treated wool, while the O/C atomic ratio increased significantly for all plasma treatments. These changes suggested a partial oxidation of the hydrocarbon chains of the fatty layer without epicuticle removal (Kan, Chan, and Yuen 1998).

Kan and Yuen (2010) investigated the decrease in contact angle after the plasma treatments. These changes might be due to the variation of the surface chemical composition or the morphology of the treated fibres (Byrne, Roberts, and Ross 1979). For measurement of contact angle, some assumptions are necessary. First, the liquid and the solid (i.e., fibre) surface should be free from contamination. To prevent contamination, the liquid used must be pure and the solid surface must be clean. Second, the solid surface must be uniform, i.e., homogeneous. The non-uniformity of the wool fibre surface was assumed to be due to the presence of surface impurities.

The reduction in contact angle meant that the liquid would spread more freely on the fibre surface and hence would help to improve the wettability of the fibre (Liu, Xiong, and Lu 2006). It was established that plasma modification of fibres resulted in oxidation and degradation of the fibre surfaces (Kan, Yuen, et al. 2010). This oxidation creates oxidised functionalities, which leads to an increase in surface energy (Sprang, Theirich, and Engemann 1995), while degradation mainly changes the surface morphology of the fibres (Liu, Xiong, and Lu 2006). SEM photographs from a study (Kan, Yuen, et al. 2010) show that the plasma treatment caused an increase of surface roughness.

Plasma-treated wool can be used in woollen spinning with favourable results being achieved both in spinnability and yarn quality. This may be attributed to the increase in both the fibre friction coefficient and the cohesion of the wool top by a factor of 2.0 to 2.2 (Rakowski 1989). When the tenacity of plasma-treated wool yarn is increased, the evenness is also improved but the hairiness is decreased (Yu and Yan 1993; Rakowski 1989). Nevertheless, after plasma treatment, the spinning behaviour of wool is changed and can be summarised as follows (Rakowski 1997):

1. The spinning aids applied to the first drawing frame must be carefully selected.
2. The rubbing intensity or twist of the slubbing should be increased.
3. Reduction in break rate at the ring-spinning frame is usually observed.
4. Increase in yarn tenacity by 10%–25% is observed for all yarns.

The high cohesion of wool top after plasma treatment can last for three to four weeks. The coefficient of variation of cohesion within a single spinning is lower after plasma treatment than before. The increased top cohesion affects the tensile strength values of both single and piled yarns to a definite extent.

The 'superwash' standard, which was introduced and promoted by the International Wool Secretariat (IWS) (now The Woolmark Company under administration of Australian Wool Innovation) as a quality concept for manufacturers, commerce and consumers, acts as a guarantee for the customer. For example, a jumper labelled with this symbol can be washed up to 100 times in a washing machine using the special programme without the alteration of its dimensions or colour.

Certain requirements related to felting resistance and colour fastness must meet before the 'superwash' label can be used. Nowadays, almost 80% of the world production of 'superwash' articles is carried out in a two-stage process on wool tops, i.e., chlorination and synthetic resin application. The pre-chlorination in this two-stage process is important because chlorination can affect the adhesion and uniformity of the resin coating on the wool fibres, thereby improving the felting resistance. On the other hand, the plasma-treated wool can be subjected to antifelting finishing more easily without using resins. Researchers have shown that the application of resin to the plasma-modified wool is more effective than that carried out on the untreated fabric (Rakowski 1989; Kan, Chan, et al. 1998a, 1998b, 1998c). In these research works, wool fabrics produced from plasma-modified tops were impregnated with two synthetic resins, namely (a) Basolan SW (polyether with reactive groups) and

(b) Synthappret BAP + Impranil DLN. For comparison, an identical fabric produced from the untreated wool was finished using the same amount of finish. The results of a felting resistance test were carried out in accordance with the IWS TM 185 Test Method, which is applicable to the 'superwash' standard.

The lowest effective resin application, i.e., the minimum effective treatment level (METL), was 25% Basolan SW for the untreated fabric and only 10% for the plasma-modified fabric. Furthermore, it was not necessary to use $Na_2S_2O_5$ with the plasma-treated fabric. The METL value determined in the same way for the open-width tubular fabric was 3%–4% Basolan SW for the untreated fabric and only 1.5%–2% for the plasma-treated knitted fabric. Lower METL values are better not only because of the reduced processing costs, but also due to less change in the fabric handle and reduced soiling during the washing process.

The improved resin application may be due to the enhanced surface tension of wool fabric after plasma treatment (Wang 2002; Kan, Chan, et al. 1997), which in turn increases the adhesion property of the fabric during coating process (Kan, Chan, et al. 1998a, 1998b, 1998c). Hence, the application of resin to the plasma-treated wool fibre surface becomes easier and more effective. On the whole, fabrics made from the plasma-modified wool top can meet the 'superwash' requirements with regard to felting resistance even without applying resins.

5.2.2 Application of Plasma Pretreatment for Silk Fibre

Silk fibre, called the 'queen' of fibres, is a high-quality and high-performance natural material, widely used in the textile industry due to the unique advantages of its fabrics, such as its typical shiny appearance, special silk scroop, soft handle and elegant drape, etc. (Freddi, Mossotti, and Innocenti 2003). Silk fibre is produced by silkworms while spinning a cocoon. An original silk fibre reeled from a cocoon, also called raw silk fibre, is a composite material. It includes two filaments, a fibroin core and a fibre-cementing gum, called *sericin*, for gluing and holding the filaments together and keeping the structure of cocoon (Freddi, Mossotti, and Innocenti 2003). The fibroin synthesised at the back gland of the silkworm is highly oriented and become fibrous during spinning. It is a semicrystalline material possessing a famous b-sheet crystal structure, and is insoluble in hot water (Freddi, Mossotti, and Innocenti 2003). On the other hand, the sericin synthesised and secreted at the middle silk gland is a fibre glue surrounding the fibroin filaments, mainly consisting of polar and branched-chain amino acids, with high concentration of serine, and possessing a low- or non-oriented and low-crystalline structure (Freddi, Mossotti, and Innocenti 2003). Therefore, the properties of sericin are different from the fibroin filament in solubility, hydrophilicity and stickiness. As described in Komatsu's (2003) typical sericin theory, sericin consists of four layers from the outmost to the inner layer close to the fibroin core, layered according to their aqueous solubility. In general, the sericin in the outmost layer readily dissolves in hot water, and then it becomes more difficult to dissolve as it approaches the fibroin core.

Usually, raw silk fibre or fabric has poor functional properties, such as a dull appearance, harsh and stiff handle, etc., hardly meeting the consumer's needs. Therefore, degumming (or sericin removal) is a necessary and key point for gaining good functional properties of fabrics in the silk-making process (Freddi, Mossotti, and Innocenti 2003). Theoretically, degumming of raw silk fibre/fabric is carried out to take advantage of the differences between the fibroin filament core and the surrounding sericin in physical and chemical structures and properties. The surrounding sericin can be removed in hot aqueous solution containing soap, alkali, or synthetic detergents, etc. (Freddi, Mossotti, and Innocenti 2003). However, the fibroin core is water-insoluble due to its structure. To date, various methods have been developed for silk fibre/fabric degumming, such as a soap-alkali process, an organic acid or synthetic detergent method, a hot-water degumming process, and even a biological protease method, among others (Freddi, Mossotti, and Innocenti 2003; Jiang et al. 2006). Among all the methods, the soap-alkali process, as a standard conventional method, is still widely employed in practice, and this usually needs to be followed with a wet-chemical process in mild conditions (called the *redegumming process*) to remove the remaining sericin at the inner layers, especially for that attached closely to the fibroin core. However, all the methods described so far are thermal- and wet-chemical processes that often cause environmental concern due to wastewater discharge or increase the cost due to effluent treatment. Furthermore, these methods are involved in water resource management and are highly energy intensive (Freddi, Mossotti, and Innocenti 2003). Therefore, efforts to find or develop effective, environmentally friendly and resource-conserving methods or processes for degumming of silk fibres/fabrics have been the focus of much research in recent years.

Low-temperature plasma treatment has been widely studied in recent years due to its rapidity as well as the opportunity for resource conservation, as it is a water- and chemical-free process (Li, Ye, and Mai 1997; Vohrer, Müller, and Oehr 1998; Borcia, Anderson, and Brown 2006; Yip et al. 2002), though it is not yet really used or well established in industry. According to the definition of plasma, it consists of negatively discharged electrons, positively discharged ions, neutral atoms and molecules, various types of radicals and ultraviolet photons (Li, Ye, and Mai 1997; Yip et al. 2002). It is an electrically conducting and neutral medium due to the equilibrium of the electrons and positively and negatively charged excited ions. These excited heavy particles, electrons and photons, etc., have energies higher than that of covalent and other bonds, and thus can initiate various chemical reactions (Vallon, Drévillon, and Poncin-Epaillard 1997; Wertheimer, Fozza, and Hollander 1999). Furthermore, through physical bombardments, these energetic heavy particles can produce an etching effect on the substrate surface by fluting and/or grooving the general material surfaces (Yuan et al. 1992; Anand, Cohen, and Baddour 1981; Smiley and Delgass 1993), even on some special and hard materials (Jeong et al. 2001). Therefore, low-temperature plasma treatment is widely applied in surface modifications in various industries, especially in polymer and textiles, such as plasma polymerization, ablation, plasma grafting, surface cleaning and etching, and even for poly(vinyl alcohol) desizing, among other applications (Chaivan, Pasaja, and Boonyawan 2005; Canal et al. 2007; Matthews, McCord, and Bourham 2008).

A new degumming method was developed for silk yarns/fabric using a low-pressure and glow-discharge argon plasma in pretreatment in an effort to improve degumming efficiency. The plasma variables such as argon pressure, discharge power, exposure time were investigated. The main properties of degummed silk fabric by the proposed method were measured and compared with the conventional degumming method (Long et al. 2008). Long et al. (2008) investigated the effect of plasma pressure (20, 40, 60, 80 and 100 Pa) on raw silk fabric weight loss by pretreatment, degumming improvement and fabric properties at 60 W of glow-discharge power for 10 min. The results revealed that fabric weight loss by plasma pretreatment decreased with plasma pressure from 20 to 100 Pa; the weight loss by degumming increased with pressure up to 80 Pa, and then it decreased at higher gas pressure. This indicates that a plasma physical etching played a role on the raw silk fabric and effectively improved the degumming efficiency. Theoretically, excited and energetic heavy argon molecules and atoms, etc., in a plasma region can bombard sericin layers and break the gum effectively, and the etched sericin particles are readily ablated or carried away by the argon plasma flow, which results in the fabric weight loss. Therefore, degumming agents will diffuse more easily into the broken sericin layers. This theory is confirmed by the fact that the degumming efficiency was improved. The increase of gas pressure probably leads to an increased density of the excited particles and causes harder plasma etching on fabrics, which results in the improvement of fabric degumming. The decrease of fabric weight loss in pretreatment might be because fewer sericin particles are ablated or carried away from the fabric surface. However, the decrease of degumming efficiency at a plasma pressure higher than 80 Pa was probably due to a cross-linking reaction induced by plasma species between sericin chains (Long et al. 2008).

In addition to studying the improved degumming efficiencies, Long et al. (2008) studied the fabric whiteness and capillary-rise values in warp and weft directions, which also increased with gas pressure up to 80 Pa by the proposed method, and achieved comparable or even higher values at 80 Pa than that of the conventional method. A tolerable strength loss within 10% of degummed fabric was also achieved in both directions of fabric by the proposed method, and comparable bending rigidity was observed at 60- and 80-Pa pressure. Reasonably, the improved degumming efficiency by plasma pretreatment in an appropriate gas-pressure range leads to an easier removal of sericin and impurities (such as sizing agents, etc.) in the subsequent wet-chemical process. Therefore, the main properties of fabric were improved by the plasma treatment. However, the decrease of fabric strength, especially for warp yarns, was probably due to the penetration of excited plasma species into the fibroin core. An 80-Pa argon plasma pressure seems optimum in pretreatment of raw silk fabric.

The variations of glow-discharge power for raw silk fabric pretreatment were investigated by Long et al. (2008) at 80-Pa pressure for 10 min. The results showed that the fabric weight loss increased gradually with discharge power from 20 to 120 W, which indicates that probably more excited and energetic species were generated in argon plasma region at a higher discharge power, resulting in more effective physical bombardment or chemical modification on sericin layers. The curve of fabric weight loss due to degumming versus discharge power shows that the degumming efficiency was improved notably by the plasma pretreatment as discharge power increased from 20 to 60 W. However, a further increase of the discharge power up to 120 W resulted in a decrease of the degumming loss. This is probably attributed to the increased cross-linking reactions between sericin macromolecules induced by argon plasma at an overly high discharge power, resulting in a problem for sericin removal in the subsequent mild degumming process.

After plasma treatment, the main properties of degummed silk fabric, such as fabric whiteness and capillary-rise height, were also improved with the increased discharge power up to 60 W, and then decreased. However, a slight decrease in degummed fabric strength was obtained with discharge power, especially for warp yarns. This is probably due to a harder physical etching effect and/or plasma chemical reactions on fibres (Long et al. 2008).

Comparable bending rigidity results were obtained at 60- and 90-W discharge power compared with the conventional method, which indicates that a soft handle of degummed fabric was also achieved by the proposed method. A 60-W discharge power was adopted in raw silk fabric plasma pretreatment to maintain strength variation within 10% and other comparable properties (Long et al. 2008).

Theoretically, by prolonging the exposure time in argon plasma, the physical etching function and plasma chemical reactions on sericin layers or even on the fibroin core will be continued, leading to more breakage of sericin chains and removal of sericin particles. Indeed, the fabric loss increased gradually, and the degumming efficiency improved at an appropriate exposure time range. However, overlong exposure time readily results in greater cross-linking between sericin chains and layers, thereby making it more difficult to remove sericin (Long et al. 2008).

In the case of weighting, a weighting agent such as stannic acid can effectively be filled into the plasma-treated degummed silk fibre. The filling rate of stannic acid was increased from 2% to 6% for the untreated and plasma-treated degummed silk fibre, respectively. It was found that the filling rate was increased with prolonged plasma treatment time up to 10 min. The improved filling process was due to the formation of microvoids on the surface and inner part of the silk fibre, which allows more sites for aggregation of the weighting agent (Ren et al. 2004; Y. Chen et al. 2004).

5.3 PLASMA PRETREATMENT FOR SYNTHETIC FIBRES

Synthetic fibres do not contain natural impurities, but they do contain added impurities such as sizing material and oil stain. Consequently, pretreatment processes for synthetic fibres are simpler than for natural fibres. However, synthetic fibres such as polyester and acrylic have poor wettability, dyeability and antistatic behaviour. Plasma treatment can significantly improve the wettability of synthetic fibres by physically altering the fibre surface and introducing hydrophilic functional groups to the fibre surface. Much work has been devoted to the modification of polyester textile materials, and good results have recently been obtained (Morent et al. 2008).

5.3.1 Application of Plasma Pretreatment for Polyester Fibre

Before colouration such as dyeing and printing, the impurities present in the synthetic fibre should be removed to improve its wettability such that good dyeing and printing results can be achieved. During the plasma treatment, high-energy plasma species bombard the fibre surface, resulting in an etching effect such that the surface impurities, e.g., oil stains and sizing materials, are either partially degraded and removed or become soluble in water. Generally speaking, the plasma-treated polyester has improved wettability property which can subsequently improve the dyeing and printing process. In addition, the etching effect will result in weight loss and reduce the fibre fineness, comparable with the results obtained with conventional alkaline treatment (Morent et al. 2008).

For those samples that were treated with oxygen plasma, ripple-like patterns oriented in a direction perpendicular to the fibre axis were observed. For those polyester samples that were exposed to prolonged plasma (60 min, 5 Pa, 20 W) (Figure 5.2), the extent of etching was much greater compared with polyester that was treated for only 10 min. The granular substances developed on the plasma-treated polymer surface are believed to be "self-deposits" from the polymer, developed during plasma etching and re-adhered onto the polymer due to the vacuum environment of the plasma treatment. The smaller weight loss shown in Table 5.9 can be attributed to this self-deposit effect (W. Wong et al. 2000).

As shown in Table 5.7, there was a significant change in the weight of all the samples after plasma treatment (10 min, 100 W, 5 Pa). The effect of plasma treatment on the weight loss of polyester was less prominent than that observed following laser treatment, which can be explained by the self-deposit effect (see Figure 5.2). However, the weight reduction is more prominent for fibres with higher draw ratio

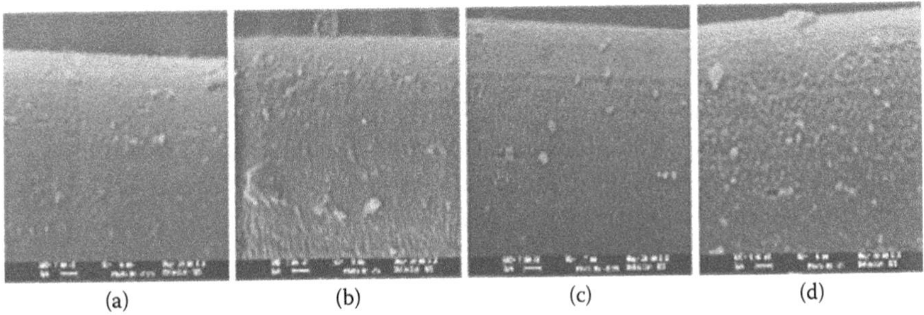

FIGURE 5.2 Morphological study of the LTP-treated PET: (a) draw ratio = 1.4, draw temperature = 200°C, energy = 10 min, 5 Pa, 100 W; (b) draw ratio = 1.4, draw temperature = 200°C, energy = 60 min, 5 Pa, 20 W; (c) draw ratio = 1.8, draw temperature = 200°C, energy = 10 min, 5 Pa, 100 W; (d) draw ratio = 1.8, draw temperature = 200°C, energy = 60 min, 5 Pa, 20 W. (From W. Wong et al. 2000.)

TABLE 5.7
Weight Loss of Plasma-Treated Polyester

	Weight before Treatment	Weight after Treatment	Weight Loss	Weight Loss (%)
Fibre (low draw ratio: 1.4)	65.50	65.17	0.33	0.50
Fibre (high draw ratio: 1.8)	56.88	56.49	0.39	0.69

Source: W. Wong et al. (2000).
Note: Plasma-treatment conditions: 10 min, 100 W, 5 Pa.

than for their low-draw-ratio counterparts. As the molecular orientation of the fibre is enhanced by drawing, this finding indicates that the weight loss increases with axial polymeric alignment (W. Wong et al. 2000). Schollmeyer (1995) has also experimentally verified that morphological modification due to plasma treatments occurred only when the polymer chains were drawn above a certain ratio. Further, greater weight loss for high-drawn fibre also brings about a higher etch rate, since the reduction of weight is the result of material melting/etching due to the treatments. Costa et al. (2006) revealed that polyester samples treated with plasma had a substantial improvement in their wettability when compared with the non-treated one. The samples treated with atmospheres of $N_{92\%}H_{8\%}$ and $O_{10\%}N_{83\%}H_{7\%}$ were the ones that presented the best wettability. The samples treated with an atmosphere of $N_{9\%}H_{1\%}M_{90\%}$ and $N_{83\%}H_{7\%}M_{10\%}$ (M is methane) had wettability behaviour similar to that of the non-treated sample.

The liquid penetration rate for the plasma-treated polyester threads was slower after the plasma treatment (SD, silent discharge) in nitrogen atmosphere at the frequencies of 10 and 20 kHz. It is necessary to note that the protective oil coating placed on the surface of the cord increases the cord wettability with water. The observed decrease in the liquid penetration rate is in accordance with the removal of the coating oil from the surface of the material (Janca et al. 2001). It can partially be explained by the activation of the oil residues on the surface of the cords. In contrast, the plasma treatment at 30 kHz caused an increase in the wettability of the cord with water (Janca et al. 2001). This result can be attributed to two effects: Cleaning of the sample surface is the first of them, as it is supposed to reduce the wettability. The second effect is the modification of the polymer surface itself (it increases the wettability), and in the case of the treatment provided at 30 kHz, this plasma activation is predominant. This modification can be seen through an increase in the surface energy. It is possible to observe an increase in the spreading time with the increasing frequency of plasma treatment (Janca et al. 2001). Experimental results indicate that with the increasing operating frequency, the efficiency of the plasma removal of the oil coating from the surface of the cord increases as well. In the case of the treatment made at 30 kHz, there are two effects. The first one is the plasma cleaning, as in the case of the samples treated at lower frequencies, and the second one is the activation of the material. The result is a decrease in the spreading time. In accordance with the results of wettability measurements, the effect of the plasma activation is predominant (Janca et al. 2001).

One study (J. Chen, Wang, and Wakida 1999) treated polyester films with plasma in six kinds of gases (O_2, N_2, He, Ar, H_2 and CH_4). The results of the surface tension γ_s for three components ($\gamma^a{}_s$, $\gamma^b{}_s$ and $\gamma^c{}_s$) and the critical surface tension γ_c of the polyester film are summarised in Table 5.8. From the data in Table 5.8, it is evident that the value of the surface tension γ_s of polyester film obtained by the extended Fowkes equation (Fowkes 1964) corresponds highly to γ_c of Zisman's plots (Zisman 1964) which coincides with the theory of Kitazaki and Hata (1972). The surface tensions of polyester films treated with O_2, N_2, He and Ar plasma for a short time (3 min) are all highly increased to 56–57.5 × 10^{-5} N·cm^{-1}; the surface of the film is in a high-energy state; and its hydrophilicity is improved. The surface tension of

TABLE 5.8
Surface Tension (10^{-5} N·cm^{-1}) of Polyester Film Treated with Low-Temperature Plasma

Plasma Treatment	Nonpolar Dispersion Force (γ^a_s)	Dipole Force (γ^b_s)	Hydrogen Bonding Force (γ^c_s)	Solid Surface Tension (γ_S)	Critical Surface Tension (γ_c)
Blank	36.3	1.6	4.3	42.2	42.0
O_2	14.3	2.7	40.1	57.1	57.5
N_2	17.6	1.0	38.4	57.0	57.0
He	16.8	1.1	37.2	55.1	56.0
Ar	17.6	1.2	37.2	56.0	56.0
H_2	33.0	0.9	15.4	49.3	50.0
CH_4	27.4	9.8	2.0	39.2	39.0

Source: J. Chen, Wang, and Wakida (1999).

polyester treated with CH_4 plasma is decreased to 39 × 10^{-5} N·cm^{-1}, and the surface free energy is also decreased. According to the surface tension γ_s and its three components, $\gamma^a{}_s$, $\gamma^b{}_s$ and $\gamma^c{}_s$, of the polyester film, one can conclude that the nonpolar dispersion force $\gamma^a{}_s$ is decreased by 50%–60%, but the hydrogen bonding force $\gamma^c{}_s$ is increased by nine times, and the surface free energy is increased markedly by the plasma treatments with O_2, N_2, He and Ar as compared with the untreated sample. The wettability of polyester is increased because of the interaction between the hydrogen bond and dipole–interdipole in the vertical direction of the interface (Westerdahi et al. 1974). The nonpolar dispersion force $\gamma^a{}_s$ keeps unchanging, but the hydrogen bonding force $\gamma^c{}_s$ is increased by 3.6 times in the H_2 plasma treatment, so the surface free energy is also increased. The hydrogen bonding force $\gamma^c{}_s$ of CH_4 plasma-treated polyester is decreased to 2.0 × 10^{-5} N·cm^{-1}; thus, the surface is in a low-energy state. Therefore, it is apparent that the increase in surface free energy and wettability by these plasmas is due to the increase in the hydrogen bonding force $\gamma^c{}_s$ (J. Chen and Wakida 1997).

The wettability of the sample is tightly related to the presence of a particular functional group that resides in the outermost surface layer (Rossmann 1956). The study showed that an acting force was produced in the vertical direction of the interface by introducing oxygen or nitrogen polar functional groups into the surface layer of a fibrogenic superpolymer, which can improve the wettability of polymer. The relationship between the surface chemical structure and the surface wettability of a plasma-treated polyester was characterised by ESCA. The contact angle to water θ_{H2O} and the relative content of the surface elements on the polyester film treated with plasma in six kinds of gases are listed in Table 5.9 (J. Chen, Wang, and Wakida 1999). It is evident that the surface wettability of polyester treated with plasma in O_2, N_2, He and Ar is improved greatly, and the contact angle to water is decreased to 24°–28°. The surface wettability of H_2 plasma-treated polyester is improved slightly, and the contact angle to water is decreased to 47°. The surface

TABLE 5.9
Contact Angles to Water (θ_{H2O}) and ESCA Intensity of Surface Element of Polyester Film Treated with Low-Temperature Plasma

Plasma Treatment	Blank	O_2	N_2	He	Ar	H_2	CH_4
θ (°)	70.00	24.40	25.00	28.00	28.00	47.20	74.00
C (%)	73.11	64.32	67.91	68.97	68.14	77.70	92.37
O (%)	26.89	34.29	26.69	30.02	31.68	22.30	7.63
N (%)	0.00	1.39	2.40	1.01	0.00	0.00	0.00
(O+N)/C	36.78	55.47	47.25	44.99	46.75	28.71	8.26

Source: J. Chen, Wang, and Wakida (1999).

wettability of CH_4 plasma-treated polyester is dropped, and the contact angle to water is increased to 74°. From the element content on the surface of polyester shown in Table 5.9, it is apparent that the O_2, N_2, He and Ar plasma treatments lead to an increase in oxygen intensity and a decrease in carbon intensity. In addition, a small amount of nitrogen is introduced in the surface layer of the O_2, N_2 and He plasma-treated polyester. The ratios of (O + N)/C for the O_2, N_2, He and Ar plasma-treated polyester are increased to 55.47, 47.25, 44.99 and 46.75, respectively, while the untreated sample is 36.78. The H_2 plasma treatment leads to a decrease in oxygen intensity and an increase in carbon intensity. On the other hand, the oxygen intensity is decreased greatly, and the carbon intensity is increased markedly by CH_4 plasma treatment, and the ratio of (O + N)/C is only 8.26 (J. Chen, Wang, and Wakida 1999).

5.3.2 Application of Plasma Pretreatment for Nylon Fibre

The first and second wetting adhesion tension hysteresis cycles of untreated and plasma-treated nylon 6 were investigated by Canal et al. (2004). It is known that the concentration of functional groups introduced on a polymer surface by plasma treatment may change as a function of the time elapsed after the treatment due to a possible reorganization process (Nakamatsu et al. 1999; Kang et al. 1996). For this reason, the adhesion tension measurements on plasma-treated samples were carried out as soon as possible after the treatment (2–12 h). The advancing adhesion tension values (force divided by perimeter [*F*/*p*]) for the second cycle of untreated, air and nitrogen plasma-treated rods were lower than the first one, which could be attributed to the dissolution of some hydrophilic material in the water (wetting liquid), thus increasing the hydrophobicity of the surface. According to Wang et al. (1994), the discrepancy between the first and second immersion cycles reflects a change in the surface state of a polymer. However, the advancing adhesion tension values for the second immersion cycle of water-vapour plasma-treated samples were slightly higher than in the first one. This could be due to the reorientation of hydrophilic groups on the polyamide surface as a consequence of its immersion in water. It also suggests that the hydrophilic groups formed are covalently bonded to the polyamide, preventing its dissolution in the water wetting liquid.

The advancing and receding contact angles and contact-angle hysteresis values for all samples are shown in Table 5.10 (Canal et al. 2004). As can be seen from Table 5.10, when air and nitrogen were used as plasma gases, an important increase in the hydrophilicity of polyamide surface was achieved, with very similar advancing contact-angle values of around 50°. The treatment with water-vapour plasma was the most effective with regards to wettability in water, reducing the advancing contact angle to 34.7° and consequently being the most effective to generate highly hydrophilic nylon 6 surfaces.

The hysteresis cycles of air, nitrogen and water-vapour plasma-treated polyamides reveal considerable wettability differences, and show a decrease in the hysteresis values from the untreated nylon 6 to that treated with air, N_2 and water vapour. This could be attributed to a decrease of the relative proportion of amide groups on the surface, since the hysteresis of polyamides tends to increase with the amide content (according to Extrand [2002]) and also to the increased chemical homogeneity of the surface, both due to the introduction of new hydrophilic groups.

Previous studies (Tusek et al. 2001; Molina et al. 2002; Nakamatsu et al. 1999; Molina et al. 2003) on different natural and synthetic polymers indicate that surface modifications by plasma are not permanent and vary progressively, depending on the storage conditions of the sample. This ageing process implies an evolution of contact-angle values with time elapsed after the plasma treatment. Canal et al. (2004) showed an increase in hydrophobicity of the plasma-treated rods as a function of the time elapsed after the plasma treatment, indicating that the concentration of hydrophilic groups on the surface decreases, which could be due to the reorientation of these groups toward the bulk phase of nylon 6 rods during their storage in an air environment. The contact-angle values for nitrogen plasma-treated samples reveal a fast ageing immediately after the treatment, suggesting that most hydrophilic groups disappear in contact with air. Air and water-vapour plasma-treated samples show approximately the same evolution with time. This suggests the formation of the same type of hydrophilic groups on the nylon 6 surface when the plasma treatment is carried out in air or water vapour. It is known that in comparison to N_2 plasma (Camacho 2003), air (Lee and Pavlath 1975) and water vapour (Camacho 2003), plasmas contain similar species, such as H_2, O_2, H, OH, H_3O^+ and OH^-.

TABLE 5.10
Advancing (θ_{adv}) and Receding (θ_{rec}) Contact Angles and Contact Angle Hysteresis ($\Delta\theta$) of Untreated and Nylon 6 Treated with Different Plasma Gases

Treatment	θ_{adv} (°)	θ_{rec} (°)	($\Delta\theta$) (°)
Untreated	71.4	15.6	55.8
Air	53.4	7.3	46.2
N_2	49.7	8.4	41.3
H_2O	34.7	8.6	26.1

Source: Canal et al. (2004).

Nakamatsu et al. (1999) have studied the ageing of differently plasma-treated polymers and have suggested that it is due to the active sites remaining on the surface, which are prone to undergo post-reactions (Gengenbach and Griesser 1999), and to the restructuring of the surface due to the tendency of the polymer to achieve an energetically favourable state by restructuring and diffusion processes (Morra, Occhiello, and Garbassi 1993), thereby reducing the free energy of the surface. The reorientation of the polar groups on the surface toward the bulk phase of the polymer is favoured by the possible breakage of the original polymer chains by the plasma, consequently increasing their mobility. Even without a decrease of the polymer molecular weight on the surface, bond rotation can produce the same effect of hiding the polar groups away from the surface, resulting in a new unmodified surface from the subsurface phase of the polymer. Van der Mei et al. (1991) suggested that ageing appears to be greatly dependent on the oxidation conditions and concluded that polymer surface dynamics is the major factor causing ageing. It is thought that reorientation of the hydrophilic groups toward the bulk phase plays the most important role in the ageing process of both air and water-vapour plasma-treated nylon 6. However, other factors such as surface contamination from the environment, post-treatment oxidation, or migration of contaminants to the surface may also play a role. Nevertheless, even though the contact-angle values increase, none of the plasma-treated nylon 6 surfaces reach the contact-angle values of the untreated polymer, even after 77 days of storage, so the hydrophilicity properties still remain improved (Canal et al. 2004).

The surface tension of the polyamide fabrics increased after the plasma exposure to acetylene and nitrogen, indicating that polar groups were grafted onto the surface (Pappas et al. 2006). In the nitrogen case, an activation process is expected, with mild surface etching. As the dissociation of molecular nitrogen is not favourable, it is expected to impose Penning ionisation to other molecules present (Penning 1927). In the acetylene treatment case, a thin carbon film is expected to be formed, meaning that the coated surface would become more hydrophobic compared to the untreated material. The role of helium is to remove any surface-residing impurities and, through energy-transfer mechanisms, to cause chain scission and the formation of cross-linked layers on the polymer surfaces (Hansen and Schonhorn 1966; Gheorgiu et al. 1997). Cross-linked layers provide stability to the material and act as a barrier to surface changes (Hall et al. 1969; J. Chen, Wang, and Wakida 1999).

As mentioned previously, the introduction of polar groups is responsible for the improved wettability of the plasma-modified nylon films. Sometimes, the surface does not show a stable behaviour with time, and the water contact angles tend to increase, an indication that an ageing process called hydrophobic recovery has been initiated. The polar groups grafted on the surface can be described as mobile and immobile (Gengenbach et al. 1994; Xie, Gengenbach, and Griesser 1992). The mobile polar groups tended to move toward the bulk of the polymer, reducing its wettability. One month after the plasma treatments, the lowest observed angle was for the sample treated the longest (9.6 s) and was smaller than that of the control film, confirming that a long-lasting improvement of the hydrophilicity of the surface was achieved.

X-ray photoelectron spectroscopy (XPS) was employed to identify the polar groups attached after the treatment. The C1s core-level spectra of the nylon control film and plasma-treated nylon films were deconvoluted with five components, as seen in Figure 5.3 (Pappas et al. 2006). In terms of binding energy, the peak C1 at 284.6 eV is attributed to C in the C–C chain CH_2 groups. The C2 peak at 285.3 eV can be associated to the amido-carbonyls [–(C=O)]. The peak C3 at 286.2 eV represents the carbon atoms neighbouring the amido nitrogen [–C–NH(C=O)–] and that at 287.9 labelled C_4 is assigned to the amide carbonyl carbons [–NH(C=O)–] (Pappas et al. 2006).

After the plasma modification of the nylon films, a fifth peak C_5 was observed, and this is believed to have formed due to the oxidation of the methylene carbons immediately adjacent to the amido carbonyls (Cui et al. 2005). Moreover, we observed a decrease of the C1 signal, which is typical of such treatments due to the chain scission mechanisms caused by the plasma active species and mainly due to the action of helium. An increase of peak C3 can be attributed to a low-level oxidation of the methylene carbons promoted by dielectric barrier discharge (DBD) processing in air. Peak C3 has also been assigned to the presence of newly formed C–OH functional groups (Oh, Kim, and Kim 2001) possibly formed by the interaction of the active surface with atmospheric air (Pappas et al. 2006).

The oxidation was also confirmed by the atomic concentration of oxygen on the surface that increased from 9.88% for the untreated nylon film to 13.49% for the film that was exposed to N_2/He plasma for 4.8 s. The ratio N1s/C1s remained almost constant for all the samples treated under different treatment times. Similar results were observed for nylon 6,6 films treated with nitrogen (Foerch and Hunter 1992) and for low-pressure treatments of polyamide fibres with oxygen plasma (Yip et al. 2002). The O1s/C1s ratio

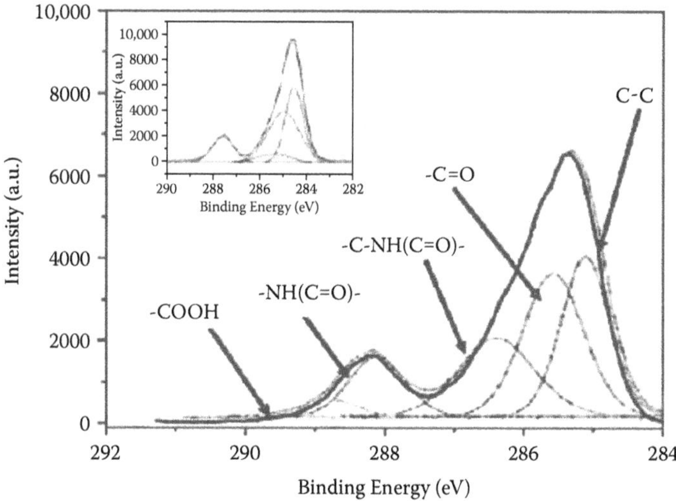

FIGURE 5.3 High-resolution XPS spectra of the C1s binding energy region of nylon films before (inset) and after plasma treatment. The positions of different functional groups are indicated. (From Pappas et al. 2006.)

increased from 0.115 for the control film to 0.177 for the nylon film treated for 4.8 s. It is suggested that the plasma treatment induced the formation of carboxylic species on the surface, either in the hydrocarbon or carbonyl groups, which finally enhances the hydrophilicity of the polymer. The oxygen uptake can be attributed to (a) the presence of atomic oxygen during the process, resulting from the dissociation of atmospheric O_2 and/or (b) the reaction of the resulting "active" surface obtained after the plasma modification (Upadhyay, Anderson, and Brown 2004). It is known that the plasma treatment is responsible for chain scission on the polymer surface and thus can react with the environment prior to reaching equilibrium (Pappas et al. 2006).

A twofold increase of the surface roughness was observed by confocal microscopy (Pappas et al. 2006), as the average roughnesses were measured to be 0.876 μm for the control film and 1.769 μm for the nitrogen–helium plasma-treated sample after a treatment time of 4.8 s, leading to an enhancement of the surface area.

5.3.3 Application of Plasma Pretreatment for Acrylic Fibre

A substantially increased surface roughness can be seen in the case of plasma-treated acrylic substrates. Liu, Xiong and Lu (2006) clearly demonstrated that plasma etching causes a significant increase in the specific surface area. The increase of the specific surface area contributes to moisture adsorption and transportation on the fibre surface. Apart from the changes in the surface structure, changes in the surface composition were also investigated. XPS analysis was performed to determine the chemical composition and the nature of the chemical bonds, and C1s and N1s data show that about 46.51% of the cyanogen and 23.73% of the ester groups were reduced with 3-min plasma treatment. While a substantial loss of cyanogen and ester groups occurred, no change was noticeable for methylene ($–CH_2–$) and methenyl (–CH=) groups. Clearly, the cyanogen and ester groups are much less stable than the methylene and methenyl groups. The O1s data indicate that plasma treatment leads to significant introduction of O atoms. The introduced O concentration accounts for 46.91% of the total O1s peak area, attributable exclusively to the introduction of the amide and carboxyl groups.

Contact angle and moisture regain were used to evaluate surface wettability. The results are shown in Table 5.11 (Liu, Xiong, and Lu 2006). The original untreated acrylic sample was quite hydrophobic (moisture regain was only 0.85 and contact angle was 77.80°). However, surface wettability was significantly increased following plasma treatment. Under the given plasma conditions, a time of 3 min was long enough for good wettability (i.e., the highest moisture regain and the smaller contact angle). With increasing treatment time, wettability decreased slightly and then remained at a relatively constant level, which seems to suggest equilibrium between introducing and removing functional groups during plasma treatment.

The half-decay time of the treated acrylic samples is also shown in Table 5.11. Here, the plasma treatment has caused a drastic reduction in the half-decay time of the fibres. The half-decay time was found to decrease from 9.57 s for the untreated sample to 1.35 s for the 3-min plasma-treated sample. This result shows that the antistatic ability of the acrylic fibres is drastically improved by plasma treatment. Surface wettability was directly related to surface energy, with more energetically

TABLE 5.11
Wettability and Antistatic Ability of the Untreated and Plasma-Treated Acrylic Samples

Samples	Moisture Regain (%)	Contact Angle, θ (°)	Half-Decay Time (s)
Untreated	0.85	77.80	9.57
		Nitrogen plasma treated	
1 min	1.35	65.64	3.98
3 min	2.24	34.92	1.35
5 min	1.65	56.41	2.62
7 min	1.59	54.73	3.16

Source: Liu, Xiong, and Lu (2006).

stable surface results observed in less wettable surfaces. It is now established that plasma modification of the fibres results in oxidation and degradation of the fibre surfaces. The oxidation creates oxidised functionalities, which lead to an increase in surface energy, while the degradation mainly changes the surface morphology of the fibres (Liu, Xiong, and Lu 2006). According to the Wenzel equation, the roughness of the surfaces influences the contact angle (Sprang, Theirich, and Engemann 1995):

$$\cos\theta_{rough} = r\cos\theta_0 \tag{5.2}$$

where θ_{rough} is the contact angle on a sample surface, θ_0 is the thermodynamic contact angle on the smooth surface, and r is the roughness (ratio of the actual area of the interface to the geometric surface area). Equation (5.2) indicates that for the surface having a lower contact angle than 90°, increasing the surface roughness probably decreases the contact angle, which will contribute to the improved surface wettability.

Water is a conductor of electricity. Therefore, the improved surface wettability will decrease the accumulation of electrostatic charges. The increase of surface roughness also induces an increase in the specific surface area. The increased specific surface area will lead to a more "moisture-rich" surface, which enhances the conductivity of the fibres. Plasma treatment not only causes the increase in surface roughness, but also introduces hydrophilic groups onto the fibre surface. XPS analyses have shown that amide and carboxyl groups are created on the fibre surface after plasma treatment. There are two possibilities of generating the polar groups. The first one is that they are generated by reacting with the ambient gas during the processing. The second is that they are generated when the samples are exposed to air after plasma processing, that is, plasma treatment produces a considerable amount of unsaturated bonds, and then the unsaturated bonds react with atmospheric oxygen to form polar groups on the sample surface. Plasma treatment introduces an amount of amide and carboxyl groups on the acrylic fibre surface.

These polar groups incorporate with moisture through hydrogen bonding and help moisture penetration and binding on the fibre surface. Under the action of water molecules, these polar groups also generate ionization and lead to a structural layer that can conduct electricity on the fibre surface, which enhances the electrostatic dissipation. Therefore, the half-decay time of the fibres decreases after plasma treatment. A 3-min plasma treatment is enough for effective surface modification. As the processing time increases further, some weakly attached polar groups will be removed from the fibre surface, which causes the slight increase in the half-decay time (Liu, Xiong, and Lu 2006).

5.4 CONCLUSIONS

This chapter reviewed the application of plasma treatment for pretreating cellulosic, protein and synthetic fibres. Different types of plasma setups have been used for treating textile fibres. Generally speaking, plasma treatment is an effective tool to remove surface impurities from textile fibres. In addition, with the use of different plasma gases, the textile surface chemical properties as well as physical properties can be modified and designed to achieve desired effects. One important property that was introduced by plasma treatment to the fibre surface was the wettability, which will in turn affect the subsequent colouration process. The effect of plasma treatment on textile colouration is discussed in Chapter 6.

REFERENCES

Anand, M., R. E. Cohen, and R. F. Baddour. 1981. Surface modification of low density polyethylene in a fluorine gas plasma. *Polymer* 22(3): 361–70.

Bhat, N. V., A. N. Netravali, A. V. Gore, M. P. Sathianarayanan, G. A. Arolkar, and R. R. Deshmukh. 2012. Surface modification of cotton fabrics using plasma technology. *Textile Research Journal*. DOI: 10.1177/0040517510397574.

Borcia, G., C. A. Anderson, and N. M. D. Brown. 2006. Surface treatment of natural and synthetic textiles using a dielectric barrier discharge. *Surface and Coatings Technology* 201: 3074–81.

Byrne, K. M., M. W. Roberts, and J. R. H. Ross. 1979. The critical surface tension of wool. *Textile Research Journal* 49: 34–40.

Cai, Z. S., Y. J. Hwang, Y. C. Park, C. Y. Zhang, M. McCord, and Y. P. Qiu. 2002. Preliminary investigation of atmospheric pressure plasma-aided desizing for cotton fabrics. *AATCC Review* 2(12): 18–21.

Camacho, J. 2003. Estudio mediante espectrometría cuadrupolar de masas (QMS) de los efectos de plasmas en muestras de lana. *Revista de Química Textil*. 163: 10–31.

Canal, C., F. Gaboriau, R. Molina, P. Erra, and A. Ricard. 2007. Role of the active species of plasmas involved in the modification of textile materials. *Plasma Processes and Polymer* 4: 445–54.

Canal, C., R. Molina, E. Bertran, and P. Erra. 2004. Wettability, ageing and recovery process of plasma-treated polyamide, 6. *Journal of Adhesion Science and Technology* 18(9): 1077–89.

Carr, C. M., I. H. Leaver, and A. E. Hughes. 1986. X-ray photoelectron spectroscopic study of the wool fibre surface. *Textile Research Journal* 56: 457–61.

Chaivan P., N. Pasaja, and D. Boonyawan. 2005. Low-temperature plasma treatment for hydrophobicity improvement of silk. *Surface and Coatings Technology* 193: 356–60.

Chan, C. M. 1993. *Polymer surface modification and characterisation*. New York: Hanser Publishers.

Chen, J. R. 2005. *Application of plasma technology in textile dyeing and finishing* (Chinese). Beijing: China Textile and Apparel Press.

Chen, J. R., and T. Wakida. 1997. Studies on the surface free energy and surface structure of PTFE film treated with low temperature plasma. *Journal of Applied Polymer Science* 63: 1733–39.

Chen, J. R., X. Y. Wang, and T. Wakida. 1999. Wettability of poly(ethylene terephthalate) film treated with low-temperature plasma and their surface analysis by ESCA. *Journal of Applied Polymer Science* 72: 1327–33.

Chen, Y. Y., H. Lin, Y. Ren, H. W. Wang, and L. J. Zhu. 2004. Study on *Bombyx mori* silk treated by oxygen plasma. *Journal of Zhejiang University SCIENCE* 5(8): 918–22.

Costa, T. H. C., M. C. Feitor, C. Alves Jr., P. B. Freire, and C. M. de Bezerra. 2006. Effects of gas composition during plasma modification of polyester fabrics. *Journal of Materials Processing Technology* 173: 40–43.

Cui, N., D. Upadhyay, C. Anderson, and N. Brown. 2005. Study of the surface modification of a nylon-6,6 film processed in an atmospheric pressure air dielectric barrier discharge. *Surface and Coatings Technology* 192: 94–100.

Douthwaite, F. J., D. M. Lewis, and U. Schumacher-Hamedat. 1993. Reaction of cystine residues in wool with peroxy compounds. *Textile Research Journal* 63: 177–83.

Extrand, C. W. 2002. Water contact angles and hysteresis of polyamide surfaces. *Journal of Colloid and Interface Science* 248: 136–42.

Foerch, R., and D. Hunter. 1992. Remote nitrogen plasma treatment of polymers: Polyethylene, nylon 6,6, poly(ethylene vinyl alcohol), and poly(ethylene terephthalate). *Journal of Polymer Science, Part A: Polymer Chemistry* 30: 279–86.

Fowkes, F. M. 1964. Dispersion forces constitutions to surface and interfacial tensions, contact angles, and heat of immersion. *Advances in Chemistry Series*. 43: 99–111.

Freddi, G., R. Mossotti, and R. Innocenti. 2003. Degumming of silk fabric with several proteases. *Journal of Biotechnology* 106: 101–12.

Gengenbach, T. R., and H. J. Griesser. 1999. Post-deposition ageing reactions differ markedly between plasma polymers deposited from siloxane and silazane monomers. *Polymer* 40: 5079–94.

Gengenbach, T., X. Xie, R. Chatelier, and H. Griesser. 1994. Evolution of surface composition and topography of perfluorinated polymers following ammonia-plasma treatment. *Journal of Adhesion Science and Technology* 8: 305–28.

Gheorgiu, M., F. Arefi, J. Amouroux, G. Placinta, G. Popa, and M. Tatoulian. 1997. Surface cross linking and functionalization of poly(ethylene terephthalate) in a helium discharge. *Plasma Sources Science and Technology* 6: 8–19.

Gomez, N., M. R. Julia, I. Munoz, M. R. Infante, A. Pinazo, A. Naik, and P. Erra. 1994. Wool treatments with mixtures of sulphite and amphiphilic cationic protein hydrolysate. *Journal of Textile Institute* 85: 215–24.

Hall, J., C. Westerdahl, A. Denne, and M. Bodnar. 1969. Activated gas plasma surface treatment of polymers for adhesive bonding. *Journal of Applied Polymer Science* 13: 2085–96.

Hansen, R., and H. Schonhorn. 1966. A new technique for preparing low surface energy polymers for adhesive bonding. *Polymer Letters* 4: 203–9.

Janca, J., P. Stahel, J. Buchta, D. Subedi, F. Krcma, and J. Pryckova. 2001. A plasma surface treatment of polyester textile fabrics used for reinforcement of car tires. *Plasmas and Polymers* 6(1–2): 15–26

Jeong, B. Y, M. S. Hwang, C. Lee, and M. Kim. 2001. Roughness of the surface layers of plasma oxidation-treated ductile cast iron. *Surface and Coatings Technology* 135: 279–85.

Jiang, P., H. Liu, C. Wang, L. Wu., J. Huang, and C. Guo. 2006. Tensile behavior and morphology of differently degummed silkworm (*Bombyx mori*) cocoon silk fibres. *Materials Letters* 60: 919–25.

Kan, C. W. 2007. Surface morphological study of low temperature plasma treated wool: A time dependence study. In *Modern research and educational topics in microscopy*. Vol. 2, ed. A. Mendez-Vilas and J. Diaz, 683–89. Badajoz, Spain: Formatex.

Kan, C. W., K. Chan, and C. W. M. Yuen. 1998. Surface analysis of plasma treated wool. *Research Journal of Textiles and Apparel* 2(1): 63–73.

Kan, C. W., K. Chan, and C. W. M. Yuen. 2000a. Application of low temperature plasma (LTP) on wool, Part II: Dyeing and felting properties. *The Nucleus* 37(1–2): 22–33.

Kan, C. W., K. Chan, and C. W. M. Yuen. 2000b. Application of low temperature plasma on wool, Part IV: Surface morphological study. *The Nucleus* 37(1–2): 161–72.

Kan, C. W., K. Chan, C. W. M. Yuen, and M. H. Miao. 1997. Physico-chemical study on the surface properties of physically and chemically treated wool fibre. *Journal of Hong Kong Institution of Textile and Apparel* 1(1): 33–47.

Kan, C. W., K. Chan, C. W. M. Yuen, and M. H. Miao. 1998a. Effect of low temperature plasma, chlorination, and polymer treatments and their combination on the properties of wool fibres. *Textile Research Journal* 68: 814–20.

Kan, C. W., K. Chan, C. W. M. Yuen, and M. H. Miao. 1998b. Plasma modification on wool fibre: Effect on the dyeing properties. *Journal of the Society of Dyers and Colourists* 114: 61–65.

Kan, C. W., K. Chan, C. W. M. Yuen, and M. H. Miao. 1998c. A study of the shrinkage properties of wool substrates following plasma and chemical treatments. *Journal of Federation of Asian Professional Association* 4: 23–32.

Kan, C. W., K. Chan, C. W. M. Yuen, and M. H. Miao. 1999. Low temperature plasma on wool substrate: The effect of nature of gas. *Textile Research Journal* 69(6): 407–16.

Kan, C. W., and C. W. M. Yuen. 2007. Plasma technology in wool. *Textile Progress* 39(3): 121–87.

Kan, C. W., and C. W. M. Yuen. 2010. Effect of nature of gas on some surface physico-chemical properties of plasma-treated wool fiber. *Journal of Adhesion Science and Technology* 24(1): 99–111.

Kan, C. W., C. W. M. Yuen, and O. N. Hung. 2013. Improving the pilling property of knitted wool fabric with atmospheric pressure plasma treatment. *Surface and Coatings Technology* 228: S588–92.

Kan, C. W., C. W. M. Yuen, W. Y. I. Tsoi, and T. B. Tang. 2010. Plasma pretreatment for polymer deposition: Improving anti-felting properties of wool. *IEEE Transactions on Plasma Science* 38(6): 1505–11.

Kang, E. T., K. L. Tan, K. Kato, Y. Uyama, and Y. Ikada. 1996. Surface modification and functionalization of polytetrafluoroethylene films. *Macromolecules* 29(21): 6872–79.

Kitazaki, Y., and T. Hata. 1972. Surface-chemical criteria for optimum adhesion. *Journal of Adhesion* 4(2): 123–32.

Komatsu, K. 2003. The formation and structure of silk. In *Chemical and structural characteristics of silk sericin*, ed. N. Hojo, 379–415. Nagano, Japan: Shinkyo Printing.

Lee, K. S., and A. E. Pavlath. 1975. The effect of afterglow, ultraviolet radiation, and heat on wool in an electric glow discharge. *Textile Research Journal* 46: 779–85.

Leung, K. T., M. T. Lo, and K. W. Yeung. 1996. *Knowledge of materials, II*. Hong Kong: Institute of Textiles and Clothing, Hong Kong Polytechnic University.

Li, R., L. Ye, and Y. W. Mai. 1997. Application of plasma technologies in fibre-reinforced polymer composites: A review of recent developments. *Composites Part A: Applied Science and Manufacturing* 28(1): 73–86.

Lindberg, B. J., K. Hamrin, G. Johansson, U. Gelius, A. Fahlman, C. Nordling, and K. Siegbahn. 1970. Molecular spectroscopy by means of ECSA. *Physica Scripta* 1: 286–98.

Liu, Y. C., Y. Xiong, and D. N. Lu. 2006. Surface characteristics and antistatic mechanism of plasma-treated acrylic fibers. *Applied Surface Science* 252: 2960–66

Long, J. J., H. W. Wang, T. Q. Lu, R. C. Tang, and Y. W. Zhu. 2008. Application of low-pressure plasma pretreatment in silk fabric degumming process. *Plasma Chemistry and Plasma Processing* 28: 701–13.

MacLaren, J. A., and A. Kirkpatrick. 1968. Partially oxidised disulphide groups in oxidised wool: Reaction with thiols. *Journal of Society of Dyers and Colourists* 84(11): 564–67.

Matthews, S. R., M. G. McCord, and M. A. Bourham. 2008. Poly(vinyl alcohol) desizing mechanism via atmospheric pressure plasma exposure. *Plasma Processes and Polymers* 2(9): 702–8.

Molina, R., P. Jovancic, F. Comelles, E. Bertran, and P. Erra. 2002. Shrink-resistance and wetting properties of keratin fibres treated by glow discharge. *Journal of Adhesion Science and Technology* 16: 1469–85.

Molina, R., P. Jovancic, D. Jocic, E. Bertran, and P. Erra. 2003. Surface characterization of keratin fibres treated by water vapour plasma. *Surface and Interface Analysis* 35(2): 128–135.

Morent, R., N. De Geyter, J. Verschuren, K. De Clerck, P. Kiekens, and C. Leys. 2008. Non-thermal plasma treatment of textiles. *Surface and Coatings Technology* 202: 3427–49.

Morra, M., E. Occhiello, and F. Garbassi. 1993. The effect of plasma-deposited siloxane coatings on the barrier properties of HDPE. *Journal of Applied Polymer Science* 48(8): 1331–40.

Nakamatsu, J., L. F. Delgado-Aparicio, R. Da Silva, and F. Soberón. 1999. Ageing of plasma-treated poly(tetrafluoroethylene) surfaces. *Journal of Adhesion Science and Technology* 13: 753–61.

Oh, K., S. Kim, and E. Kim. 2001. Improved surface characteristics and the conductivity of polyaniline–nylon 6 fabrics by plasma treatment. *Journal of Applied Polymer Science* 81(3): 684–94.

Pappas, D., A. Bujanda, J. D. Demaree, J. K. Hirvonen, W. Kosik, R. Jensen, and S. McKnight. 2006. Surface modification of polyamide fibers and films using atmospheric plasmas. *Surface and Coatings Technology* 201(7): 4384–88.

Peng, S. J., and Y. P. Qiu. 2009. Size removal of PVA with various moisture content via atmospheric plasma jet. *Journal of Xi'an Polytechnic University* 23(2): 413–17.

Penning, F. M. 1927. Über ionisation durch metastabile Atome. *Naturwissenschaften.* 15(40): 818.

Rakowski, W. 1989. Plasma modification of wool under industrial conditions. *Melliand Textilberichte/International Textile Report* 70: E334–37.

Rakowski, W. 1997. Plasma treatment of wool today, Part I: Fibre properties, spinning and shrinkproofing. *Journal of Society of Dyers and Colourists* 113(9): 250–55.

Ren, Y., Y. Y. Chen, H. Lin, and H. W. Wang. 2004. Weighting properties of mulberry silk after the plasma treatment. *Journal of Textile Research* 25(5): 32–33.

Rossmann, K. 1956. Improvement of bonding properties of polyethylene. *Journal of Polymer Science* 19(91): 141–44.

Roy Choudhury, A. K. 2006. *Textile preparation and dyeing*. Enfield, NH: Science Publishers, Edenbridge.

Schollmeyer, E. 1995. Structuring of polymer surfaces by UV-laser irradiation. *AATCC Book of Papers* 10: 353–58.

Shishoo, R. 2007. *Plasma technologies for textiles.* Cambridge, UK: Woodhead Publishing.

Smiley, R. J., and W. N. Delgass. 1993. AFM, SEM and XPS characterization of PAN-based carbon fibres etched in oxygen plasma. *Journal of Materials Science* 28: 3601–11.

Sprang, N., D. Theirich, and J. Engemann. 1995. *Plasma and ion beam surface treatment of polyethylene. Surface and Coatings Technology* 74–75: 689–95.

Tokino, S., T. Wakida, H. Uchiyama, and M. Lee. 1993. Laundering shrinkage of wool fabric treated with low temperature plasmas under atmospheric pressure. *Journal of Society of Dyers and Colourists* 109: 334–35.

Tusek, L., N. Nitschke, C. Werner, K. Stana-Kleinschek, and V. Ribtisch. 2001. Surface characterisation of NH_3 plasma treated polyamide 6 foils. *Colloids and Surfaces A: Physicochemical and Engineering Aspects* 195: 81–95.

Upadhyay, D., C. Anderson, and N. Brown. 2004. A comparative study of the surface activation of polyamides using an air dielectric barrier discharge. *Colloids and Surfaces A: Physicochemical and Engineering Aspects* 248(1–3): 47–56.

Vallon, S., B. Drévillon, and F. Poncin-Epaillard. 1997. In situ spectro-ellipsometry study of the crosslinking of polypropylene by an argon plasma. *Applied Surface Science* 108: 177–85.

Van der Mei, H. C., I. Stokroos, J. M. Schakenraad, and H. J. Busscher. 1991. Aging effects of repeatedly glow-discharged polyethylene: Influence on contact angle, infrared absorption, elemental surface composition, and surface topography. *Journal of Adhesion Science and Technology* 5: 757–69.

Vohrer, U., M. Müller, and C. Oehr. 1998. Glow-discharge treatment for the modification of textiles. *Surface and Coatings Technology* 98(1–3): 1128–31.

Wakida, T., S. Tokino, S. Niu, H. Kaeamura, Y. Sato, M. Lee, H. Uchiyama, and H. Inagaki. 1993. Surface characteristics of wool and poly(ethylene terephthalate) fabrics and film treated with low temperature plasma under atmospheric pressure. *Textile Research Journal* 63: 433–38.

Wang, H. 2002. Effect of corona discharge treatment on surface property of wool fabric. *Journal of Textile Research* 23(2): 30–32.

Wang, J. H., P. M. Claesson, J. L. Parker, and H. Yasuda. 1994. Dynamic contact angles and contact angle hysteresis of plasma polymers. *Langmuir* 10: 3887–97.

Ward, T. L., and R. R. Benerito. 1982. Modification of cotton by radiofrequency plasma of ammonia. *Textile Research Journal* 52(4): 256–63.

Wenzel, R. N. 1936. Resistance of solid surfaces to wetting by water. *Industrial and Engineering Chemistry*. 28(8): 988–994.

Wertheimer, M. R., A. C. Fozza, and A. Hollander. 1999. Industrial processing of polymers by low-pressure plasmas: The role of VUV radiation. *Nuclear Instruments and Methods in Physics Research Section B: Beam Interactions with Materials and Atoms* 151: 65–75.

Westerdahi, C. A. L., J. R. Hall, E. C. Schramn, and D. W. J. Levi. 1974. Gas plasma effects on polymer surfaces. *Journal of Colloid and Interface Science* 47: 610–20.

Wong, K. K., X. M. Tao, C. W. M. Yuen, and K. W. Yeung. 1999. Low temperature plasma treatment of linen. *Textile Research Journal* 69(11): 846–55.

Wong, K. K., X. M. Tao, C. W. M. Yuen, and K. W. Yeung. 2000. Effect of plasma and subsequent enzymatic treatments of linen fabric. *Journal of Society of Dyers and Colourists* 116: 208–13.

Wong, W., K. Chan, K. W. Yeung, Y. M. Tsang, and K. S. Lau. 2000. Surface structuring of poly(ethylene terephthalate) fibres with a UV excimer laser and low temperature plasma. *Journal of Materials Processing Technology* 103: 225–29.

Xa, D. 1997. An application of plasma technique to ramie fabric pretreatment. *Dyeing and Finishing* 23(4): 10–11, 25.

Xie, X., T. Gengenbach, and H. Griesser. 1992. Changes in wettability with time of plasma-modified perfluorinated polymers. *Journal of Adhesion Science and Technology* 6(12): 1411–31.

Yan, H. J., and W. Y. Guo. 1989. A study on change of fibre structure caused by plasma action. In *Proceedings of the Fourth Annual International Conference of Plasma Chemistry and Technology*, ed. H. V. Boenig, 181–88. Lancaster, PA: Technomic.

Yeung, K. W., K. Chan, Q. Zhang, and S. Y. Wang. 1997. Surface modification of polyester by low temperature plasma treatment and its effect on coloration. *Journal of Hong Kong Institution of Textile and Apparel* 1: 10–17.

Yip, J., K. Chan, K. M. Sin, and K. S. Lau. 2002. Low temperature plasma-treated nylon fabrics. *Journal of Materials Processing Technology* 123(1): 5–12.

Yu, W. D., and H. J. Yan. 1993. Application of plasma etching technique to modification of wool. *Journal of China Textile University (Eng. Ed.)* 10(3): 17–22.

Yuan, L. Y., C. S. Chen, S. S. Shyu, and J. Y. Lai. 1992. Plasma surface treatment on carbon fibers, Part 1: Morphology and surface analysis of plasma etched fibers. *Composites Science and Technology* 45(1): 1–7.

Zisman, W. A. 1964. Relation of the equilibrium contact angle to liquid and solid constitution. *Advances in Chemistry Series*. 43: 1–51.

6 Application of Plasma Treatment in the Dyeing of Textiles

Morent et al. (2008) summarised some previous research studies related to application of plasma treatment in the dyeing of textiles. Plasma treatment modifies or removes the fibre's hydrophobic outer layer, improving dye–fibre interaction and increasing the flux of dye molecules through the fibre surface into the fibre bulk. An improvement of the dyeing properties of textiles has many benefits: (a) increase in dyeing rate, (b) increased dye-bath exhaustion and (c) improved dyeing homogeneity. Improvements in dyeing properties have been reported on all fibre types. Two types of plasma processing can be distinguished for this purpose: depositing plasmas or non-depositing plasmas. In the former, the required functional groups are part of a coating that is plasma polymerised *in situ* at the fibre surface. Non-depositing plasmas (e.g., O_2 plasma) introduce functional groups that interact with an (ionic) dye molecule directly at the fibre polymer surface by chemical reaction, or alter the hydrophobic character of the fibre surface to improve diffusion of (ionic) dye molecules (Morent et al. 2008).

Öktem et al. (1999) introduced carboxylic acid groups at a polyester fibre surface with a 13.56-MHz low-pressure plasma treatment in two ways. In the first case, fabrics were directly treated in an acrylic acid plasma, while in the other case they were first treated in an argon plasma and then immersed in an aqueous acrylic acid bath. The argon plasma creates a large amount of surface radicals that initiate the polymerisation reaction. Dyeing with a basic dye caused the dyeability to increase from 0.34 K/S for the untreated fabric to 0.82 K/S for a 5-min *in situ* plasma polymerization. A value of 1.48 K/S was obtained for a 15-min plasma treatment in argon followed by a 5-min incubation time in an acrylic acid solution. Öktem et al. (2002) also reported on the treatment of polyester/cotton fabrics with acrylic and water plasma, and the treated fabrics showed a much higher dyeability than the untreated polyester/cotton fabrics.

Özdogan et al. (2002) treated scoured and bleached cotton jersey knitted fabric in a 13.56-MHz ethylene diamine (EDA) or triethylene tetramine (TETA) plasma. Another treatment was done in argon, after which the fabric was immersed in a solution of EDA or TETA. The purpose of these treatments was the introduction and protonation of amine functional groups at the fibre surface, resulting in an increasing affinity of anionically charged reactive dye molecules for the cotton fibres. An increase of colour strength K/S between 6.6% and 29.0% for EDA and between 0.3% and 33.9% for TETA was observed, with the best results obtained after a 15-min treatment in a TETA plasma (Morent et al. 2008).

Ferrero et al. (2004) tested the dyeability of polyester (PET), polyamide (PA) and polypropylene (PP) woven fabrics after plasma polymerisation of acrylic acid. The overall colour strength was significantly increased. However, while the wash fastness was acceptable on polyamide, it was unsatisfactory on polyester and polypropylene fabrics, probably due to the lack of penetration of the acrylic acid monomer in the fibre. Scanning electron microscopy and Fourier transform infrared spectroscopy (FTIRS) confirmed that grafting of polyacrylic acid had taken place only on the surface of polyester and polypropylene, but in the case of polyamide the interior of the fibre was also modified. Similar experiments on nylon-6, with a plasma pretreatment and a subsequent acrylic acid or 2-hydroxyethyl methacrylate grafting process, were also reported (Liao et al. 2005).

The impact of a nitrogen plasma treatment on the dyeing properties of wool fabric was studied (El-Zawahry, Ibrahim, and Eid 2006). The non-thermal plasma treatment resulted in etching of the fibre surface, enhancing the hydrophilicity and wettability of the treated wool along with creating and introducing new active sites onto the wool surface. The treatment also improved the dyeing rate and shortened the time to reach dyeing equilibrium. The nitrogen plasma introduced new $-NH_2$ groups onto the wool surface, thereby enhancing the extent of dye exhaustion. Wakida et al. (1993a, 1993b, 1998) showed that atmospheric-pressure plasma treatment was effective for ameliorating the dyeing properties of wool and nylon and that the effectiveness of the treatment regarding dyeing rate strongly depended on the specific dye used.

El-Nagar et al. (2006) performed a low-pressure treatment of PET resulting in significant improvements in water absorbance and dyeability. Recently, Raffaele-Addamo et al. (2006) applied a radio-frequency (RF) plasma in air or argon to PET fibres, and their results showed an increased colour depth upon dyeing. Iriyama et al. (2002) generated low-pressure plasmas in different gases (O_2, N_2 and H_2) for deep dyeing of silk fabrics. All plasma-treated silk fabrics showed increased K/S values, and the best results were obtained with the O_2 plasma. The plasma-treated fabrics dyed in lower dye concentration, as low as 6%, gained greater K/S values compared with untreated ones dyed in a 10% dye solution, which could lead to great saving of dyes. Radetic et al. (2000) did a similar study on knitted wool fabrics with different gases (O_2, air and Ar) and also achieved the best results with an oxygen plasma. The effect of the plasma treatment on the K/S values increased in the following order: argon plasma < air plasma < oxygen plasma. The results on K/S of plasma-treated fabrics were compared with a traditional chlorination process and showed that even short plasma treatments yielded considerable improvements in the printing properties of wool fabric, making plasma technology a viable alternative for chlorination.

Jin and Dai (2003) reported an increased dyeing uptake of nitrogen glow-discharge treated wool. In another study, Jocic et al. (2005) compared the effect of plasma treatment and chitosan treatment on wool dyeing with acid red 27. They observed that wool treated with a RF air plasma had improved dyeing properties.

Another publication described the grafting of $SiCl_x^+$ cations from a 30-kHz low-pressure $SiCl_4$ plasma onto additive-free acetone-extracted PET fibres (Sarmadi, Denes, and Denes 1996). The highly reactive $SiCl_x$ species implanted on the surface are hydrolysed to $Si(OH)_x$ in a post-plasma reaction with atmospheric humidity. Untreated samples showed a 0.4 K/S value. Optimum surface modification was reached after a 1-min plasma treatment, after which K/S values of up to 1.6 were noted following 30 s of dyeing with a basic blue dye.

Polypropylene nonwoven fabrics were activated by an atmospheric-pressure plasma treatment using a surface dielectric barrier discharge in N_2 and ambient air by Cernakova et al. (2005). Subsequently, the plasma-activated samples were grafted using a catalyst-free water solution of acrylic acid. The grafted nonwoven exhibited improved water transport and dyeing properties, and the plasma activation in nitrogen was more efficient than in air.

Covalently bonded coatings were deposited on CCl_4 and acetone-extracted PP fabrics in an argon/acrylonitrile 13.56-MHz cold plasma by Sarmadi, Ying and Denes (1993). The fabrics showed both increased water uptake and increased uptake of acid dye: K/S increased from 0.6 for the untreated fabric up to 4.25 for the treated fabric. Sarmadi and Kwon (1993) studied the O_2 and CF_4 plasma treatment of polyester fabrics. The O_2 plasma improved both the water uptake and the surface dyeability, while CF_4 plasma resulted in excellent water repellency and unexpected improved surface dyeability. They suggested that this behaviour could be due to intense fluorination reactions and simultaneous formation of unsaturated bonds and trapped free radicals. Ageing analysis showed insignificant post-plasma modifications under open laboratory conditions. However, as in many other plasma–textile applications, ageing of the plasma effect often plays a role in dye uptake, as shown by Malek and Holme (2003) for cotton fibres.

Another possible explanation for the positive influence of a plasma treatment on the dyeing kinetics of polyester fibres was suggested by Urbanczyk, Lippsymonowicz and Kowalska (1983). They treated a polyester filament yarn in a 6.5-MHz air plasma at low pressure and determined several dyeing properties. The diffusion coefficient of a disperse dye increased up to 28%. An explanation for the improved dyeing properties was given by considering that the outer molecular layers of polyester fibres are markedly more ordered than the bulk polymer, creating a barrier for molecules diffusing into the fibre. A plasma treatment alters or removes these outer layers, thus increasing diffusion rates of dye and solvent molecules.

In contrast to the previous studies, a plasma treatment can also decrease diffusion rates, as it has been shown that a plasma selectively etches the amorphous domains of the fibre surface polymer structure, thus increasing its crystalline fraction. Because dye molecules progress through the fibres' amorphous polymer fraction, dye rates can be decreased, as was noted when PET and nylon 6.6 were treated in an air or oxygen plasma (Okuno, Yasuda, and Yasuda 1992).

6.1 EFFECT OF PLASMA TREATMENT ON DYEING OF CELLULOSIC FIBRES

Plasma treatment has been shown to be a useful and suitable technique to modify a polymer surface, especially natural polymers like cellulose (Ward et al. 1978, 1979; Pavlath and Slater 1971; Riccobono and Rolden 1973; Stone and Barrett 1962; Jung, Ward, and Benerito 1977) in a dry and pollution-free system (Shishoo 1996). A plasma is, on average, an electrically neutral gas consisting of free particles, in which the potential energy of a typical particle due to its nearest neighbour is much smaller than its kinetic energy (Lieberman and Lichtenberg 1994). A simple way to produce a plasma discharge without the risk of damaging or deforming the solid material is by using strong electric or magnetic fields (Rakowski 1982; Marsh 1987; Venugopalan 1971). This will create highly energetic electrons which can generate new chemically active species of atoms, ions, and free radicals (Benerito et al. 1981). The highly energetic particles produced in the plasma chamber are only able to affect the surface of the polymer being treated (Ward and Benerito 1982).

6.1.1 Dyeing Behaviour of Plasma-Treated Cotton Fibre

Over the past three decades, low-temperature plasma technology has been the focus of much research for improving the surface properties of polymeric materials without changing the bulk properties (Hollahan and Bell 1974; Clark and Feast 1978; H. Yasuda 1981, 1985; d'Agustino 1991). Research has also been carried out to improve wettability, water repellency, soiling, soil release, printing, dyeing and other finishing processes of textile fibres and fabrics by using plasma technology (Byrne and Brown 1972; Wrobel, Kryszewski, and Gazicki 1978; Jahagirdar and Venkatarkrishnan 1990; Okuno, Yasuda, and Yasuda 1992; Sarmadi and Kwon 1993). In most of these studies, two major types of discharges have been considered, i.e., high-frequency discharge (low-pressure plasma) and low-frequency discharge (corona discharge). The most recent type of discharge to be investigated is low-temperature plasma under atmospheric pressure, which was first applied by Wakida et al. (1993a, 1993b).

Plasma treatments for the surface modification of textiles are usually performed by applying two main procedures, i.e., depositing or non-depositing plasmas. In the case of depositing plasmas, the plasma is usually applied by using saturated and unsaturated gases such as fluoro- and hydro-carbons or vapours (monomers) such as acetone, methanol, allylamine and acrylic acid. Several reactive etching (Ar, He, O_2, N_2, F_2) or non-polymerisable gases (H_2O, NH_3) are used in the non-depositing plasmas. Almost all cotton dyes are anionic during the dyeing process, but due to the negative zeta-potential of cotton in water, anionic dyes have a medium-to-poor affinity for cotton (Yang and Liy 1994). Many studies over the last decade have involved the modification of cotton fibres in order to enhance their dyeability (Cardamone et al. 1996; Wu and Chen 1992, 1993a, 1993b; Burkinshaw et al. 1990; Lei and Lewis 1990; Clipson and Roberts 1989, 1994; Kamel, Youssef, and Shokry 1998; Chhagani, Iyer, and Shenai 2000), the basis of which involves changing the charge of the fibre from anionic (in aqueous medium) to cationic. The resulting attraction between the anionic dyestuffs and the fibres therefore increases, resulting in more deeply dyed fibres. As a result, compared to untreated fabrics, the creation of amine groups on the surface of cotton samples means that less dyestuff is required to produce equivalent levels of dyeings (Ozdogan et al. 2002).

Reactive dyes are often used for application with cellulose fibres, as they provide a complete colour range and are easily applied, particularly in exhaust dyeing. However, reactive dyes have only a moderate affinity for cotton fibre. Several previous attempts have been made to overcome this impairment. One of the most feasible methods is to enhance the interaction between the dyestuff and fibre (Hauser 2000). Consequently, since reactive dyes carry an anionic charge, cationically charged cotton should be expected to have a higher affinity for these dyestuffs.

In a study, low-temperature plasma methods were used to modify the cotton fibres prior to dyeing. Plasma refers to a partially ionised gas that consists of ions, electrons and neutral particles (Radetic et al. 2000). Plasma processing is a clean, simple and multifunctional process that meets today's strict economic and ecological demands. As noted in many previous studies, the main advantages of plasma treatments are a shortened processing time and the non-existence of water in the medium (Grzegorz et al. 1983; Öktem et al. 1999, 2000).

Dyeing of desized, scoured, and plasma-treated fabrics was carried out using reactive red dye, direct dye, and also a natural dye (Bhat et al. 2011). In the dyeing studies, the samples of cotton fabrics were exposed to air plasma and then dyed using three different types of dyes: reactive, direct, and natural. The dye absorption was determined spectroscopically by measuring the absorption-band maximum of the dye-bath solution before starting and after exhaustion. In addition, the amount of dye uptake was also determined from the colour-matching instrument. Table 6.1 gives the comparative values of the colour strength of different dyes in cotton fabric.

From the data in Table 6.1, it can be seen that the changes in dye absorption are dependent on the type of dye used. When reactive dye was used, the colour strength increased after the plasma treatment, whereas for the direct dye there was a decrease in the colour strength. In the case of the natural dye, it was found that the colour strength increased after the plasma treatment. Thus it is seen that the type of dye is important (Bhat et al. 2011).

The increase in the dyeing in the initial stage could be due to the fact that plasma etches out the surface of the cotton fibres, removing the waxy layer and creating a rougher surface with irregularities, as discussed previously. This etching of the surface has been confirmed by other researchers (Ozdogan et al. 2009; Karahan et al. 2009). The removal of the waxy layer and the creation of a rougher surface make the effective surface area larger per unit volume, thus facilitating the interaction and diffusion of the dye molecules. In addition, with the prolonged plasma treatment, the glycoside bonds (–O–) may break, creating COOH groups at these sites that make the fibres hydrophilic. The chemical changes in the cotton fibre surface can lead to the possibility of the formation of free radicals on the cellulose chains and the subsequent formation of carbonyl and carboxyl groups (Bhat et al. 2011).

Bhat et al. (2011) discovered that the amount of dye absorbed depended on the time of plasma treatment. For the reactive dye, there are two SO_3^- groups at the benzene ring, and these groups can bond at the OH site by abstraction of H due to

TABLE 6.1
Comparative Study of Dye Uptake of Different Dyes for Cotton Fabric

Type of Dye	Colour Strength (untreated)	Colour Strength (plasma treated)
Reactive dye	100	102.7
Direct dye	100	97.9
Natural dye	100	103.2

Source: Bhat et al. (2011).

plasma interaction. Upon prolonged exposure to plasma, additional sites of –C–O– are created where dye molecules can interact and dyeing is enhanced. Although there is no direct proof of these suggested interactions at the moment, this seems to be one of the possibilities. Additional experiments are planned to verify this aspect.

In the case of direct dye, the mechanism of dyeing is mainly through the diffusion phenomenon. For direct dye, there appears to be a slight decrease in the dye uptake, which can be due to cross-linking on the surface of the cellulose fibre. This will not allow the diffusion of the dye into the fibre structure, and thus the decrease could occur. It is also important to mention that the dyeing process mainly occurs through the amorphous regions. The etching away of the amorphous regions during the plasma treatment can lead to lowering of the dye uptake, which has indeed been observed. This is in conformity with previous studies (Bhat et al. 2011).

For the natural dye, it appears that the chemical bonding with the cellulose chain can occur through the COOH group. Since the chain is long, the diffusion into the fibre may be difficult. From Table 6.1, it appears that the uptake of different types of dyes was different on the same type of plasma-treated cotton fabric. It appears that the fibre–dye pair should be such that it is the fibre and the dye that are compatible with each other to enhance the dye uptake. This is on account of the different types of interacting groups on the dye molecule that will have affinity to the sites on the fibre. For example, if the C–H, C–N, or C–O groups of the cellulose get split during the plasma processing, then the reactive dye can react at these sites. In the case of direct dyes, the main process of dyeing could be the diffusion rather than interaction. Therefore, the colour strength for fabrics dyed with direct dye reduces the control, as diffusion is hindered on account of some cross-linking on the surface, although some reactive sites may be available on the fibre (Bhat et al. 2011).

6.1.1.1 Wicking Rate after Plasma Treatment

The fabric water-absorption property can be evaluated by wicking rate, which is related to the dyeing behaviour. Inbakumar et al. (2010) studied the effect of the plasma process parameter on the wicking rate of cotton fabric.

6.1.1.1.1 Effect of Exposure Time

In the first series of experiments, Inbakumar et al. (2010) kept the operating parameters of pressure and discharge power constant at 0.03 kPa and 4.2 W, respectively, while the exposure time is varied between 1 and 10 min. Experimental results revealed that the evolution of the average wicking height is a function of wicking time for cotton fabrics which were plasma-treated at different exposure times (Inbakumar et al. 2010). The wicking height of the untreated cotton fabric remained the same (i.e., zero) as a function of the wicking time due to its poor wetting nature. This bad wicking behaviour of the scoured untreated cotton fabrics can be explained by the significant amount of non-cellulosic components present on the scoured cellulose fibres (Inbakumar et al. 2010). However, it was shown that initially the wicking height increased quickly with increasing wicking time for plasma-treated samples before levelling off at a certain wicking time. For the upward wicking, it is obvious that if the distance travelled by the liquid becomes high enough, there will be a noticeable effect of gravity on the flow rate, leading to a reduction in the increase of wicking height. A higher wicking rate was achieved with the increase in the plasma exposure time. This observation suggests that with increasing treatment time, the cellulose fibres may become more hydrophilic, since wettability is a prerequisite for wicking: A liquid that does not wet fibres cannot wick into a fabric. Moreover, due to plasma etching effects, the effective pore size present in the plasma-treated cotton fabrics may increase and adversely reduce the capillary pressure, thus increasing the wicking ability (Sun and Stylios 2004; K. Wong et al. 2001).

6.1.1.1.2 Effect of Discharge Power

In order to study the influence of discharge power, Inbakumar et al. (2010) performed a series of experiments at constant operating pressure (0.03 kPa) and constant plasma exposure time (6 min). To easily visualise the effect of discharge power on cotton wettability, it is preferred to present the average wicking height after 5 min of wicking (H_{5min}) as a function of discharge power instead of presenting the wicking height as a function of wicking time for various discharge powers. Figure 6.1 shows the evolution of H_{5min} as a function of discharge power. Based on Figure 6.1, one can assume that increasing the discharge power leads to an enhanced wicking behaviour of the cellulose fibres. This can be explained by the fact that increasing the discharge power results in an increase of the amount of reactive plasma species. The presence of more plasma species can lead to an increased fibre wettability and/or an increased wicking ability due to a more intense bombardment of active plasma species.

6.1.1.1.3 Effect of Operating Pressure

In the final series of experiments done by Inbakumar et al. (2010), the influence of pressure on the hydrophilicity of plasma-treated cotton fabrics was studied in detail in the pressure range 0.03–0.07 kPa. The pressure variation is limited to this small range to ensure that the stability of the DC glow discharge and the discharge power and the plasma exposure time are kept constant at 4.2 W and 6 min, respectively. The experimental results revealed that increasing the plasma pressure

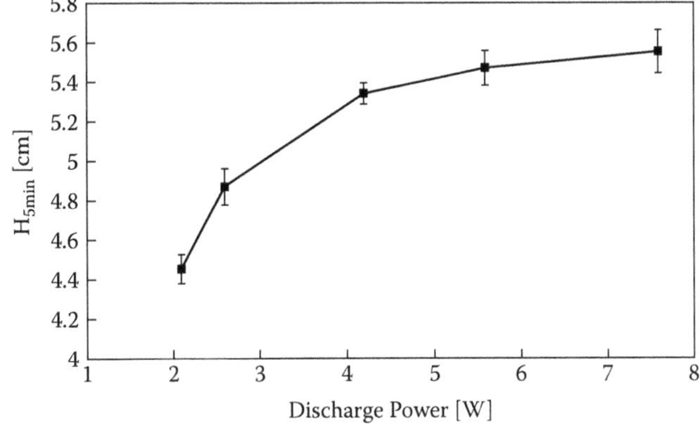

FIGURE 6.1 Average wicking height after 5 min of wicking (H_{5min}) as a function of discharge power (pressure = 0.03 kPa, treatment time = 6 min). (From Inbakumar et al. 2010.)

in the range 0.03–0.07 kPa led to an increase in cotton wicking ability. It is known that while the electron energy (and hence the average energy of the reactive plasma particles) decreases with increasing pressure, its density increases (Shi and Clouet 1992). The observed wicking behaviour is probably a result of these two antagonistic effects; however, since wicking proceeds faster at higher pressures, it is likely that the density increase dominates the energy decrease.

6.1.1.1.4 XPS Results

X-ray photoelectron spectroscopy (XPS) measurements were carried out to evaluate the chemical composition of untreated and plasma-treated cotton fabrics (Inbakumar et al. 2010). Low-resolution XPS survey scans were performed to determine the percentages of elements present at the cotton surfaces before and after plasma treatment. As expected for cellulose fibres, the main elements detected were carbon and oxygen, since hydrogen cannot be detected using XPS. Using area sensitivity factors, the oxygen-to-carbon (O/C) atomic ratios were calculated for the untreated fabric and plasma-treated fabrics at different treatment times, as shown in Table 6.2. The O/C atomic ratio of the untreated cellulose fabric is 0.30, while the value expected for pure cellulose is 0.83 (Inbakumar et al. 2010). This observation indicates that the surface of the untreated cotton fabric does not consist of pure cellulose. The low O/C ratio is due to a considerable amount of carbon atoms without oxygen neighbours, which should not be present in pure cellulose. However, it is well known (Topalovic et al. 2007; Chung, Lee, and Choe 2004; Karahan and Özdogan 2008) that laminar layers of waxes, proteins and pectin cover the natural cotton fibres, and since these layers mainly consist of unoxidised carbon atoms, one can conclude that waxes and proteins are still present on the cellulose fibres, even after NaOH pretreatment. This result is in agreement with the results of Mitchell et al. (2005), who also observed the presence of noncellulosic components on scoured cotton fibres.

Table 6.2 also shows that with increasing plasma exposure time, the O/C atomic ratio gradually increases from 0.30 for an untreated sample to 0.46 for a sample plasma-treated for 10 min. This increase suggests that new oxygen-containing groups are formed on the

TABLE 6.2
O/C Ratio of Untreated and Plasma-Treated Cellulose Fabrics

Treatment Time (min)	O/C Ratio (%)
0	0.30
2	0.38
4	0.40
6	0.42
8	0.43
10	0.46

Source: Inbakumar et al. (2010).
Note: Pressure = 0.03 kPa; discharge power = 4.2 W.

cellulose fabrics after plasma treatment in air. Since these oxygen groups have a polar character, one can conclude that plasma treatment increases the hydrophilicity of the cellulose fabrics, which can contribute to the increased wicking behaviour. It is also important to mention that the plasma modification does not lead to a significant increase in N/C atomic ratio. Therefore, no nitrogen-containing functional groups are introduced on the cotton surfaces during plasma treatment, despite the abundance of nitrogen present in the air plasma. These results underlie the extreme reactivity of oxygen species present in the air plasma compared to the nitrogen species (Inbakumar et al. 2010).

In order to visualise the plasma-induced morphological changes on the fibre surfaces, scanning electron microscopy (SEM) measurements were performed. The untreated cotton fabric consisted of cellulose fibres with a relatively smooth surface, although some grooves and cracks were already present. After plasma treatment, the surface of the cellulose fibres seemed to become rougher, with a significant amount of tiny grooves and cracks. Based on the SEM images, one may conclude that the applied plasma treatment etched the surface of the cotton fabrics. This etching effect may be due to electron and ion bombardment in addition to the contribution of oxidative reactions with activated oxygen atoms to degradation reactions (Inagaki et al. 2004). Taking into account these results, one can conclude that the plasma treatment causes both chemical and physical modifications to the cellulose fibres, which both contribute to the increased wicking behaviour after plasma treatment, which in turn affects the dyeing behaviour (Inbakumar et al. 2010).

6.1.2 Dyeing Behaviour of Plasma-Treated Hemp Fibre

Low-temperature plasma (LTP) treatment of hemp resulted in weight loss of fabric, likely due to severe interaction between different high-energy plasma particles and the fibre surface (i.e., plasma etching) (K. Wong et al. 2000a, 2001). Although the values of weight loss were low (Radetic et al. 2007), it is obvious that prolongation of treatment time caused a slight increase in weight loss. Enzyme (ENZ) treatment resulted in a remarkably higher weight loss of fabric, which was more prominent at higher enzyme concentrations. LTP treatment (10 min) prior to enzyme (5%) application induced weight loss comparable to that of the ENZ-treated sample for the same enzyme concentration (Radetic et al. 2007; K. Wong et al. 2000a, 2001).

The increase in water retention of LTP-treated fabrics can be attributed to plasma etching of the fibre surface. An even higher increase in water retention was evident in the case of ENZ-treated samples, which could be attributed to the removal of non-cellulosic hydrophobic components from the fibre surface (K. Wong et al. 2000a). LTP+ENZ-treated fabric showed a water-retention value lying between the values of LTP- and ENZ-treated samples (Radetic et al. 2007; K. Wong et al. 2000a, 2001).

SEM images of untreated, LTP-treated, ENZ-treated and LTP+ENZ-treated fibres were examined by Radetic et al. (2007). Such a study could be expected, since SEM analysis seems to be a powerful tool for assessing the surface morphology of LTP-treated bast fibres only after longer exposure times. The formation of pits and holes that can reach microscopic dimensions has been noted in the literature. After oxygen or argon plasma treatments of flax (K. Wong et al. 2000b), the existence of pits and cracks likely caused the macroscopic porosity changes in hemp fibre and an increase in specific surface area, making the hemp more accessible to water molecules, i.e., enhancing the water retention of the fibre (Radetic et al. 2007; K. Wong et al. 2000a, 2001).

Dye-exhaustion curves of untreated, LTP-treated, ENZ-treated and LTP+ENZ-treated samples are shown in Figure 6.2 (CI Acid Blue 13, AB113) and Figure 6.3 (CI Direct Red 81, DR81) (Radetic et al. 2007). LTP treatment induced a remarkable increase in the dyeing rate and final dye exhaustion. The longer the LTP treatment time, the higher the final exhaustion was. Dyeing with DR81 is considerably slower as compared to AB113. However, prolongation of LTP treatment time in the case of DR81 showed no significant influence on dye exhaustion, and obviously the dye-exhaustion curves overlapped.

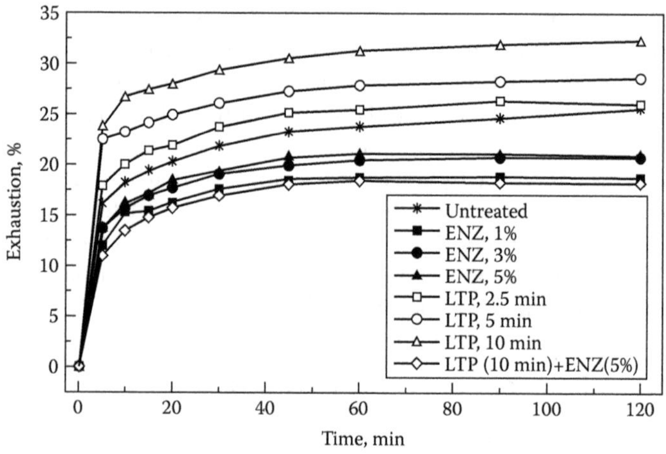

FIGURE 6.2 Dye-exhaustion curves of untreated, LTP- and/or ENZ-treated hemp fabrics (C.I. acid blue 113). (From Radetic et al. 2007.)

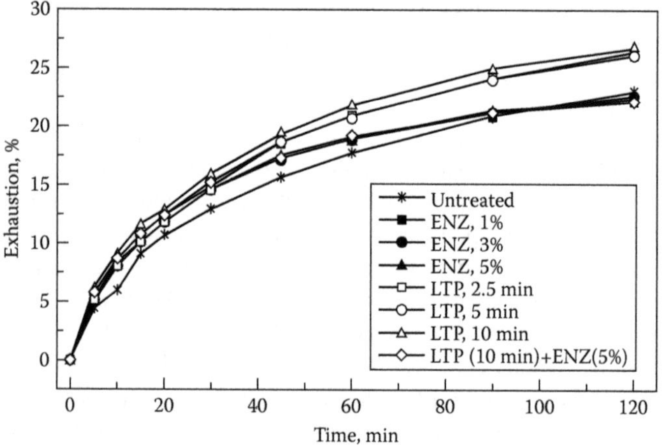

FIGURE 6.3 Dye-exhaustion curves of untreated, LTP- and/or ENZ-treated hemp fabrics (C.I. direct red 81). (From Radetic et al. 2007.)

Enhanced dye exhaustion and higher dyeing rates of LTP-treated samples are attributed to plasma etching and oxidation (K. Wong et al. 2000b). It is likely that etching increased fibre porosity and induced minor topographical changes that make hemp fibre more susceptible to dye and water molecules. Easier diffusion of dye into the fibre caused by LTP treatment is not sufficient for an increase in dye exhaustion, as this is also considerably influenced by the structure, molecular weight and state of the dye in a dyeing bath. DR81 is a dye with high substantivity and rate of diffusion, and thus it is easily bound to the active sites of the fibre. Consequently, it is poorly affected by prolongation of LTP treatment time. In contrast, acid dye exhibits low substantivity, and dyeing is mainly controlled by diffusion, which is remarkably promoted by LTP treatment (Radetic et al. 2007).

ENZ treatment induced a significant decrease in the dyeing rate and final exhaustion of AB113. The initial dyeing rate of DR81 in the case of ENZ-treated hemp was of the same order as the LTP-treated sample. After 45 min of dyeing, ENZ-treated samples exhibited reduced dyeing rates as compared to LTP-treated samples, though they were still higher when compared to untreated hemp. However, final exhaustion of the ENZ-treated sample was slightly lower in comparison with an untreated sample. The decrease in final exhaustion of both the investigated dyes is suggested to be due to the enzymatic hydrolysis of hemp and digestion of amorphous regions that are responsible for fibre dyeability (K. Wong et al. 2000a).

LTP+ENZ-treated hemp showed a dyeing pattern similar to that of the ENZ-treated samples, though the dyeing rate and final exhaustion of both dyes were much lower compared to all other investigated samples. Such behaviour can be attributed to plasma action prior to ENZ treatment. The appearance of cracks and holes makes the fibre surface more accessible to enzymes, promoting easier digestion of amorphous regions (K. Wong et al. 2000a).

The changes in hemp dyeability after LTP, ENZ and LTP+ENZ treatments are also illustrated with K/S values (Table 6.3). The K/S value presents an indirect parameter of the colour yield of the dyed sample, and it is correlated to the dye concentration

TABLE 6.3
Kubelka–Munk Values (K/S) of Dyed Hemp Samples

	K/S	
Sample	**CI Direct Red 81 (DR81)**	**CI Acid Blue 113 (AB113)**
Untreated	3.56	2.90
LTP (2.5 min)	3.74	3.27
LTP (5 min)	3.90	3.18
LTP (10 min)	3.90	3.19
ENZ (1%)	3.28	2.36
ENZ (3%)	3.37	2.71
ENZ (5%)	3.29	3.73
LTP (10 min) + ENZ (5%)	3.36	2.66

Source: Radetic et al. (2007).

TABLE 6.4
Colour Fastness of Dyed Hemp Samples

	Colour Fastness			
	CI Direct Red 81 (DR81)		CI Acid Blue 113 (AB113)	
Sample	**ΔE**	**Grey Scale Grade [a]**	**ΔE**	**Grey Scale Grade [a]**
Untreated	4.35	2–3	16.1	1
LTP (2.5 min)	3.93	3	17.4	1
LTP (5 min)	4.22	2–3	17.1	1
LTP (10 min)	4.89	2–3	16.3	1
ENZ (1%)	3.35	3	15.1	1
ENZ (3%)	3.07	3	14.5	1
ENZ (5%)	4.21	2–3	14.6	1
LTP (10 min) + ENZ (5%)	4.09	3	17.5	1

Source: Radetic et al. (2007).
[a] 1: great change, 2: considerable change, 3: noticeable change, 4: slight change, 5: no change.

on hemp fibre. K/S values are in good correlation with results corresponding to dyeing curves (Radetic et al. 2007).

The colour fastness of dyed hemp samples is presented in Table 6.4 (Radetic et al. 2007). Obviously, there is no significant difference in grey-scale grade between untreated and LTP-treated samples. However, since the cold-dyeing procedure was carried out and a specific type of dye was used, the results of colour fastness, particularly for DR81, could be considered as satisfactory. The better colour fastness of samples dyed with DR81 is likely due to higher substantivity.

6.1.3 Dyeing Behaviour of Plasma-Treated Flax (Linen) Fibre

Flax (linen) was treated with oxygen and argon plasma (K. Wong et al. 2000b). Table 6.5 shows the relative intensities of C1s and O1s, representing the chemical composition percentages of the fibre surface by the O1s/C1s ratio. The chemical states of atoms represented by the relative peak area can be obtained by wave separation of the C1s spectrum. The carbon component is further divided into four subcomponents with peak 1 at 285 eV corresponding to CH, peak 2 at 286.5 eV to CO, peak 3 at 287.9 eV to C=O and peak 4 at 289.1 eV to COOH (Ryu, Wakida, and Takagishi 1991; Sato et al. 1994). The relative peak areas of chemical component percentages are also shown in Table 6.5. The results show that the COOH and CH components increase dramatically on the fibre surface after being treated with either oxygen or argon. The increment of oxidised component by the oxygen plasma is slightly greater than that by the argon plasma. The increment of oxidised and CH components can be explained by the increase of free-radical intensity and the breakage of glucoside bonds to form activated carbonyls on the fibre surface,

TABLE 6.5
Relative Intensities of Surface Atoms and C1s and O1s Spectra of Linen Treated with Oxygen and Argon Plasma Pretreatments

Linen Sample	Relative Intensities of Chemical Composition (%)			Relative Peak Area of Chemical Component (%)			
	C1s	O1s	O1s/C1s	CH	CO	C=O	COOH
Untreated	67.65	32.35	0.48	45.6	40.9	12.4	1.1
Oxygen plasma pretreated	67.32	32.68	0.49	63.4	17.2	10.6	8.8
Argon plasma pretreated	68.85	31.15	0.45	67.1	14.2	9.9	8.8

Source: K. Wong et al. (2000a).

TABLE 6.6
Changes in Whiteness of Flax (Linen) Fabrics after Plasma Pretreatment and Recovery by Washing and Enzyme Treatment

Linen Sample	CIE Whiteness Index
Untreated	68.8
After washing	67.4
After enzyme treatment	66.0
Oxygen plasma pretreatment	46.4
After washing	44.7
After enzyme treatment	58.4
After plasma pretreatment	47.9
After washing	48.6
After enzyme treatment	56.2

Source: K. Wong et al. (2000a).

respectively (Walker and Wilson 1991; Wakida et al. 1989; Ward and Benerito 1982). Argon plasma treatments initiate reactions mainly associated with the cleavage of C1–C2 linkages, leading to the formation of C=O and O–CO–O groups, while oxygen plasma treatments are associated with more intense pyranosidic ring (C–O–C) bond-splitting mechanisms (Hua et al. 1997). The experimental results in Table 6.6 show that the fabric whiteness decreased significantly after the plasma exposure. The reduction of fabric whiteness was mainly attributed to the oxidation effect brought about by the plasma treatment on the fibre surface. However, the sources of fabric yellowing are still unknown, and further investigation is needed (K. Wong et al. 2000a, 2000b).

A significant improvement of fabric water uptake was found after exposure to both types of gas, with the argon plasma pretreatment providing a slightly higher increment. This is similar to the results from a previous study of cotton fabrics treated with plasma (Stone and Barrett 1962). As the weight loss percentage and the changes of fibre surface appearance are not so significant under the treatment conditions, the modification should largely be attributed to the increase of surface hydrophilicity. XPS results indicated that the oxidised components, especially the COOH group, increased significantly, which is in agreement with literature results (Ryu, Wakida, and Takagishi 1991).

Table 6.7 lists the values for the time of half-dyeing, final exhaustion, K/S value, colour change and colour staining for samples treated under various conditions. For the samples dyed with a direct dye of a smaller molecular size (red dye), the dyeing rates do not change significantly, but there is a slight improvement of the final exhaustion percentage. Moreover, the K/S value increases slightly, and these increments may be basically due to the porous surface structure created by the plasma etching and the increased dye adsorption. With regard to the direct dye of a larger molecular size (green dye), the dyeing rates also decrease slightly. However, the final exhaustion percentage is increased in contrast to the reduced K/S value. This may be explained by the fact that the changes of fibre size and shape after the plasma treatment greatly affect the colour reflection of the material dyed with a larger dye molecule. The previous section described how the plasma etches the fibre surface, causing the surface reactivity and fabric water uptake to increase, as demonstrated by the XPS analysis and water uptake test, respectively. Table 6.7 also confirms that the dye-fastness properties of the plasma-treated linen have been improved in most of the cases. The plasma treatments can enhance the interaction of dye molecules with the fibre surface, as reflected by the higher final exhaustion percentage and improved colourfastness (K. Wong et al. 2000a, 2000b).

It is obvious from Table 6.7 that when dyeing the plasma-pretreated samples after subsequent enzyme treatment with a small dye molecule (red dye), the change of dyeing rate is not very significant in all the cases. There is a slight reduction of the final exhaustion value in the plasma-treated samples after the enzyme treatment, but the K/S value does not undergo significant change. However, the dyeing rate, final exhaustion percentage and K/S value of the plasma-pretreated sample are greatly reduced by the enzyme treatment when dyed with a direct dye of larger molecular size (green dye). The susceptibility of the plasma-pretreated linen toward the larger dye molecule is reduced by the enzyme treatment. The effect of enzyme treatment on the dyeing performance is greater for the oxygen plasma-pretreated sample when compared with the argon plasma-pretreated sample. Plasma treatment alone apparently does not change the dyeing rate of linen significantly. The cellulase treatment reduces the dyeing rate, especially when dyeing with the dye of a larger molecule. As the cellulase digests the amorphous regions and reduces the accessibility of the larger dye molecule to the fibre, it is expected that a reduction of final exhaustion percentage and shade depth will occur. The effect is more prominent for the oxygen plasma-pretreated sample. Table 6.7 shows that the enzyme treatment did not significantly affect the colourfastness properties for either the untreated or the plasma-pretreated linen (K. Wong et al. 2000a, 2000b).

TABLE 6.7
Time of Half-Dyeing, Final Exhaustion, K/S Value and Colourfastness to Washing of Untreated, Plasma-Pretreated and Enzyme-Treated Linen Samples Dyed with CI Direct Red 81 and Green 26

	CI Direct Red 81					CI Direct Green 26				
Linen Sample	Half-Time of Dyeing (min)	Final Exhaustion (%)	K/S Value	Colour Change	Colour Staining	Half-Time of Dyeing (min)	Final Exhaustion (%)	K/S Value	Colour Change	Colour Staining
Untreated	3.0	54.4	3.4	3.7	2.2	4.2	63.6	5.1	4.5	4.5
After enzyme treatment	2.9	53.9	3.5	3.5	2.0	5.7	62.7	5.0	4.7	4.8
Oxygen plasma pretreated	2.9	54.9	3.4	4.3	2.3	4.3	66.6	4.9	4.7	4.7
After enzyme treatment	3.0	53.7	3.6	3.8	2.7	8.9	63.0	4.6	4.7	5.0
Argon plasma pretreated	2.9	57.5	3.6	4.3	2.5	4.7	66.5	4.8	4.8	4.8
After enzyme treatment	3.0	54.2	3.6	4.2	2.5	7.3	62.7	4.7	4.8	4.7

Source: K. Wong et al. (2000a).

6.2 EFFECT OF PLASMA TREATMENT ON DYEING OF PROTEIN FIBRES

This section reviews the effect of plasma on the dyeing behaviour of wool, silk and mohair.

6.2.1 Dyeing Behaviour of Plasma-Treated Wool Fibre

Dyeability modification was studied by a number of researchers (Thomas and Hoecker 1995; Kan et al. 1998a, 1998b, 1998c, 1998d; Kan, Chan, and Yuen 2000, 2001). The exhaustion curves of the plasma-treated samples with different plasma gases—namely oxygen (PO), nitrogen (PN) and nitrogen/hydrogen mixture (PM)—and the untreated samples are shown in Figure 6.4. Table 6.8 summarises the results of time of half-dyeing ($t_{1/2}$) and percentage of exhaustion at equilibrium (%E at Em) obtained from Figure 6.4. The percentage exhaustion curve shows the variation of dye-bath concentrations against time, from which the characteristics of a dyeing system, i.e., $t_{1/2}$, %E at Em and the initial rate of dyeing (strike), can be determined.

Figure 6.4 shows that the slopes of the curves representing plasma-treated fibres at the start of dyeing are steeper than that of the untreated fibre, implying that the initial dyeing rate of the plasma-treated samples is faster than the untreated fibre. This phenomenon is probably due to the fact that the diffusion rate of dye molecules becomes relatively faster for the plasma-treated fibre as a result of surface modification. In addition, the time to reach the dyeing equilibrium also become significantly shorter for the plasma-treated samples, i.e., the percentage exhaustion curve starts to flatten at a faster time than the untreated fibre. Of the three different gases used, PN has the greatest effect on the dyeing rate, followed by PO and PM. According to this sequence, it is likely that the nature of the plasma gas used could influence the dyeing behaviour of a dyeing system (Kan et al. 1998b).

A previous study showed that the nature of the plasma played an important role in altering the surface composition of the fibre (Kan, Chan, and Yuen 2000).

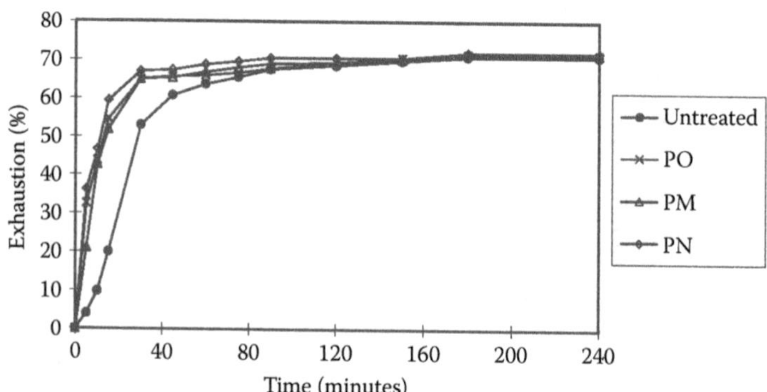

FIGURE 6.4 Percentage dye-bath exhaustion of different samples (acid dye). (From Kan et al. 1998b.)

TABLE 6.8
Time of Half-Dyeing ($t_{1/2}$) and Percentage Exhaustion at Equilibrium (%E at Em) of Control and Plasma-Treated Wool Samples

Sample	$t_{1/2}$ (min)	%E at Em (%)
Untreated	23.43	70.98
PO	6.43 (↓72.56%)	71.56 (↑0.83%)
PN	4.98 (↓78.75%)	72.02 (↑1.47%)
PM	8.39 (↓64.19%)	71.69 (↑1.00%)

Source: Kan et al. (1998b).
Note: The figures in parentheses indicate the increase or decrease in percent when the treated sample is compared with the control (↑ means increase in %, and ↓ means decrease in %).

The changes in surface composition are likely to affect the dyeing behaviour of a dyeing system. When nitrogen gas was used for the plasma treatment, it may have introduced amino groups ($-NH_2$) to the fibre (Kan et al. 1997, 1999). The induced $-NH_2$ groups might then have become the dye sites on the wool fibre, resulting in increased dye absorption. Unlike nitrogen gas, the oxygen plasma will increase the cysteic acid content on the wool fibre surface (Kan et al. 1997, 1999). The cysteic acid groups will facilitate the hydrophilic and wetting character of the wool fibre, thereby enhancing the dyeability of the wool fibre. Although the composition of gas-mixture plasma is quite similar to the nitrogen plasma, different results are still obtained for each kind of plasma. The hydrogen gas in the gas-mixture plasma becomes a very active species under the influence of electrical discharge. This hydrogen species will not only perform an etching effect on the wool fibre, but also have a strong reducing power which can generate free radicals of carbon on the fibre surface during the plasma process. The generated carbon-free radicals on the fibre surface will have the possibility of combining together to form single-bonded carbon chains resulting in the formation of cross-linkages on the fibre surface, which may present a barrier to the dye absorption. Hence, the PM-treated wool fibres show the least effect on the dyeing behaviour in the present dyeing system (Kan et al. 1997, 1999).

Furthermore, similar experimental results shown in Table 6.8 also indicate that the plasma treatment can alter the dyeing rates. The time of half-dyeing ($t_{1/2}$), defined as the time required to reach half-equilibrium, is used as an effective value to quantify the rate of dyeing in a dyeing system. The $t_{1/2}$ values of the plasma-treated wool fabrics were found to be greatly reduced, i.e., more than a 64% decrease when compared with the untreated sample. However, the change of %E at Em is not significant, i.e., ranging merely from 0.83% to 1.47%. This interesting observation may be due to the presence of a number of available dye sites in the wool fibres, which would affect the percentage of exhaustion at equilibrium. The dye sites of the fibre are generally associated with the internal structure of the fibre, so any change of the internal fibre structure can alter the number of dye sites. However, the plasma species can only penetrate to a depth of 0.1 μm (H. Yan and Guo 1989) at the fibre surface within the duration of the treatment time. This penetration of plasma species is not deep enough to alter the whole or partial internal structure of the wool fibre. As a result, most of the available dye sites will remain unchanged after plasma treatment, which has little effect on the final dye-bath exhaustion (Kan et al. 1998a).

The fabric dyeability measurement can be studied through the observation of the colour reflectance curve. The reflectance curves of different plasma-treated fabric samples, namely oxygen plasma (PO), nitrogen plasma (PN) and nitrogen/hydrogen plasma (PM), are shown in Figure 6.5.

The reflectance curve provides information about the depth of shade of the material in the visible spectrum. When the value of reflectance is large, the depth will be a pale shade, and vice versa. Figure 6.5 compares the depth of shade for different samples. It is clear that the position of the reflectance curve of the control sample is higher than that of the plasma-treated samples over the visible spectrum. This indicates that the shade of 1% depth of the untreated fabric is paler than the plasma-treated samples. For three plasma treatments with different gases—namely oxygen, nitrogen and nitrogen/hydrogen gas mixture—their colour reflectance curves nearly coincide with each other, although there are still some differences between them. The oxygen plasma gives the highest colour reflectance of the three plasma treatments, while the colour reflectance curves for both nitrogen plasma and gas-mixture plasma are almost overlapped. These differences may be due to the introduction of new functional groups such as $-NH_2$ on the fibre surface in the case of nitrogen and gas-mixture plasma treatments. In addition, the molecular chains present on the wool fibre surface might be broken into smaller molecules during the plasma etching process. Consequently, these low-molecular-weight materials are likely to be ejected from the fibre surface, leaving polar molecules on the surface and enhancing the dye absorption. Furthermore, the main factor contributing to the improved dyeability of the plasma-treated wool fibre is the apparent increase of the overall surface area, i.e., cracks and gaps (Kan et al. 1998a, 1998b, 1998c, 1998d), as a result of the morphological modification induced by plasma treatment. The increased surface area provides more opportunities for the dye to contact the fibre and thus increases the possibility for the dye to enter the fibre. The dye concentration in the fibre may then be increased, resulting in a deeper shade.

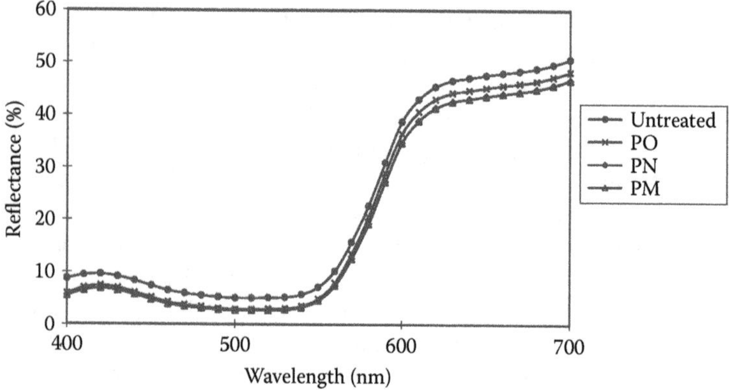

FIGURE 6.5 Colour reflectance curves of each sample. (From Kan et al. 1998a, 1998b, 1998c, 1998d.)

6.2.2 Chrome Dyeing

Microscopic studies elucidate that the plasma treatment can influence the dyeing behaviour of wool fibre (Hocker et al. 1994). Light microscopic studies demonstrate that the plasma-treated wool fibres can be easily penetrated by the dyes, which are evenly distributed over the cross-section of the fibre (Hocker et al. 1994). This phenomenon can be due to both the plasma-induced cystine oxidation in the A-layer of the exocuticle and the reduced number of cross-linkages at the fibre surface. These two changes in surface morphology obviously facilitate a transcellular dye diffusion in addition to the intercellular dye diffusion. Transmission electron microscopy investigation (Hocker et al. 1994) also shows that plasma treatment only modifies the A-layer of the cuticle to various extents due to sputtering, resulting in a partial swelling of the A-layer. In addition, the etching of the A-layer leads to the formation of grooves in this layer. Due to a partial degradation of the A-layer, which acts as a barrier to the diffusion of dyes and other chemicals, the affinity of the fibre for dyes will be increased correspondingly. The increase of dye absorption is most probably caused by the modification of the endocuticle and the neighbouring cell membrane complex, resulting in a modification of the intercellular path of diffusion (Hocker et al. 1994).

Figure 6.6 shows the behaviour of dye-bath exhaustion during the dyeing process with different types of plasma gas, namely oxygen (PO), nitrogen (PN) and nitrogen/hydrogen gas mixture (PM). The results demonstrate that the plasma treatment can influence the dyeing behaviour of wool to different extents. Table 6.9 shows the time of half-dyeing and final dye-bath exhaustion for wool fibre, derived from Figure 6.6. The curves in Figure 6.6 show that there is an increase in dyeing rate for the plasma-treated wool fabrics, but the extent of increase is different. Obviously, the PM-treated fabric shows the fastest rate of dyeing among the fabrics, followed by PN-treated, PO-treated and untreated fabrics (Kan et al. 1998a).

In Table 6.9, the results of time of half-dyeing provide good support for the determination of rate of dyeing. There is a significant change in the time of half-dyeing

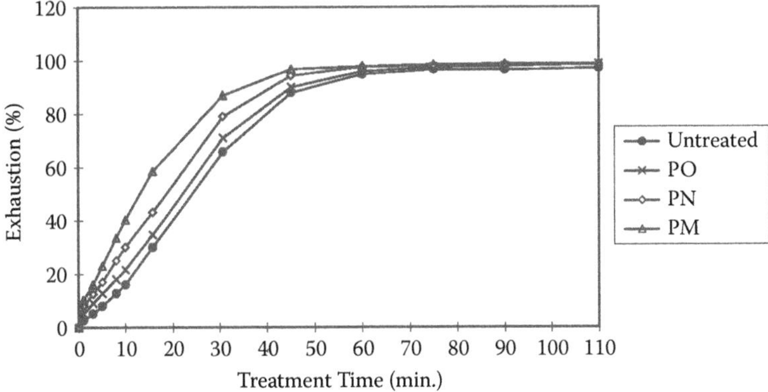

FIGURE 6.6 Dye-bath exhaustion of the plasma-treated wool sample with chrome dye (CI Mordant Black 11, 4% on weight of fabric [owf]) as compared to that of the untreated sample. (From Kan et al. 1998a, 1998b, 1998c, 1998d.)

TABLE 6.9
Time of Half-Dyeing ($t_{1/2}$) and Final Bath Exhaustion for Chrome Dyeing of Wool Fibre

Sample	$t_{1/2}$ (min)	Final Bath Exhaustion (%)
Untreated	22.53	97.16
PO	21.06 (↓6.52%)	98.73 (↑1.62%)
PN	17.34 (↓23.04%)	98.67 (↑1.55%)
PM	12.52 (↓44.43%)	98.91 (↑1.80%)

Source: Kan et al. (1998).
Note: The values in parentheses indicate the increase or decrease in percent when the treated sample is compared with the control (↑ means increase in %, and ↓ means decrease in %).

for all the plasma-treated wool fabrics, i.e., a drop from 6.52% to 44.43% for the plasma-treated fabrics when compared with the untreated wool fibre. If the nature of plasma gas is taken into consideration, the gas mixture shows the greatest effect among the other gases. The reduction in the time of half-dyeing indicates that all plasma treatments can lead to a considerable shortening of dyeing time, thereby reducing the energy consumption and hence improving the dyeing operation. Based on such results, it is suggested that the nature of plasma gas plays an important role in the alteration of the dyeing properties of the plasma-treated wool fibres. The final dye-bath exhaustions shown in Table 6.9 do not show significant changes, i.e., the changes are within 2%, with the greatest increase for the PM-treated fabric, followed by PO and PN treatment. Although PO shows a slower rate of dyeing than PN, the final dye-bath exhaustion is affected in an opposite way, i.e., the PO-treated fabric has better final dye-bath exhaustion than the PN-treated fabric. The final dye-bath exhaustion depends very much on the available dye sites of the wool fibre. Since the plasma species penetrate to a depth of about 0.1 μm (H. Yan and Guo 1989), the depth of penetration and etching are not sufficient to alter the internal structure of fibre or to induce any new dye sites in the fibre. As a result, only a small increase of the final dye-bath exhaustion is observed, and the nature of the plasma gas shows no significant alteration of the final dye-bath exhaustion. Of the three plasma gases used, the gas mixture shows the fastest rate of dyeing, followed by nitrogen and oxygen, but the sequence of nitrogen and oxygen is interchanged in the case of final dye-bath exhaustion (Kan et al. 1998a).

6.2.2.1 Hexavalent Chromium Determination

CI Mordant Black 11 requires a removal of yellowish staining by means of an ammonia after-treatment during which chromium can be extracted from the wool fabrics. This will result in a further effluent load, which can be used for the determination of hexavalent chromium. It is obvious that the hexavalent chromium concentration decreases during the after-chroming process (Kan et al. 1998a). The chromium uptake by the plasma-treated fabrics occurs more rapidly than for the untreated wool fabric. The effect is very similar to the results obtained in the rate of dyeing, i.e., the hexavalent chromium concentration is the smallest in the PM treatment dye bath followed by the PN and PO dye bath. The first effluent sample collected at 5 min after the application of potassium dichromate at 50°C shows that a large amount of the chromium has already been exhausted by the plasma-treated wool fabrics. In contrast, the untreated wool fabric still shows a relatively lower affinity for chromium at the beginning of the after-chroming process. For the first 10 min of the after-chroming process, all the plasma-treated wool fabrics (except for the PM-treated fabric) show a similar rate of chromium exhaustion as compared with the untreated wool fabric. After 10 min of the after-chroming process, it is clearly demonstrated that all the plasma-treated fabrics have a faster rate of chromium exhaustion than the untreated wool fabric. This phenomenon is maintained until the end of the after-chroming process. Although the concentration of hexavalent chromium is dropping during the after-chroming process, the decrease still reaches a state of equilibrium at the end of the after-chroming process (Kan et al. 1998a).

6.2.2.2 Trivalent Chromium Determination

The trivalent chromium concentration absorbed by different plasma-treated wool fabrics indicated an increase in the amount of absorbed trivalent chromium content throughout the after-chroming process when compared with the untreated wool fibre (Kan et al. 1998a). Such results are related closely to the amount of hexavalent chromium exhausted by the wool fibres. Similarly, the fixation of final trivalent chromium is slightly increased for all the plasma-treated wool fabrics, and this phenomenon is similar to the result of final dye-bath exhaustion. The improved trivalent chromium fixation and hexavalent chromium exhaustion will reduce the amount of effluent load discharged to the environment. The experimental results show that the plasma treatment of wool fibres will not only facilitate the uptake of chromium by the fibre, but also reduce its discharge in the effluent. In addition, the use of different plasma gases has a definite influence on the after-chroming process, although the final uptake is slightly increased. Of the three plasma gases used, the gas mixture shows the greatest effect on the after-chroming process, followed by nitrogen and oxygen (Kan et al. 1998a).

6.2.3 Dyeing Behaviour of Plasma-Treated Silk Fibres

In plasma treatments with non-polymer-forming gases, etching of substrates is not negligible. In a general study of plasma treatment of wool and silk fabrics, etching was observed by weight loss of the fabrics and examination of SEM photographs. The degradation rate (weight loss) of the plasma-treated fabric increased with increasing exposure time due to etching of a contaminant layer (Inbakumar and Anu kaliani 2010) that is responsible for the hydrophobicity of the fabric. This etching process is predominant on the amorphous region of the surface rather than the crystalline region (Blais, Carlsson, and Wiles 1972). Therefore, it is possible that the initial rate of the etching is more rapid than supposed. Once all the amorphous materials on the surface have been removed, the remaining crystalline and tightly bound amorphous regions could not be removed easily, causing a decline in the etching rate (Novak, Pollak, and Chodak 2006). After the removal of this waxy layer, the fabric acquires hydrophilicity. Etching may not be limited to just the surface, because the fibre became thinner, and the fibre treated with plasma also showed micro-pitting on the surface (Inbakumar and Anu kaliani 2010).

Plasma treatment increased the dye affinity of the silk fibres, as evidenced by the deepening of the shades of the dyed silk fabric samples (Inbakumar and Anu kaliani 2010).

Silk colour fastness to light showed good resistance to plasma treatment. In contrast, wool colour fastness ranged from 4 to 3.4 after 10 min of plasma treatment and then remained constant upon further treatment. The reduction of colour fastness in wool fibre is probably the result of physical ablation, in which the outer layer is removed, causing the wool fabric to become more vulnerable to UV rays and X-rays. The smooth surface of the silk fabric makes it less vulnerable (Inbakumar and Anu kaliani 2010).

Changes in the surface morphology of silk after air plasma treatment were investigated using SEM by Inbakumar and Anu kaliani (2010). The control sample showed a clean and smooth surface, while slight longitudinal flutes and small pits appeared on the surface of treated silk fibre. This phenomenon was the result of etching by plasma, leading to an increase in surface roughness, which promotes wettability, printability, dyeability and adhesion properties of fabrics. Many researchers have published work on this aspect of plasma treatment (e.g., Nadigar and Bhat 1985).

6.2.4 Dyeing Behaviour of Plasma-Treated Mohair Fibre

Mohair fibres are found in limited regions throughout the world and are produced in small quantities. These factors combine to make the finished products very expensive, and hence mohair fibres are termed *luxury fibres* (Öktem and Atav 2007). In the domain of colour and lustre, mohair fibres have superiority when compared to wool fibres (Öktem and Atav 2007; Atav and Öktem 2006). Mohair is famous for its strength, durability and shine. It is stronger and warmer than wool and is not subject to shrinking or wrinkling (Atav and Yurdakul 2011). Mohair can be used in many items: accessories for hats, scarves, lounging boots, slippers, throws, blankets, carpeting and rugs, wigs, paint rollers, ink-transfer pads, and children's toys (Atav and Yurdakul 2011). As these fibres are also protein based, their dyeing characteristics are similar to those of wool. However, there are some differences between them. It is well known that mohair tends to lose its lustre when dyed for prolonged periods at a boiling temperature. To preserve its lustre, it is generally necessary to use short dyeing cycles or low dyeing temperatures (Atav and Yurdakul 2011). In recent years, many attempts have been made to improve various aspects of dyeing, and new technologies have been (and are still being) developed to reduce fibre damage, decrease energy consumption and increase productivity. Conventional processes, such as chlorination, do not comply with environmental legislation due to the adsorbable organo halogens (AOX) that are generated during processing (Shao, Liua, and Carr 2001). Alternative surface modifications for improving wool dyeability are therefore being explored, one of which is plasma treatment. Plasma technology is an important alternative to wet treatments because there is no water usage: Treatment is carried out in a short-duration gas phase that does not generate industrial waste and saves energy (Atav and Yurdakul 2011).

Results of the experiments carried out to optimise the parameters of plasma treatment are given in Figures 6.7 and 6.8 (Atav and Yurdakul 2011). When Figures 6.7 and 6.8 are examined, it can be understood that colour yields obtained in dyeing increase with an increase in the power and time of plasma treatment. Comparing the gases, it can be said that the best results are obtained with argon gas (Atav and Yurdakul 2011).

For argon gas, the results obtained at 140 W are better. As for the effect of the time, it can be said that increasing the time from 30 to 60 s significantly increased the colour yield of dyeing, but a further increase from 60 to 120 s is not essential. Thus it can be concluded that the optimum conditions of plasma treatment for improving mohair fibre dyeability are treatments carried out using argon gas at 140 W for 60 s (Atav and Yurdakul 2011).

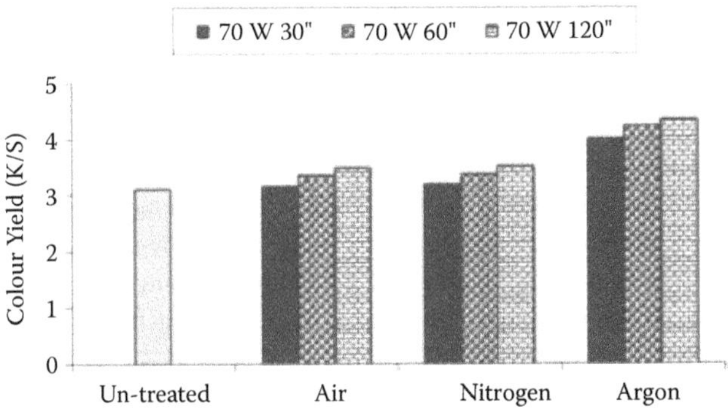

FIGURE 6.7 Effect of gas type and time of plasma treatments carried out at 70 W on the colour yield (K/S at 630 nm) of mohair dyed with 2% owf Telon Blue M-RLW. (From Atav and Yurdakul 2011.)

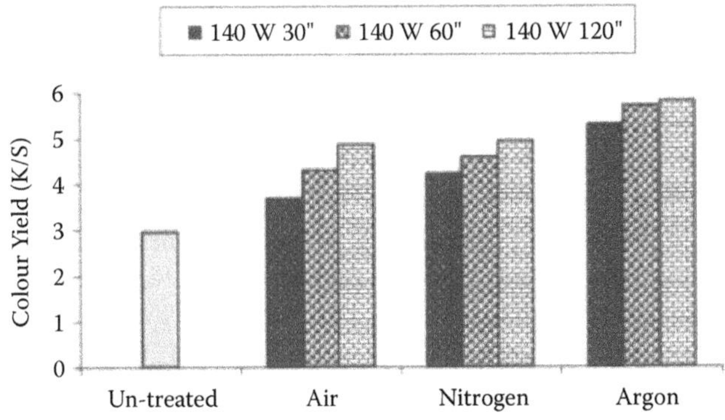

FIGURE 6.8 Effect of gas type and time of plasma treatments carried out at 140 W on the colour yield (K/S at 630 nm) of mohair dyed with 2% owf Telon Blue M-RLW. (From Atav and Yurdakul 2011.)

The increase in colour yield for plasma-treated fibres can be attributed to the oxidation effect brought about by the plasma treatment of the fibre surface, including the creation of polar groups by oxidation and removing surface lipids. The formation of hydrophilic groups in the hydrocarbon chains of the lipid layer and/or its elimination makes the fibre surface even more hydrophilic. Plasma pretreatment more or less removes the surface barrier present on the fibre surface, meaning that dyes can enter the treated fibres more easily in comparison with untreated fibre (Jocic et al. 2005). Furthermore, although the wool fibre itself contained amino groups ($-NH_2$), further introduction of amino groups by plasma treatment may have enhanced the absorption of anionic dye during the dyeing process (Atav and Yurdakul 2011).

Colour yields of plasma-treated and untreated mohair fibres dyed at various temperatures and times were studied by Atav and Yurdakul (2011). Experimental results revealed that the plasma-treated fibres were dyed darker than the untreated ones, but for acid-levelling dye there was no difference between them in colour yield; and also the dyeing temperature was not important. Acid-levelling acid dyes are small molecular dyes that do not need high energy for diffusion, and for this reason they can also be exhausted by fibre at lower temperatures than the boiling point. As a result, it can be suggested that acid dyes of low molecular weight should be preferred for preventing colour yield losses during low-temperature dyeings; however, this will not be a good solution when high wet-fastness values are desired, as these dyes have fairly low wet-fastness properties. However, if low-temperature dyeing can be achieved with high-molecular-weight dyes without causing any decrease in colour yield, it would result in improved dyeing characteristics, such as wet fastness, of the final product (Atav and Yurdakul 2011).

The increase in colour yield for plasma-treated fibres dyed with milling acid, 1:2 metal complex, and reactive dye can be attributed to the oxidation effect brought about by plasma treatment of the fibre surface, including the creation of polar groups by oxidation and removing surface lipids. The formation of hydrophilic groups in the hydrocarbon chains of the lipid layer and/or its elimination makes the fibre surface more hydrophilic. Plasma pretreatment more or less removes the surface barrier present on the fibre surface, meaning that dyes can enter the treated fibres more easily in comparison with untreated fibre (Jocic et al. 2005). Furthermore, although the wool fibre itself contained amino groups ($-NH_2$), further introduction of amino groups by plasma treatment may have enhanced the absorption of anionic dye during the dyeing process (Atav and Yurdakul 2011).

While the colour yields of plasma-treated mohair fibres reactively dyed at 90°C are approximately equal to untreated fibres dyed at a boiling temperature, the colour yields of plasma-treated fibres dyed with milling acid dye or 1:2 metal complex dye at 90°C were lower than untreated fibres dyed at a boiling temperature. The reason for these differences is that even if the fibre structure becomes more diffusible to dyes because of the decomposition of the cuticle layer (Atav and Yurdakul 2011), for large molecular dyes, such as milling acid and 1:2 metal complex, at lower temperatures (such as 90°C), it is difficult to reach and diffuse into the fibres because of the low kinetic energy in the medium. From these results, it can generally be concluded that plasma-treated mohair fibres can be dyed at lower temperatures (90°C) over shorter times (1 h instead of 1.5 h) with reactive dye without causing any decrease in colour yield.

CIE L*a*b* values of the dyed fibres (untreated and plasma treated) were studied (Atav and Yurdakul 2011). If the L* values are examined, it can be seen that the L* values of untreated fibres are higher except for acid-levelling dye. The L* value is that of lightness-darkness, and an increase in L* shows that the colour is getting lighter. Furthermore, the L* values for plasma-treated fibres dyed at 90°C with reactive dye are similar to those for untreated fibres dyed at 100°C, which means that the fibres can be dyed at 90°C with reactive dye without any decrease in colour yield. From this point of view, the results obtained are parallel with the K/S values. Generally speaking, the differences in the a* and b* values of the colour obtained for untreated and plasma-treated fibres are smaller compared to the differences in the L* values. The a* values are higher for the plasma-treated samples, which means the colours are redder. The b* values of the plasma-treated fibres are generally lower compared to those of untreated ones, which means the colours are bluer. If the washing and light fastness values of the dyed samples are taken into consideration, it can be seen that there are not any big differences in washing fastness among the untreated and plasma-treated fibres. However, the light fastness of plasma-treated fibres is higher except for acid-levelling dye, the reason for which is that plasma-treated fibres were dyed darker. Because the dye amount which is damaged by the effect of light is consistent, and if the dyeing shade is higher, the dye percentage damaged will decrease; hence light fastness values will be higher (Atav and Yurdakul 2011).

6.3 EFFECT OF PLASMA TREATMENT ON DYEING SYNTHETIC FIBRES

This section reviews the effect of plasma on the dyeing behaviour of polyester, nylon, acrylic and Tencel fibres.

6.3.1 Dyeing Behaviour of Plasma-Treated Polyester Fibre

Plasma, which contains ions, excited molecules and energetic photons, can be used to modify the surface properties of polymers such as polyester (PET), which is an important material for textile use. However, because of its highly hydrophobic nature, the comfort of PET garments is rather poor. Also, due to the high crystallinity and compact structure, dyeing of the fibre can be quite problematic. Much research (H. Yan and Guo 1987; Wakida et al. 1993a; Bhat and Benjamin 1989) has been conducted to study the changes in the structure, birefringence and wettability of PET after plasma etching.

The effect of plasma treatment on the dyeing properties of the PET fabrics is illustrated in the exhaustion curves of three selected disperse dyes: (a) Dispersol C-2R (low-energy disperse dye), (b) Palanil Blue RM (medium-energy disperse dye) and (c) Celliton Blue FFR (high-energy disperse dye) (Yeung et al. 1997). It was shown that the plasma-treated PET fabrics had a higher equilibrium dye uptake and a faster rate of dyeing. At equilibrium, the dye uptake on the treated PET fabrics, on average, is 10% greater than that of the untreated ones.

In addition, it can be observed from Table 6.10 that the improvement in dyeability (both initial rate and equilibrium exhaustion) is more significant for the high-energy

TABLE 6.10
Initial Rate and Equilibrium Exhaustion of Different Disperse Dyes on PET

	Initial Rate (gradient)			Final Exhaustion (%)		
Type of Disperse Dye	**Plasma-Treated**	**Untreated**	**Increase (%)**	**Plasma-Treated**	**Untreated**	**Increase (%)**
High energy	1.53	1.32	15.9	83	70	18.57
Medium energy	1.20	0.73	64.4	85	81	4.94
Low energy	1.17	0.67	74.6	87	83	4.8

Source: Yeung et al. (1997).

disperse dyes. This phenomenon could probably be due to the comparatively large molecular volumes of the high-energy disperse dyes. After the surface of the PET fabric was etched by plasma treatment, the diffusion of the high-energy dyes into the available dye sites of the fibre would become easier, resulting in a higher initial rate of dyeing and equilibrium dye uptake. After the plasma treatment, it was found that the surface morphological structure was changed, resulting in increased free volume of the polyester during the dyeing process (Yeung et al. 1997). Thus more dyes can enter the plasma-treated polyester, and finally the dye uptake at equilibrium was increased. As the diffusion rate of low- and medium-energy disperse dyes is always higher than that of the high-energy ones, the benefit of plasma treatment in terms of dyeability for the former types of dyes is less significant. It is postulated that such improvement is a direct result of the change of the internal molecular structure of PET, including crystallinity and glass-transition temperature (Yeung et al. 1997).

The crystallinity and glass-transition temperature of the plasma-treated and untreated PET fabrics were examined by Yeung et al. (1997). After plasma treatment, the degree of crystallinity of PET was greatly reduced (by 20%), and the glass-transition temperature was also lowered by 5°C. The reduction in crystallinity would mean a loosening of the compact structure of PET and also an increase of dye sites for disperse dyes. As a result, both the rate of diffusion and equilibrium dye uptake would be improved. On the other hand, the reduction in glass-transition temperature would mean an easier movement of the molecular chain at a low temperature and, hence, an increase in the rate of dyeing (Yeung et al. 1997).

The SEM images shown in Figure 6.9 suggest that plasma treatment is a very effective tool to dramatically increase the roughness of tested samples in comparison to untreated ones, which confirmed our theoretical predictions and results obtained recently on similar polymer systems (Lehocky and Mracek 2006). The surface energy, surface charge, streaming potential, adhesive properties as well as amount of polar groups on the surface were increased by plasma treatment without a change in the bulk properties, which are very important for the subsequent dyeing process and for the mechanical properties of the fibre body (Yeung et al. 1997).

Wettability is a sensitive parameter indicating changes in the surface properties of a polymeric fibre. Wettability is measured in terms of changes in the contact angle

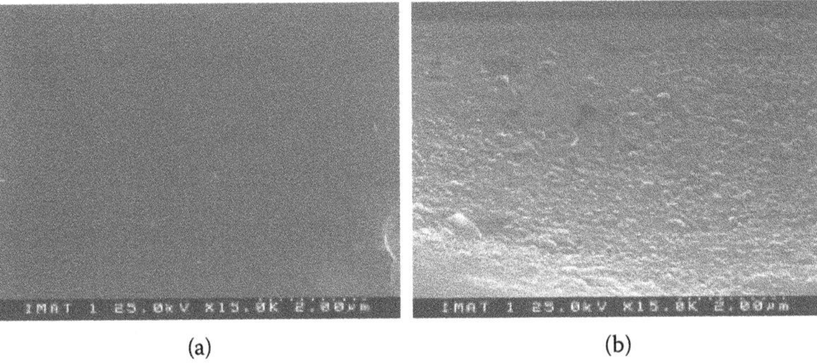

(a) (b)

FIGURE 6.9 SEM images of PET: (a) untreated and (b) plasma-treated. (From Lehocky and Mracek 2006.)

of a liquid on the surface. Untreated PET surfaces are usually hydrophobic, and therefore they have contact angles between 60° and 90° using water as the liquid. In this case, the liquid does not wet the solid, and the liquid drop moves easily on the surface. Complete wetting of the surface is observed when a contact angle of 0° is measured. This is a limiting extreme only in a geometric sense. The effect of the surface treatment was illustrated by comparing the contact angles of water on the surfaces before and after plasma treatment (Lehocky and Mracek 2006). It can be seen that plasma treatment of PET fabric had a strong impact on surface wettability. The initial water contact angle was nearly 80°, which decreased to 18° after the surface treatment. Surface treatment longer than 20 min had only a slight influence on the wetting properties (Lehocky and Mracek 2006).

XPS was used to determine the surface chemical composition of plasma-treated polyester. It is known that air plasma treatment increases the amount of polar oxygen and nitrogen groups. Therefore, the ratios of oxygen content/carbon content (O/C) and nitrogen content/carbon content (N/C) are useful in quantitatively describing chemical surface composition (Lehocky and Mracek 2006). The air plasma treatment produced treatment-time-dependent results, which were quite similar to one another. The changes observed in their relative intensities were a function of treatment time. Rapid initial increase of relative intensities was followed by a less rapid increase and final saturation. A longer treatment process led to conversions of functionalities but not to further incorporation of oxygen or nitrogen. Thus, XPS data confirmed previous wettability results showing that polar hydrophilic groups are present due to the plasma treatment. It was found by visible reflectance spectrometry that the dye adsorption on natural and synthetic polymer fibres treated by low-temperature plasma proceeded much more efficiently, approximately 10%–15% in comparison to the untreated ones. It is important to note that only a relatively short plasma-treatment exposure time was needed, which was a maximum of 5 min. Following a prolonged plasma treatment, the increase of the matrix polarity was no more effective, and the effect on increased dye adsorption was minimal (Lehocky and Mracek 2006).

W. Wong et al. (2000) investigated plasma-treated PET samples and observed only a slight improvement colour fastness and staining to washing over the untreated control. However, the improved dyeability, in terms of faster dyeing and greater dye uptake, due to plasma treatment is also expected to produce a fabric that can withstand frequent laundering.

6.3.2 Dyeing Behaviour of Plasma-Treated Nylon Fibre

The SEM results of control and plasma-treated nylon 6 fibres under different moisture regains (1.23%, 5.19% and 9.70% MR) were examined by Zhu, Wang, and Qiu (2007). The surface of control fibres was very smooth, while a few micro-pits appeared at the surface of plasma-treated samples with 1.23% MR. In comparison, a more roughened surface was observed for the treated 5.19% MR fibres. The most aggressive etching took place at the surface of the 9.70% MR samples after plasma treatment. The surfaces of the nylon fibres were partially peeled off, and only small fragments remained on the surface. The etching of so much material from the surface of a fibre or polymer in such a short time (3 s) under the influence of moisture has not been previously reported in the literature. This is an important finding which could lead to a new method for selective etching of polymer surfaces if a surface of hygroscopic and hydrophilic polymer alloy or block copolymers is etched when the hygroscopic polymer is saturated with water (Zhu, Wang, and Qiu 2007).

Figure 6.10 shows the rate of exhaustion as a function of dyeing time for control and helium plasma-treated fibres (Zhu, Wang, and Qiu 2007). Compared to the control specimens using acid dye, the treated fibres had an overall higher rate of exhaustion. Furthermore, the dyeing of treated fibres needed less time to reach equilibrium. Active species in plasma are known to be capable of breaking primary chemical bonds and induce chain scission (Hancock 1995). The increased amount of amine end groups induced by plasma treatment may play an important role in enhancing the formation of ionic linkages with acid dyes for plasma-treated nylon fibres, thereby speeding up the dye adsorption and increasing the rate of dye exhaustion at equilibrium. In contrast to the 5.19% and 9.70% MR groups, it is obvious that the 1.23% MR group had a higher final rate of exhaustion (Zhu, Wang, and Qiu 2007).

In the case of dispersive groups, both the rate of exhaustion and the total amount of dye adsorption were increased after plasma treatment, as shown in Figure 6.10b. The curves for three treated groups show almost no differences, although the exhaustion for the 1.23% MR group is slightly higher. The increased surface energy for treated fibres may be responsible for the improved main linkages (hydrogen bonds and van der Waals forces) between the dye and polyamide fibres in the surface layer (Ginns, Silkstone, and Nunn 1979). In addition, it is possible that the rougher surface resulting from plasma etching provided a convenient pathway for both acid and dispersive dyes to diffuse into the fibre structure (Kan et al. 1998b, 1998c, 1998d).

Figure 6.11 presents spectral reflectance curves of both acid and dispersive dyes (Zhu, Wang, and Qiu 2007). These curves are almost the same for the control and the treated fibres. This looks contrary to the results obtained in Figure 6.10, since a higher amount of total dye uptake should produce a darker colour. Nevertheless, it should be noted that as a surface modification technique, plasma treatment has a penetration depth of normally less than 500 Å if no plasma ablation occurs on the polymer surface (Li, Ye, and Mai 1997). The available dye sites may be increased

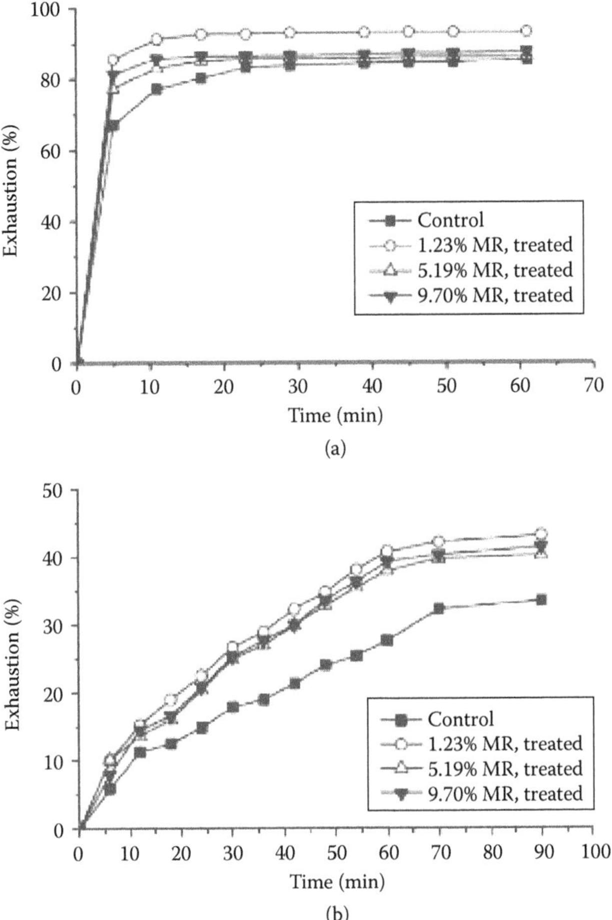

FIGURE 6.10 Dye-exhaustion curves for control and plasma-treated nylon 6 fibres with different moisture regains: (a) acid dye; (b) dispersive dye. (From Zhu, Wang, and Qiu 2007.)

primarily on the fibre surface, while the dye sites related to the internal structure were hardly altered. Therefore, the difference may not be significant enough to make a difference in spectral reflectance curves (Zhu, Wang, and Qiu 2007).

In an effort to gain insight into the interaction between plasma and fibre surface, XPS has been applied to analyze the surface chemical composition of nylon 6 fibres (Zhu, Wang, and Qiu 2007; Borcia, Dumitrascu, and Popa 2005; Yip et al. 2003). The photoelectron peaks at 532, 400 and 285 eV correspond to the O1s, N1s and C1s orbits, respectively. Compared to the control, the intensities of O1s peaks after plasma treatment were raised at the expense of the C1s intensities. The O/C ratios increased dramatically when the fibres with 1.23% and 5.19% MR were exposed to atmospheric-pressure plasma, whereas a slight increase of O/C ratio was found for 9.70% MR fibres after plasma treatment. This finding is consistent with what was

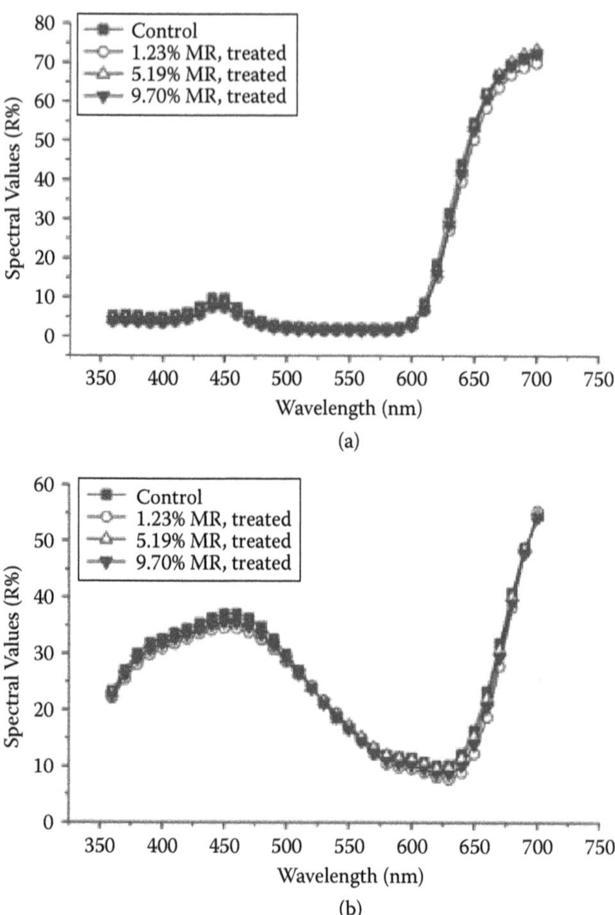

FIGURE 6.11 Spectral reflectance curves for control and plasma-treated nylon 6 fibres with different moisture regains: (a) acid dye; (b) dispersive dye. (From Zhu, Wang, and Qiu 2007.)

reported in the literature (Borcia, Dumitrascu, and Popa 2005; Pappas et al. 2006; Oh, Kim, and Kim 2001; Yip et al. 2004; Upadhyay et al. 2004). The reduced chemical composition change for the 9.70% MR group could be a result of severe etching by the plasma which removed the modified surface quickly, thus leaving a relatively fresh fibre surface with little alteration of its chemical composition. The N/C ratios showed only small fluctuations for different groups after plasma modification (Zhu, Wang, and Qiu 2007).

Deconvolution analyses of Cls peaks were performed and are illustrated in Figure 6.12. The Cls spectra of the control specimen are fitted to four parts: (a) an adventitious carbon (C_{adv}) appeared at a binding energy of 282.4 eV, (b) a major peak at 284.6 eV assigned to aliphatic carbon atoms, (c) a peak at 285.5 eV corresponding to the carbon atoms in C–N bonds and (d) a peak at 287.5 eV arising from the carbon atoms in amide group (–CONH–). The advent of C_{adv} should be induced by sample

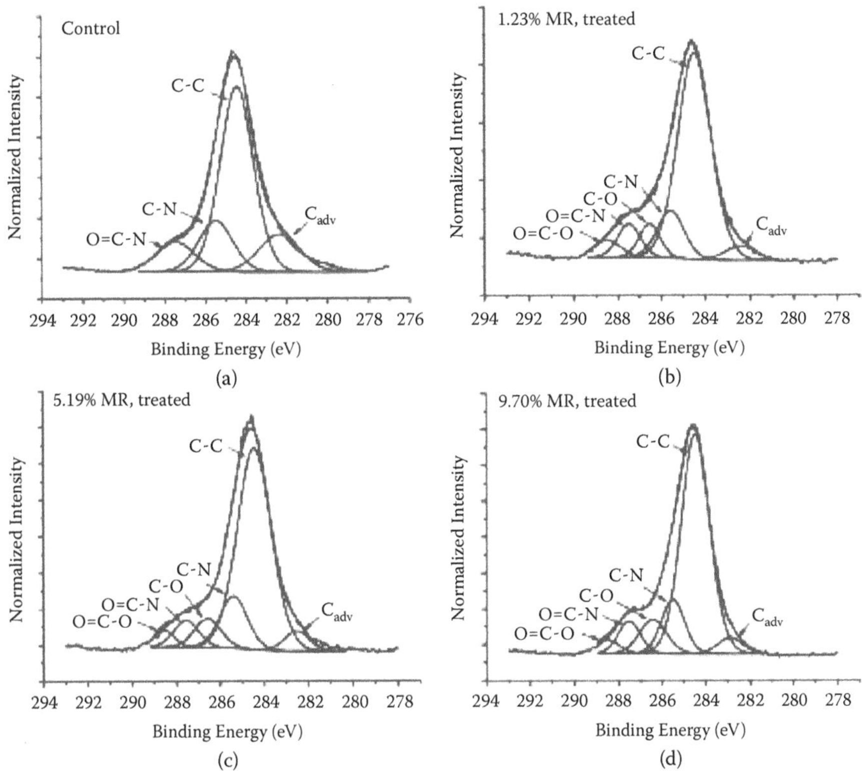

FIGURE 6.12 Deconvoluted XPS C1s core level spectra for nylon fibres: (a) untreated (control), (b) treated with 1.23% MR, (c) treated with 5.19% MR and (d) treated with 9.70% MR. (From Zhu, Wang, and Qiu 2007.)

charging and, therefore, in principle the C_{adv} can be regarded as the chemical shift of aliphatic carbon atoms (Brack et al. 1999; Molina et al. 2003). The C_{adv} was greatly reduced after the plasma treatments, which could result from an increase of the surface electrical conductivity due to more polar groups that were introduced onto the fibre surface. Table 6.11 gives the corresponding contributions of the different carbon-containing chemical bonds. Relative to the repeating unit of nylon 6 molecules, theoretically the ratio of carbon atoms in C–C, C–N and CONH is approximately equivalent to 4:1:1. From the experimental results, on the other hand, the calculated ratio of (C_{adv} + C–C), C–N and CONH in control samples is 6.6:1.5:1, in contrast to the theoretical ratio. This provides evidence that a surface layer of a polymer displays a different property from the bulk. After plasma treatment, it can be seen that the newly oxidised carbon functional groups are imparted to the fibre surface. They are primarily assigned to the C–O bonds and carboxyl groups, referring to the most likely formed oxidised groups in plasma treatment (Allred and Schimpf 1994; Tan, Xiao, and Suib 1991). From the results shown in Table 6.11, for the treated groups, the amount of carbon atoms bonded to oxygen (C–O) in the fibres with 5.19% and 9.70% MR were larger than that in

TABLE 6.11
Data of Deconvolution of C1s Peaks for Nylon 6 Fibres Untreated and Treated with Plasma

	Relative Area Corresponding to Different Chemical Bonds (%)					
Sample	**C_{adv} (282.4 eV)**	**–C–C– (284.6 eV)**	**–C–N– (285.5 eV)**	**–C–O– (286.5 eV)**	**–O=C–NH (287.5 eV)**	**–O=C–OH (288.5 eV)**
Control	14.04	58.48	16.37	0	11.11	0
1.23% MR, treated	3.94	65.79	11.84	6.58	7.24	4.61
5.19% MR, treated	4.55	64.93	13.64	7.14	6.49	3.25
9.70% MR, treated	3.82	63.69	12.74	9.55	7.01	3.19

Source: Zhu, Wang, and Qiu (2007).

the samples with 1.23% MR. This means that probably more hydroxyl groups are present in these two groups (Zhu, Wang, and Qiu 2007). The higher concentration of hydroxyl groups in the 5.19% and 9.70% MR fibres may be attributed to the enhanced interaction between the active particles and the water molecules present in the fibres during the plasma treatment (Zhu, Wang, and Qiu 2007).

After plasma treatment, the advancing contact angles of fibres decreased significantly from 66° to around 54°. Among the three different treatment groups, no significant difference was observed for the advancing contact angles. A significant decrease of about 9°–19° of the receding contact angles was achieved in plasma-treated samples, while that for the 1.23% MR group was lowered less than the other two treated groups. In combination with the analysis of surface morphology and chemical composition of nylon 6 fibres, atmospheric-pressure plasma treatment resulted in considerable changes on the fibre surface, such as increased roughness and more functional groups. It can be seen from XPS analysis that part of the hydrophobic groups in nylon fibre, such as C–C and C–N bonds, are transformed to hydrophilic groups after plasma treatment. The increase in polar functional groups is responsible for the improvement of the wettability of the fibres. In addition, the greater hysteresis between advancing and receding contact angles observed in the treated fibres with higher moisture regain suggests an increased surface roughness (Wiele et al. 2006).

The scanning electron micrographs of untreated and tetrafluoromethane plasma-treated nylon 6 filaments were examined by Yip et al. (2002). Small patches were observed on the filament surface after treatment with the tetrafluoromethane plasma. It is known that tetrafluoromethane is a non-polymerising gas; however, it tends to form thin films on surfaces subjected to glow discharge (T. Yasuda, Gazicki, and Yasuda 1984). The patches are believed to be the formation of such films.

The atomic ratios for untreated and tetrafluoromethane plasma-treated samples were also examined by Yip et al. (2002). The oxygen to carbon ratio decreased after plasma treatment, while the fluorine to carbon and fluorine to oxygen ratios increased from 0.00 to 0.97 and 5.74, respectively. This indicates a substantial incorporation of fluorine atoms into the fabric surface after the tetrafluoromethane plasma treatment. Chemical states of atoms, represented by relative peak areas, can be obtained by the wave separation method. The carbon component of untreated nylon 6 can be divided into four peaks at 285.00, 285.31, 286 and 288.01 eV. These peaks are assigned as C1 (CH_2), C2 ($CH_2C=O$), C3 (CH_2NH) and C4 (NHC=O), respectively (Beamson 1992). After the plasma treatment, these peaks were fluorine-shifted (that is, they had higher binding energies), and their relative intensities were altered. Three additional peaks were evident at 289.10, 290.70 and 292.80 eV, and can be assigned as C5 (CHF), C6 (CF_2) and C7 (CF_3), respectively (T. Yasuda, Gazicki, and Yasuda 1984; Beamson 1992). From the relative peak areas of the chemical components of the treated sample, the relative contents of the CF, CF_2 and CF_3 groups were 4.3%, 18% and 9.87%, respectively. Since CF_2 and CF_3 are non-polar groups, an increase in the number of these groups on the polyamide surface enhances the hydrophobicity of the fabric (Yip et al. 2002).

Nylon 6 contains equal numbers of negatively charged carboxylic acid and positively charged amine end groups. Therefore, either anionic or cationic dyes can be used to dye the material (Ginns, Silkstone, and Nunn 1979). Acid dyes, which are anionic in nature, are the most important group for the coloration of nylon. The dye is initially adsorbed onto the fibre surface, then slowly diffuses into the fibre and forms an ionic bond with the amine end groups. Since nylon 6 has only a small number of amine end groups, it is easily saturated with acid dyes, resulting in poor uniformity. The main linkages between disperse dyes and polyamide are hydrogen bonding and van der Waals forces. Excellent uniformity can be achieved due to the high affinity of disperse dyes (Yip et al. 2002).

Figure 6.13 shows the rate of exhaustion of the dyes on untreated and tetrafluoromethane plasma-treated nylon 6 fabrics (Yip et al. 2002). For the acid dye, the initial rate of exhaustion is much lower for the plasma-treated sample than for the untreated one, and it takes longer to absorb the dye onto the fibre surface. The results can be attributed to the hydrophobic fabric surface caused by the plasma treatment. Acid dyes are hydrophilic and can be absorbed by nylon fabric in a short time under acidic conditions. Tetrafluoromethane treatment produces a hydrophobic fabric surface and thus repels the dye at the initial dyeing stage. However, it is interesting to find that both untreated and plasma-treated fabrics have the same amount of dye exhaustion at equilibrium. The results suggest that tetrafluoromethane plasma treatment does not affect the amount of amine end groups of nylon. Therefore, the resultant electrostatic attraction between acid dyes and the treated material remains unchanged and does not affect the amount of absorption (Yip et al. 2002).

For disperse dyes, the rate of exhaustion and the total amount of dye absorption by tetrafluoromethane plasma-treated fabrics were higher than those of the untreated ones (Figure 6.13b). The results can be explained by the chemical modification caused by tetrafluoromethane plasma treatment. The disperse dye used in this study had a nonionic structure, so it was hydrophobic and less soluble in water. Non-polar CF_2 and

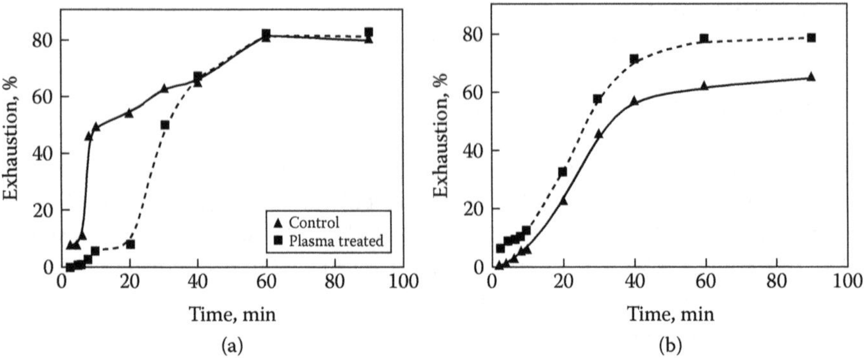

FIGURE 6.13 Dye-bath exhaustion study of untreated and tetrafluoromethane plasma-treated nylon 6 fabric dyed with (a) acid and (b) disperse dyes. (From Yip et al. 2002.)

CF_3 groups are introduced to the fabric surface after tetrafluoromethane plasma treatment. Therefore, the hydrophobic attraction between nylon and the disperse dye was increased, leading to increases in the rate and amount of dye absorption. Spectral reflectance can be used to verify the colour depth of the dyed samples. In the case of the reflectance spectra for both acid- and disperse-dyed samples, for acid-dyed fabrics, there was little change in the reflectance upon treatment with tetrafluoromethane plasma. The spectral reflectance results agree well with those of the dye-bath exhaustion for the appropriate dye. For the disperse-dyed fabrics, the reflectance of the plasma-treated fabrics is slightly lower, as a deeper dyeing was achieved. The increase in depth of shade is explained by the increased amount of dye absorbed due to the chemical changes induced in the fabric surface (Yip et al. 2002).

After CF_4 plasma treatment, the wettability of the undyed nylon fabric decreased considerably (Yip et al. 2002). This result is attributed to the incorporation of non-polar groups onto the fabric surface. Wettability of all the dyed samples increased due to the nature of the dyes, since the dyes used in the study contained hydrophilic groups. However it was found that the pretreated samples retained their hydrophobicity after the dyeing process. This indicates that stable water-repellent surfaces were achieved after the plasma treatment. The CF_4 plasma-treated samples showed a slightly improved fastness to staining during washing when compared with the control, especially for the staining of nylon fabrics (Yip et al. 2002). Although the improvement was small, it is considered that the improved dyeability gained by the CF_4 plasma treatment is permanent.

6.3.3 Dyeing Behaviour of Plasma-Treated Acrylic Fibre

Conventional dyeing of acrylic fibre is based on the use of basic dye. However, with the use of low-temperature plasma, acid dyes can be used for dyeing acrylic fibres (X. Yan, Liu, and Lu 2005). Under the influence of nitrogen plasma at a pressure of 25 Pa and discharge power from 100 to 300 W, experimental results revealed that the moisture regain of acrylic fibre increased gradually with the increment of discharge power. The reason is that, after plasma treatment, the surface area of the acrylic fibre increased, and hydrophilic carboxyl and amide groups were formed on the fibre surface. As a result, when dyeing with acid dyes, the exhaustion rate of acid dye on nitrogen-plasma modified acrylic fibres increased accordingly. On the other hand, with the increment of plasma treatment time from 1 to 5 min, the exhaustion rate of nitrogen plasma-treated acrylic fibre reached a maximum at 3 min and decreased afterwards. This is because prolonged treatment time enhanced the surface etching effect and reduced the amount of hydrophilic groups in the acrylic fibre surface. Therefore, the exhaustion rate no longer increased with prolonged treatment time (Liu, Yan, and Lu 2005).

6.3.4 Dyeing Behaviour of Plasma-Treated Tencel Fibre

The reflectance curves of argon and oxygen plasma-treated Tencel standard fabric samples are shown in Figures 6.14a and 6.14b, respectively (Mak et al. 2006). The results showed that the short treatment duration, i.e., 1 or 2 min, did not give significant improvement on the dyeability of the Tencel fabric; therefore, a 5-min interval was set during the experiment. The dye-bath exhaustion curves of argon and oxygen plasma-treated Tencel fabric samples are shown in Figure 6.15 (Mak et al. 2006). A prolonged treatment time of 60 min was also used to study the effect of prolonged treatment time on the dyeability of Tencel fibre. Experimental results showed that even at prolonged treatment time, the result of dyeability was not as good as that obtained at a short treatment time of 5 min. This may be because the longer the exposure to plasma treatment, the greater was the degradation of the Tencel fabric, with an accompanying breakdown of the reactive dyeing sites on the fabric. Therefore, the longer exposure time of plasma treatment would not be directly proportional to an increase of the dye uptake (Mak et al. 2006).

Small values of reflectance indicated the deep shade of the Tencel dyed fabric. Figures 6.14a and 6.14b show that the position of the reflectance curves of the untreated Tencel was higher than all the plasma-treated samples over the visible spectrum. This indicated that the shade of the untreated Tencel fabric was lighter than all the plasma-treated samples. The plasma treatments on the Tencel fabric (a) induced the generation of more surface area, thereby providing more space for the dye diffusion to occur, and (b) produced more hydroxyl groups on the fibre surface, thereby increasing the number of sites to react with reactive dyes by means of chemical bonding. These changes improved the dye affinity on Tencel fabric in a manner similar to that observed for cotton (Bogaerts et al. 2002). In fact, the chemical structure of Tencel is the same as cotton, but the length of cellulose molecules is different. In terms of degree of polymerization (DP), cotton is 2000, while Tencel is around 600–800. In both fibres, the cellulose molecules form crystalline and non-crystalline regions. The crystallinity in cotton is about 80% and in Tencel it is 60%. The chain orientation is very high in cotton and somewhat lower in Tencel. In addition, plasma treatment of cellulosic fibres leads to an improvement in the wetting properties with water. Hence, an increase in affinity for dyestuff was achieved during the dyeing process (Bhat and Benjamin 1989; T. Yasuda, Gazicki, and Yasuda 1984).

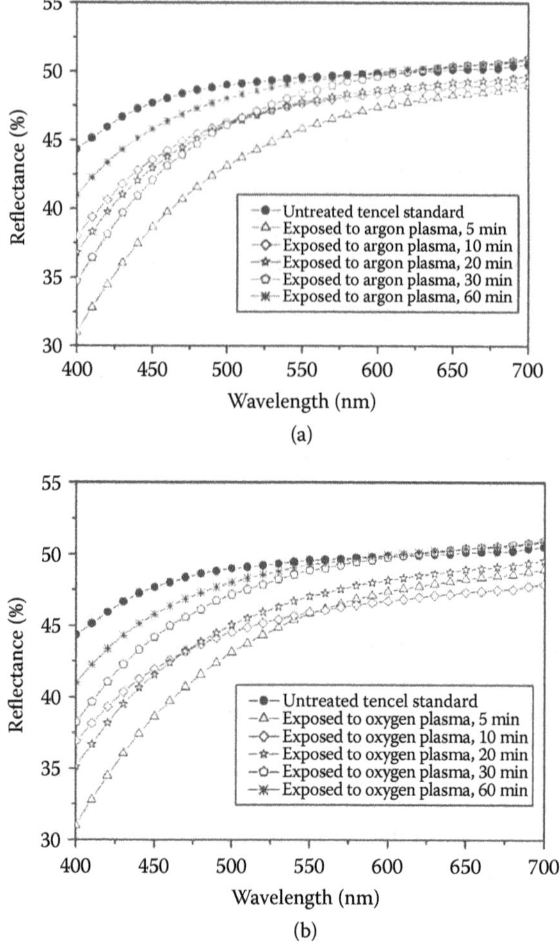

FIGURE 6.14 Reflectance curves of Tencel fabric exposed to different time intervals of plasma treatment: (a) argon plasma treatment; (b) oxygen plasma treatment. (From Mak et al. 2006.)

Another important factor contributing to the improved dyeability of plasma-treated Tencel fabric is the apparent increase in the overall surface area as a result of the morphological modifications induced by the plasma treatment. The increase in surface area provides more opportunity for the dye to contact with the fibre surface and increases the possibility of the dye to enter the fibre, resulting in a deeper shade. According to the reflectance curves of Figures 6.14a and 6.14b, 5-min exposure of Tencel fabric to argon or oxygen plasma obtained the lowest reflectance, i.e., the deepest shade. This confirmed that a short exposure time of 5 min was sufficient to increase the dye shade of Tencel fabric. The dye-bath exhaustion curves for the untreated and plasma-treated Tencel samples are shown in Figure 6.15. Table 6.12 summarises the results of time of half-dyeing ($t_{1/2}$) and percentage of exhaustion at equilibrium obtained from Figure 6.15 (Mak et al. 2006).

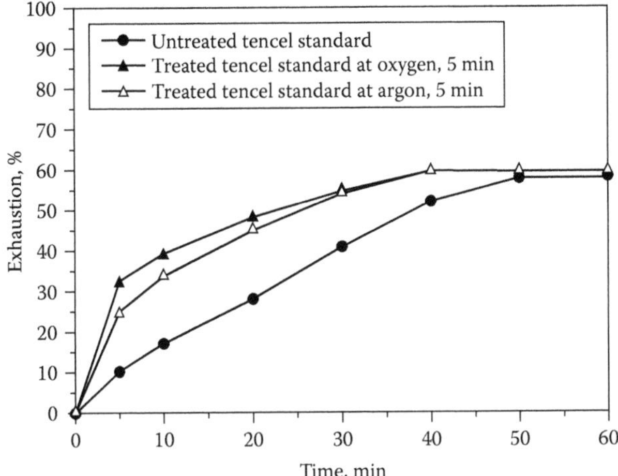

FIGURE 6.15 Dye-bath exhaustion of Tencel fabric treated with different plasma gases for 5 min. (From Mak et al. 2006.)

TABLE 6.12
Time of Half-Dyeing and Final Dye-Bath Exhaustion of Tencel Fabric

Sample	$t_{1/2}$ (min)	Percentage Exhaustion at Equilibrium (%E)
Untreated Tencel standard	21.4	58
Tencel standard treated by oxygen plasma for 5 min	4.61 (↓78.46%)	59 (↑1.72%)
Tencel standard treated by argon plasma for 5 min	7.76 (↓63.74%)	60 (↑3.45%)

Source: Mak et al. (2006).
Note: The figures in parentheses indicate the increase or decrease in percent when the treated sample is compared with the control (↑ means increase in %, and ↓ means decrease in %).

From the percentage–exhaustion curves in Figure 6.15, it is obvious that the slopes of the curves representing plasma-treated fabric at the start of dyeing are steeper than those of the untreated fabric, indicating a high initial dyeing rate. This might be due to dye molecules diffusing relatively quickly into the treated fibres as a result of surface modification. In addition, the time to reach the dyeing equilibrium was also shorter for plasma-treated samples. Finally, the time of half-dyeing was significantly altered. As time of half-dyeing is an indicator for determining the rate of dyeing, a shorter time of half-dyeing may be interpreted as a faster rate of dyeing (Mak et al. 2006).

After the plasma treatment using oxygen and argon, more hydroxyl groups were generated inside the polymer chain of Tencel fibre, as confirmed by the FTIRS spectra near the wave number of 1100 cm^{-1} (Mak et al. 2006). On the basis of this phenomenon, the dye affinity of Tencel fabric was also increased, as more hydroxyl groups of Tencel could bond with the dye molecules by means of increased hydrogen bonding. However, this increase in hydroxyl groups could only assist the dye uptake, as the cellulosic fibres already had lots of hydrogen bonding capacity. Therefore, this increment of hydroxyl groups could only assist the dye-uptake effect, but the increment of surface area due to the plasma treatment was playing the major role in the dye uptake effect. After 5-min plasma treatment using argon or oxygen, the surface of Tencel fabric experienced different extents of damage. The SEM results showed that Tencel experienced various degrees of damage after plasma treatment (Mak et al. 2006). Argon plasma represented the sputtering effect, while oxygen plasma represented the etching effect on the fabric surface, which is a phenomenon similar to that of cotton (Rakowski et al. 1982; J. Yan and Yu 1985). In addition, oxygen plasma caused the micro-deformations of voids and cracks on the fibre surface of Tencel, which might increase fabric weight loss compared to treatment with argon plasma. Morphologically, more surface areas were revealed by the oxygen plasma treatment than those revealed by the argon plasma treatment. As a result, this enlargement of surface area would allow more dyestuffs to approach and adhere to the fibre surface, thus promoting dye absorption (Mak et al. 2006).

6.4 CONCLUSION

This chapter reviewed the effects of plasma treatment on the dyeing properties of commonly used textile fibres. Generally speaking, the dyeing properties were improved due to the modification of the fibre surface. The modification can either be chemical or physical or both. However, depending on the plasma process parameters—such as the nature of the plasma gas used, the treatment time, discharge power, etc.—the extent of surface modification is varied but controllable.

REFERENCES

Allred, R. E., and W. C. Schimpf. 1994. CO_2 plasma modification of high-modulus carbon fibers and their adhesion to epoxy resins. *Journal of Adhesion Science and Technology* 8(4): 383–94.

Atav, R., and T. Öktem. 2006. Structural properties of mohair (angora goat) fibres. *Tekstil ve Konfeksiyon* 16(2): 105–9.

Atav, R., and A. Yurdakul. 2011. Low temperature dyeing of plasma treated luxury fibres, Part I: Results for mohair (angora goat). *Fibres and Textiles in Eastern Europe* 19(2): 84–89.

Beamson, G. 1992. *High resolution XPS of organic polymers: The Scienta ESCA300 database*. Chichester, UK: Wiley.

Benerito, R. R., T. L. Ward, D. M. Soignet, and O. Hinojosa. 1981. Modifications of cotton cellulose surfaces by use of radiofrequency cold plasma and characterisation of surface changes by ESCA. *Textile Research Journal* 51: 224–32.

Bhat, N. V., and Y. N. Benjamin. 1989. Modification of PET by direct current plasma: Effect on fine structure and textile properties. *Indian Journal of Textile Research* 14(1): 1–8.

Bhat, N. V., A. N. Netravali, A. V. Gore, M. P. Sathianarayanan, G. A. Arolkar, and R. R. Deshmukh. 2011. Surface modification of cotton fabrics using plasma technology. *Textile Research Journal.* DOI: 10.1177/0040517510397574.

Blais, P., D. J. Carlsson, and D. M. Wiles. 1972. Surface changes during polypropylene photo-oxidation: A study by infrared spectroscopy and electron microscopy. *Journal of Polymer Science Part A: Polymer Chemistry* 10(4): 1077–92.

Bogaerts, A., E. Neyts, R. Gijbels, and J. V. D. Mullen. 2002. Gas discharge plasma and their applications. *Spectrochimica Acta Part B: Atomic Spectroscopy* 57:609–58.

Borcia, G., N. Dumitrascu, and G. Popa. 2005. Influence of helium-dielectric barrier discharge treatments on the adhesion properties of polyamide-6 surfaces. *Surface and Coatings Technology* 197(2–3): 316–21.

Brack, N., R. N. Lamb, D. Pham, T. Phillips, and P. Turner. 1999. Effect of water at elevated temperatures on the wool fibre surface. *Surface and Interface Analysis* 27(12): 1050–54.

Burkinshaw, S. M., X. P. Lei, D. W. Lewis, J. R. Easton, B. Parton, and D. A. S. Philips. 1990. Modification of cotton to improve its dyeability, Part 2: Pretreating cotton with a thiourea derivative of polyamide–epichlorohydrin resins. *Journal of Society of Dyers and Colourists* 106(10): 307–15.

Byrne, G. A., and K. C. Brown. 1972. Modifications of textiles by glow-discharge reactions. *Journal of Society of Dyers and Colourists* 88(3): 113–17.

Cardamone, M. J., W. N. Marmer, J. Blanchard, A. H. Lambert, and J. B. Brady. 1996. Pretreatment of wool/cotton for union dyeing, Part 1: Resins plus choline chloride. *Textile Chemist and Colorist* 28:19–23.

Cernakova, L., D. Kovacik, A. Zahoranova, M. Cernak, and M. Mazur. 2005. Surface modification of polypropylene non-woven fabrics by atmospheric-pressure plasma activation followed by acrylic acid grafting. *Plasma Chemistry and Plasma Processing* 25(4): 427–37.

Chhagani, R. R., V. Iyer, and V. A. Shenai. 2000. Modifying cotton for dyeing. *Colourage* 47(2): 27–32.

Chung, C., M. Lee, and E. K. Choe. 2004. Characterization of cotton fabric scouring by FT-IR ATR spectroscopy. *Carbohydrate Polymer* 58: 417–20.

Clark, D. T., and W. J. Feast. 1978. *Application of plasmas to the synthesis and surface modifications of polymers*. New York: John Wiley and Sons.

Clipson, J. A., and G. A. F. Roberts. 1989. Differential dyeing cotton, Part 1: Preparation and evaluation of differential dyeing cotton yarn. *Journal of Society of Dyers and Colourists* 105(4): 158–62.

Clipson, J. A., and G. A. F. Roberts. 1994. Differential dyeing cotton. Part 2: Stoichiometry of interaction with acid and direct dyes. *Journal of Society of Dyers and Colourists* 110(2): 69–73.

d'Agustino, R. 1991. *Plasma deposition, treatment and etching of polymers*. New York: Academic Press.

El-Nagar, K., M. A. Saudy, A. I. Eatah, and M. M. Masoud. 2006. DC pseudo plasma discharge treatment of polyester textile surface for disperse dyeing. *Journal of the Textile Institute* 97(2): 111–17.

El-Zawahry, M. M., N. A. Ibrahim, and M. A. Eid. 2006. The impact of nitrogen plasma treatment upon the physical-chemical and dyeing properties of wool fabric. *Polymer-Plastics Technology and Engineering* 45(10): 1123–32.

Ferrero, F., C. Tonin, R. Peila, and F. R. Pollone. 2004. Improving the dyeability of synthetic fabrics with basic dyes using *in situ* plasma polymerisation of acrylic acid. *Coloration Technology* 120(1): 30–34.

Ginns, P., K. Silkstone, and D. M. Nunn. 1979. *The dyeing of synthetic-polymer and acetate fibre*. London: Worshipful Company of Dyers and Society of Dyers and Colourists.

Grzegorz, H., W. Urbanczyk, B. Lipp Symonowicz, and S. T. Kowalska. 1983. Einfluss von Nieder temperatur Plasma anf Feinstructur und Anfarbbarkeit von polyester fasern. *Melliand Textilberichte* 64: 838–40.

Hancock, G. 1995. Diagnostics of active species in plasmas. *Surface and Coatings Technology* 74–75: 10–14.

Hauser, P. J. 2000. Reducing pollution and energy requirements in cotton dyeing. *Textile Chemist and Colorist and American Dyestuff Reporter* 32(6): 44–48.

Hocker, H., H. Thomas, A. Kuster, and J. Herrling. 1994. Dyeing of plasma treated wool. *Melliand Textilberichte/International Textile Report* 75: E131–33.

Hollahan, J. K., and A. T. Bell. 1974. *Techniques and applications of plasma chemistry*. New York: John Wiley and Sons.

Hua, Z. Q., R. Sitaru, F. Denes, and R. A. Young. 1997. Mechanisms of oxygen- and argon-RF-plasma-induced surface chemistry of cellulose. *Plasmas and Polymers* 2(3): 199–224.

Inagaki, N., K. Narushim, N. Tuchida, and K. Miyazaki. 2004. Surface characterization of plasma-modified poly(ethylene terephthalate) film surfaces. *Journal of Polymer Science. Part B: Polymer Physics* 42:3727–40.

Inbakumar, S., and A. Anu kaliani. 2010. Effect of plasma treatment on surface of protein fabrics. *Journal of Physics: Conference Series* 208:012111.

Inbakumar, S., R. Morent, N. De Geyter, T. Desmet, A. Anukaliani, P. Dubruel, and C. Leys. 2010. Chemical and physical analysis of cotton fabrics plasma-treated with a low pressure DC glow discharge. *Cellulose* 17: 417–26.

Iriyama, Y., T. Mochizuki, M. Watanabe, and M. Utada. 2002. Plasma treatment of silk fabrics for better dyeability. *Journal of Photopolymer Science and Technology* 15(2): 299–306.

Jahagirdar, C. J., and S. Venkatarkrishnan. 1990. Antisoiling of polyester (PET) by a novel method of plasma treatments and its evaluation by color measurement. *Journal of Applied Polymer Science* 41(1–2): 117–28.

Jin, J. C., and J. J. Dai. 2003. Dyeing behaviour of nitrogen low-temperature glow discharge treated wool. *Indian Journal of Fibre and Textile Research* 28(4): 477–79.

Jocic, D., S. Vilchez, T. Topalovic, R. Molina, A. Navarro, P. Jovancic, M. R. Julia, and P. Erra. 2005. Effect of low-temperature plasma and chitosan treatment on wool dyeing with acid red 27. *Journal of Applied Polymer Science* 97(6): 2204–14.

Jung, H. Z., T. L. Ward, and R. R. Benerito. 1977. The effect of argon cold plasma on water absorption of cotton. *Textile Research Journal* 47: 217–22.

Kamel, M., B. M. Youssef, and G. M. Shokry. 1998. Dyeing of cationised cotton with acid dyes. *Journal of Society of Dyers and Colourists* 114(3): 101–4.

Kan, C. W., K. Chan, and C. W. M. Yuen. 2000. Application of low temperature plasma (LTP) on wool, Part II: Dyeing and felting properties. *The Nucleus* 37(1–2): 22–33.

Kan, C. W., K. Chan, and C. W. M. Yuen. 2001. The effect of low temperature plasma treated on different types of wool fibre. In *Proceedings of the 6th Asian Textile Conference*. Hong Kong: Hong Kong Institution of Textiles and Clothing.

Kan, C. W., K. Chan, C. W. M. Yuen, and M. H. Miao. 1997. An FTIR spectroscopic study of low temperature plasma treated wool fabric. *Journal of China Textile University (English Edition)* 14(3): 34–40.

Kan, C. W., K. Chan, C. W. M. Yuen, and M. H. Miao. 1998a. The effect of low temperature plasma on the chrome dyeing of wool fibre. *Journal of Materials Processing Technology* 82: 122–26.

Kan, C. W., K. Chan, C. W. M. Yuen, and M. H. Miao. 1998b. Plasma modification on wool fibre: Effect on the dyeing properties. *Journal of the Society of Dyers and Colourists* 114: 61–65.

Kan, C. W., K. Chan, C. W. M. Yuen, and M. H. Miao. 1998c. Effect of low temperature plasma, chlorination, and polymer treatments and their combination on the properties of wool fibres. *Textile Research Journal* 68: 814–20.

Kan, C. W., K. Chan, C. W. M. Yuen, and M. H. Miao. 1998d. Surface properties of low-temperature plasma treated wool fabrics. *Journal of Materials Processing Technology* 83(1–3): 180–84.

Kan, C. W., K. Chan, C. W. M. Yuen, and M. H. Miao. 1999. Low temperature plasma on wool substrate: The effect of nature of gas. *Textile Research Journal* 69(6): 407–16.

Karahan, H. A., and E. Ozdogan. 2008. Improvements of surface functionality of cotton fibres by atmospheric plasma treatment. *Fibres and Polymers* 9: 21–26.

Karahan, H. A., E. Ozdogan, A. Demir, H. Ayhan, and N. Seventekin. 2009. Effects of atmospheric pressure plasma treatments on certain properties of cotton fabrics. *Fibers and Textiles in Eastern Europe* 73: 19–22.

Lehocky, M., and A. Mracek. 2006. Improvement of dye adsorption on synthetic polyester fibers by low temperature plasma pre-treatment, *Czechoslovak Journal of Physics* 56 (Supplement B): B1277–82.

Lei, X. P., and D. M. Lewis. 1990. Modification of cotton to improve its dyeability, Part 3: Polyamide–epichlorohydrin resins and their ethylenediamine reaction products. *Journal of Society of Dyers and Colourists* 106(11): 352–56.

Li, R., L. Ye, and Y. W. Mai. 1997. Application of plasma technologies in fibre-reinforced polymer composites: A review of recent developments. *Composites Part A: Applied Science and Manufacturing* 28(1): 73–86.

Liao, J. D., C. Y. Chen, Y. T. Wu, and C. C. Weng. 2005. Hydrophilic treatment of the dyed nylon-6 fabric using high-density and extensible antenna-coupling microwave plasma system. *Plasma Chemistry and Plasma Processing* 25(3): 255–73.

Lieberman, M. A., and A. J. Lichtenberg. 1994. *Principle of plasma discharges and materials processing*. New York: John Wiley.

Liu, Y. C., X. Yan, and D. N. Lu. 2005. Effects of nitrogen plasma treatment on surface properties of acrylic fiber. *Dyeing and Finishing* 31(1): 8–10.

Mak, C. M., C. W. M. Yuen, S. K. A. Ku, and C. W. Kan. 2006. Low-temperature plasma treatment of Tencel. *Journal of the Textile Institute* 97(6): 533–40.

Malek, R. M. A., and I. Holme. 2003. The effect of plasma treatment on some properties of cotton. *Iranian Polymer Journal* 12(4): 271–80.

Marsh, D. E. 1987. Plasma torch cutting of textiles. *Melliand Textilberichte: International Textile Report* 68: 558–60.

Mitchell, R., C. M. Carr, M. Parfitt, J. C. Vickerman, and C. Jones. 2005. Surface chemical analysis of raw cotton fibres and associated materials. *Cellulose* 12: 629–39.

Molina, R., P. Jovancic, D. Jocic, E. Bertran, and P. Erra. 2003. Surface characterization of keratin fibres treated by water vapour plasma. *Surface and Interface Analysis* 35(2): 128–35.

Morent, R., N. De Geyter, J. Verschuren, K. De Clerck, P. Kiekens, and C. Leys. 2008. Non-thermal plasma treatment of textiles. *Surface and Coatings Technology* 202: 3427–49.

Nadigar, G. S., and N. V. Bhat. 1985. Effect of plasma treatment on the structure and allied textile properties of mulberry silk. *Journal of Applied Polymer Science* 30(10): 4127–35.

Novak, I., V. Pollak, and I. Chodak. 2006. Study of surface properties of polyolefins modified by corona discharge plasma. *Plasma Processes and Polymers* 3(4–5): 355–64.

Oh, K. W., S. H. Kim, and E. A. Kim. 2001. Improved surface characteristics and the conductivity of polyaniline–nylon 6 fabrics by plasma treatment. *Journal of Applied Polymer Science* 81(3): 684–94.

Öktem, T., and R. Atav. 2007. The usage possibilities of plasma technology in the textile usage of plasma technology in natural fibers. *Tekstil ve Konfeksiyon* 17(1): 9–14.

Öktem, T., H. Ayhan, N. Seventekin, and E. Piskin. 1999. Modification of polyester fabrics by *in situ* plasma or post-plasma polymerisation of acrylic acid. *Journal of Society of Dyers and Colourists/Coloration Technology* 115(9): 274–79.

Öktem, T., H. Ayhan, N. Seventekin, and E. Piskin. 2000. Modification of polyester and polyamide fabrics by different *in situ* plasma polymerization methods. *Turkish Journal of Chemistry* 24: 275–85.

Öktem, T., N. Seventekin, H. Ayhan, and E. Piskin. 2002. Improvement in surface-related properties of poly(ethylene terephthalate)/cotton fabrics by glow-discharge treatment. *Indian Journal of Fibre and Textile Research* 27(2): 161–65.

Okuno, T., T. Yasuda, and H. Yasuda. 1992. Effect of crystallinity of PET and nylon 66 fibers on plasma etching and dyeability characteristics. *Textile Research Journal* 62(8): 474–80.

Özdogan, E., A. Demir, H. A. Karahan, H. Ayhan, and N. Seventekin. 2009. Effects of atmospheric plasma on the printability of wool fabrics. *Tekstil ve Konfeksiyon* 19: 123–27.

Özdogan, E., R. Saber, H. Ayhan, and N. Seventekin. 2002. A new approach for dyeability of cotton fabrics by different plasma polymerisation methods. *Coloration Technology* 118(3): 100–3.

Pappas, D., A. Bujanda, J. D. Demaree, J. K. Hirvonen, W. Kosik, R. Jensen, and S. McKnight. 2006. Surface modification of polyamide fibers and films using atmospheric plasmas. *Surface and Coatings Technology* 201(7): 4384–88.

Pavlath, A. E., and R. F. Slater. 1971. Low-temperature plasma chemistry, I: Shrink proofing of wool. *Journal of Applied Polymer Science, Applied Polymer Symposium* 18: 1317–24.

Radetic, M., D. Jocic, P. Jovancic, R. Trajkovic, and Z. L. Petrovic. 2000. The effect of low-temperature plasma pretreatment on wool printing. *Textile Chemist and Colorist and American Dyestuff Reporter* 32(4): 55–60.

Radetic, M., P. Jovancic, D. Jocic, T. Topalovic, N. Puac, and Z. L. J. Petrovic. 2007. The influence of low-temperature plasma and enzymatic treatment on hemp fabric dyeability. *Fibres and Textiles in Eastern Europe* 15(4): 93–96.

Raffaele-Addamo, A., E. Selli, R. Barni, C. Riccardi, F. Orsini, G. Poletti, L. Meda, M. R. Massafra, and B. Marcandalli. 2006. Cold plasma-induced modification of the dyeing properties of poly(ethylene terephthalate) fibers. *Applied Surface Science* 252(6): 2265–75.

Rakowski, W., M. Okoniewski, K. Bartos, and J. Zawadzki. 1982. Plasma treatment of textiles: Potential applications and future prospects, *Melliand Textilberichte: International Textile Report* 63: 301–13.

Riccobono, P. X., and L. Rolden. 1973. Plasma treatment of textiles: A novel approach to the environmental problems of desizing. *Textile Chemist and Colorist* 5: 239–48.

Ryu, J., T. Wakida, and T. Takagishi. 1991. Effect of corona discharge on the surface of wool and its application to printing. *Textile Research Journal* 61(10): 595–601.

Sarmadi, A. M., A. R. Denes, and F. Denes. 1996. Improved dyeing properties of $SiCl_4$ (ST)-plasma treated polyester fabrics. *Textile Chemist and Colorist* 28(6): 17–22.

Sarmadi, A. M., and Y. A. Kwon. 1993. Improved water repellency and surface dyeing of polyester fabrics by plasma treatment. *Textile Chemist and Colorist* 25(12): 33–40.

Sarmadi, A. M., T. H. Ying, and F. Denes. 1993. Surface modification of polypropylene fabrics by acrylonitrile cold plasma. *Textile Research Journal* 43(12): 697–705.

Sato, Y., T. Wakida, S. Tokino, S. Niu, M. Ueda, H. Mizushima, and S. Takekoshi. 1994. Effect of crosslinking agents on water repellency of cotton fabrics treated with fluorocarbon resin. *Textile Research Journal* 64(6): 316–20.

Shao, J., J. Liua, and C. M. Carr. 2001. Investigation into the synergistic effect between UV/ozone exposure and peroxide pad: Batch bleaching on the printability of wool. *Coloration Technology* 117(5): 2–275.

Shi, M. K., and F. Clouet. 1992. Study of the interactions of model polymer surface with cold plasmas, II: Degradation rate versus pressure and gas flow rate. *Journal of Applied Polymer Science* 46: 2063–74.

Shishoo, L. R. 1996. Plasma treatment: Industrial applications and its impact on the C&L industry. *Journal of Industrial Textiles* 26(1): 26–35.

Stone, R. B., Jr., and J. R. Barrett Jr. 1962. U.S.D.A. study reveals interesting effects of gas plasma radiations on cotton yarn. *Textile Bulletin* 88(1): 65–68.

Sun, D., and G. K. Stylios. 2004. Effect of low temperature plasma treatment on the scouring and dyeing of natural fabrics. *Textile Research Journal* 74: 751–56.

Tan, B. J., Y. O. Xiao, and S. L. Suib. 1991. Effect of microwave nitrogen plasma treatment on Nicalon fibers. *Chemistry of Materials* 3(4): 652–60.

Thomas, H., and H. Hoecker. 1995. Plasma treatment of wool with special regard to the dyeing properties. In *Proceedings of the 9th International Wool Textile Research Conference 4*, 351–58. Biella, Italy: Città degli Studi Biella.

Topalovic, T., V. A. Nierstrasz, L. Bautista, D. Jocic, A. Navarro, and M. M. C. G. Warmoeskerken. 2007. XPS and contact angle study of cotton surface oxidation by catalytic bleaching. *Colloids and Surfaces A: Physicochemical and Engineering Aspects* 296:76–85.

Upadhyay, D. J., N. Y. Cui, C. A. Anderson, and N. M. D. Brown. 2004. A comparative study of the surface activation of polyamides using an air dielectric barrier discharge. *Colloids and Surfaces A: Physicochemical and Engineering Aspects* 248(1–3): 47–56.

Urbanczyk, G. W., B. Lippsymonowicz, and S. Kowalska. 1983. Influence of low-temperature plasma on the fibre structure and dyeability of polyester fibres. *Melliand Textilberichte: International Textile Report* 64: 838–40.

Venugopalan, M. 1971. *Reactions under plasma conditions*. Vol. I. New York: John Wiley.

Wakida, T., S. Cho, S. Choi, S. Tokino, and M. Lee. 1998. Effect of low temperature plasma treatment on color of wool and nylon 6 fabrics dyed with natural dyes. *Textile Research Journal*. 68(11): 848–53.

Wakida, T., K. Takeda, I. Tanaka, and T. Takagishi. 1989. Free radicals in cellulose fibers treated with low temperature plasma. *Textile Research Journal* 59(1): 49–53.

Wakida, T., S. Tokino, S. Niu, H. Kawamura, Y. Sato, M. Lee, H. Uchiyama, and H. Inagaki. 1993a. Surface characteristics of wool and poly(ethylene terephthalate) fabrics and film treated with low temperature plasma under atmospheric pressure. *Textile Research Journal* 63(8): 433–38.

Wakida, T., S. Tokino, S. H. Niu, M. Lee, H. Uchiyama, and M. Kaneko. 1993b. Dyeing properties of wool treated with low-temperature plasma under atmospheric pressure. *Textile Research Journal* 63(6): 438–42.

Walker, L. P., and D. B. Wilson. 1991. Enzymatic-hydrolysis of cellulose: An overview. *Bioresource Technology* 36(1): 3–14.

Ward, T. L., and R. R. Benerito. 1982. Modification of cotton by radio frequency plasma of ammonia. *Textile Research Journal* 52: 256–63.

Ward, T. L., H. Z. Jung, O. Hinojosa, and R. R. Benerito. 1978. Effect of RF cold plasmas on polysaccharides. *Surface Science* 76(1): 257–73.

Ward, T. L., H. Z. Jung, O. Hinojosa, and R. R. Benerito. 1979. Characterization and use of RF plasma-activated natural polymers. *Journal of Applied Polymer Science* 23: 1987–2003.

Wiele, L. C. V., M. Ostenson, P. Gatenholm, and A. J. Ragauskas. 2006. Surface modification of cellulosic fibers using dielectric-barrier discharge. *Carbohydrate Polymers* 65(2): 179–84.

Wong, K. K., X. M. Tao, C. W. M. Yuen, and K. W. Yeung. 2000a. Effect of plasma and subsequent enzymatic treatments on linen fabrics. *Journal of Society of Dyers and Colourists* 116: 208–14.

Wong, K. K., X. M. Tao, C. W. M. Yuen, and K. W. Yeung. 2000b. Topographical study of low temperature plasma treated flax fibers. *Textile Research Journal* 70(10): 886–93.

Wong, K. K., X. M. Tao, C. W. M. Yuen, and K. W. Yeung. 2001. Wicking properties of linen treated with low temperature plasma. *Textile Research Journal* 71: 49–56.

Wong, W., K. Chan, K. W. Yeung, and K. S. Lau. 2000. Pulsed UV laser and low temperature plasma modification on polyester microfiber: Effect of dyeing properties. *Journal of Textile Engineering* 46(2): 32–41.

Wrobel, A. M., M. Kryszewski, and M. Gazicki. 1978. Mechanism of polysilazane thin film formation during glow discharge polymerization of hexamethylcyclotrisilazane. *Polymer* 17(8): 673–77.

Wu, T. S., and K. M. Chen. 1992. New cationic agents for improving the dyeability of cellulose fibres, Part 1: Pretreating cotton with polyepichlorohydrin-amine polymers for improving dyeability with direct dyes. *Journal of Society of Dyers and Colourists* 108(9): 388–94.

Wu, T. S., and K. M. Chen. 1993a. New cationic agents for improving the dyeability of cellulose fibres, Part 2: Pretreating cotton with polyepichlorohydrin-amine polymers for improving dyeability with reactive dyes. *Journal of Society of Dyers and Colourists* 109(4): 153–58.

Wu, T. S., and K. M. Chen. 1993b. New cationic agents for improving the dyeability of cellulosic fibres, Part 3: The interaction between direct dyes and polyepichlorohydrin-dimethylamine. *Journal of Society of Dyers and Colourists* 109(11): 365–68.

Yan, H. J., and W. Y. Guo. 1987. Study on change of fibre structures caused by plasma action. *Journal of China Textile University (English Edition)* 4(2): 1–9.

Yan, H. J., and W. Y. Guo. 1989. A study on change of fibre structure caused by plasma action. In *Proceedings of the Fourth Annual International Conference of Plasma Chemistry and Technology*, 181–88. Lancaster, PA: Technomic.

Yan, J., and W. Yu. 1985. Application of plasma etching technique to the investigation of fibre structures. *Journal of East China Institute of Textile Science and Technology* 2: 10–21.

Yan, X., Y. C. Liu, and D. N. Lu. 2005. Dyeing behaviour of low temperature plasma modified acrylic fibres with acid dyes. *Dyeing and Finishing* 31(8): 8–9.

Yang, Y., and S. Li. 1994. An unusual application of a usual crosslinking agent: Dyeing trimethylolmelamine pretreated cotton without added salt. *Textile Research Journal* 64(8): 433–39.

Yasuda, H. 1981. Glow discharges polymerization. *Journal of Polymer Science: Macromolecule Review* 16: 199–293.

Yasuda, H. 1985. *Plasma polymerization*. Orlando, FL: Academic Press.

Yasuda, T., M. Gazicki, and H. Yasuda. 1984. Effects of glow discharges on fibres and fabric. *Journal of Applied Polymer Science: Applied Polymer Symposium* 38: 201–14.

Yeung, K. W., K. Chan, Q. Zhang, and S. Y. Wang. 1997. Surface modification of polyester by low temperature plasma and its effect on coloration. *Journal of Hong Kong Institution of Textile and Apparel* 1(1): 10–17.

Yip, J., K. Chan, K. M. Sin, and K. S. Lau. 2002. Study of physico-chemical surface treatments on dyeing properties of polyamides, Part 1: Effect of tetrafluoromethane low temperature plasma. *Coloration Technology* 118: 26–30.

Yip, J., K. Chan, K. M. Sin, and K. S. Lau. 2003. Study on the surface chemical properties of UV excimer laser irradiated polyamide by XPS, ToF-SIMS and CFM. *Applied Surface Science* 205(1–4): 151–59.

Yip, J., K. Chan, K. M. Sin, and K. S. Lau. 2004. Comprehensive study of polymer fiber surface modifications, Part 2: Low-temperature oxygen-plasma treatment. *Polymer International* 53(6): 634–39.

Zhu, L., C. Wang, and Y. Qiu. 2007. Influence of the amount of absorbed moisture in nylon fibers on atmospheric pressure plasma processing. *Surface and Coatings Technology* 201: 7453–61.

7 Application of Plasma Treatment in the Printing of Textiles

This chapter reviews and discusses the application of plasma treatment and its effect on the printing of cellulosic, protein and synthetic fibres.

7.1 EFFECT OF PLASMA TREATMENT ON PRINTING OF CELLULOSIC FIBRES

In the research of applying plasma treatment on cellulosic fibre, especially in cotton printing, most studies were focused on digital ink-jet printing because the plasma treatment can enhance the application of pretreatment paste coating on the cotton fabric before digital ink-jet printing (Yuen et al. 2004). Figure 7.1 shows that the colour yield of the plasma-treated cotton fabrics subjected to digital ink-jet printing with reactive dye under different plasma treatment time was higher than that of the untreated fabric (Yuen and Kan 2007). The colour yield increased gradually with plasma treatment time and reached the maximum at 2 min of treatment time. Further increase of plasma treatment time did not give further enhancement of the colour yield, but instead resulted in a gradual reduction in colour yield.

Urea is generally used in print paste, as it can swell the cotton fibres during the steaming process, especially superheated steaming, so that dye can penetrate into them rapidly (Achwal 2002; W. Chen, Wang, and Bai 2002). Urea holds some of the water very strongly, and the mixture of urea and water can provide the solvent required for the reaction to occur inside the fibres. Hence, urea acts as a solvent for reactive dyes because it performs like a moisture-absorbing agent in the pretreatment paste to increase moisture regain during steaming, thereby accelerating the migration of dye from the thickener film, i.e., the pretreatment paste containing sodium alginate, into the cotton fibres (W. Chen, Wang, and Bai 2002; Miles 1994).

A reactive dye with vinylsulphone chemical structure was used in a study done by Yuen and Kan (2007), and thus it had a tendency to be deactivated in the presence of urea due to the thermal decomposition of urea to biuret and ammonia. This would result in the conversion of the vinylsulphone dye to an inactive aminoethylsuphone rather than a direct reaction between dye and urea, leading to a decrease in the colour yield of the plasma-treated fabrics. On the other hand, with a longer exposure time under plasma using oxygen gas, the amount of hydrophilic groups present in the cellulosic fibre such as –CHO–, –C=O and –COOH would increase correspondingly (Wong et al. 1999; Inbakumar et al. 2010). The increased amount

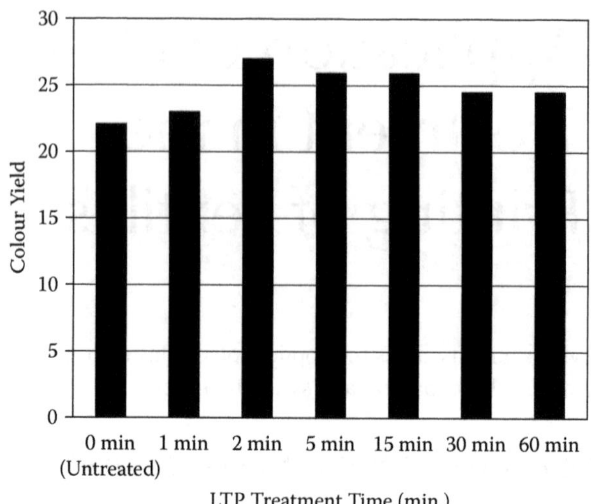

FIGURE 7.1 Colour yield of the ink-jet-printed cotton fabrics with plasma treatment. (From Yuen and Kan 2007.)

of hydrophilic groups coupled with the urea present in the pretreatment paste could hold more moisture during the steaming process. As a result, hydrolysis of reactive dye might occur and reduce the colour yield. In addition, the changes of the hydroxyl groups in the cellulosic fibre into aldehyde and carboxyl groups also decrease the reaction between cellulosic fibre and the reactive dyes, subsequently reducing the colour yield (Yuen and Kan 2007).

Scanning electron microscopy (SEM) was used to examine the surface changes of plasma-treated fabric samples for digital ink-jet printing. It was observed that the untreated cotton fibre surface had a smooth surface, whereas the plasma-treated cotton fibre surface became wrinkled and roughened. The changes in the appearance of the cotton fibre surface might be due to the localised oxygen plasma ablation of the surface layer causing damage at the fibre surface. During the oxygen plasma ablation, the fibre surface was subjected to certain degree of etching. The presence of micropores on the fibre surface indicated this predominant effect of the interaction of oxygen plasma (chemical etching) with the fibre surface. The differential etching of crystalline and amorphous regions might be the origin of the roughness. This process led to an almost complete breakdown of a relatively small number of molecules on the fibre surface into very low-molecular-weight components, which would eventually vaporise in the low-pressure system. As a result, development of cracks was found along the fibre axis (Yasuda, Gazicki, and Yasuda 1984). When untreated cotton fabric surface was padded with pretreatment paste coating, the coating adhered on the fibre in the form of a web. In the case of plasma-treated cotton fabric, the padded pretreatment paste coating had a better surface smoothness when compared with the untreated sample. The surface appearance revealed that during the padding of pretreatment paste, the cracks that had formed on the fibre surface during the plasma treatment were filled up and encased by the pretreatment paste.

In addition, with the formation of cracks on the fibre surface, the surface area of fibre would be increased accordingly, thereby facilitating more dyes to approach the fibre surface and, consequently, to increase the uptake of dye (Yuen and Kan 2007).

The colourfastness results to light, washing, and crocking of the digital ink-jet-printed cotton fabrics with plasma pretreatment were studied by Yuen and Kan (2007). The results showed that the colourfastness of ink-jet-printed cotton fabric with the plasma pretreatment achieved better colourfastness properties than the untreated cotton fabric. Of the different plasma treatment exposure times being studied, it was observed that better maximum colourfastness could be achieved at 2 and 5 min of plasma exposure times when compared with the prolonged exposure times, i.e., 15–60 min. The changes in colourfastness with respect to prolonged plasma exposure time might be due to the improvement of moisture-absorption properties of cotton fabric, i.e., enhancement of hydrophilic groups in the cotton fibre (Wong et al. 1999), resulting in more moisture being absorbed during the steam fixation process. This would eventually cause the hydrolysis of reactive dye in the printing ink, leading to a slight reduction in the colourfastness results. Although the colourfastness properties of 2- and 5-min plasma-treated fabrics were the same, the colour yield of the 2-min plasma-treated fabric was better than that of the fabric treated for 5 min (Yuen and Kan 2007).

The outline sharpness of the ink-jet-printed pattern was measured by optical analysis (Yuen and Kan 2007). Obviously, the ink-jet-printed patterns in the warp direction were thicker than those in the weft direction for both the untreated and plasma-treated fabrics. This might be due to the differential wicking effect caused by the warp and weft yarns. When comparing the width of the printed patterns, the patterns printed on the plasma-treated cotton fabrics were narrower than the untreated fabric in both the warp and weft directions. This could be attributed to the reduced spreading of the printed reactive inks as a result of the strong fibre and dye attraction, i.e., formation of covalent bonding between the hydroxyl group of fibre (Wong et al. 1999) and the reacting system of the reactive dye present in the printing ink. Consequently, the plasma treatment on cotton fabric could enhance the outline sharpness of the ink-jet prints.

7.2 EFFECT OF PLASMA TREATMENT ON PRINTING OF PROTEIN FIBRES

7.2.1 Printing Properties of Plasma-Treated Wool Fabrics

Figures 7.2 and 7.3 show the Kubelka–Munk (K/S) values of plasma-treated (air or argon) and untreated wool fabrics printed with and without wetting agents (Özdogan et al. 2009). It can be seen from the K/S values that atmospheric plasma treatment had a positive effect on the printability of wool fabrics. There are mainly two reasons for this: Firstly, atmospheric plasma treatment increases wettability of wool fabric, which increases penetration of any material like printing paste (Demir et al. 2008; Ryu, Wakida, and Takagishi 1991; Radetic, Jocic, et al. 2000; Poletti et al. 2003). Secondly, it decreased the fibre fuzziness, as can be seen in Figure 7.4 (Özdogan et al. 2009). *Fabric fuzziness* refers to the severity of hairiness of a fabric

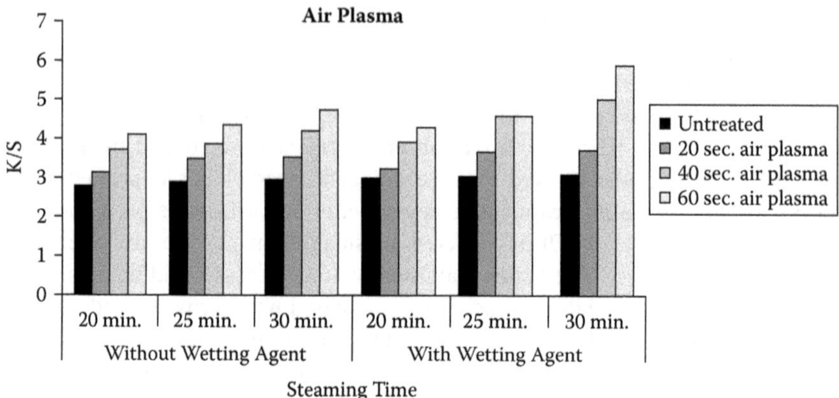

FIGURE 7.2 K/S values of untreated and air plasma-treated wool fabrics. (From Özdogan et al. 2009.)

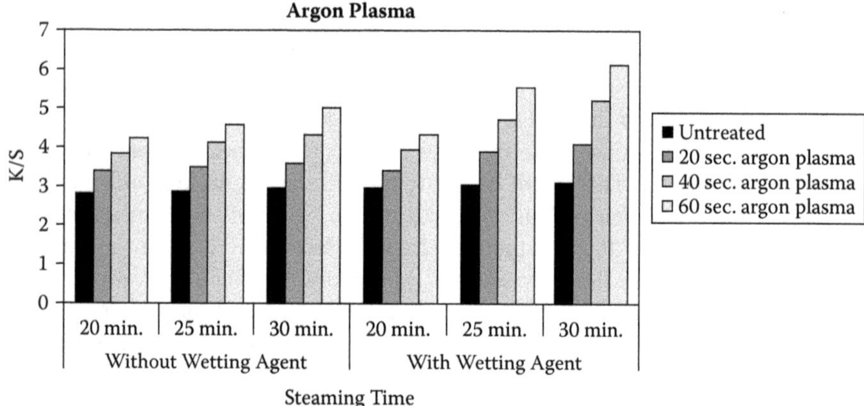

FIGURE 7.3 K/S values of untreated and argon plasma-treated wool fabrics. (From Özdogan et al. 2009.)

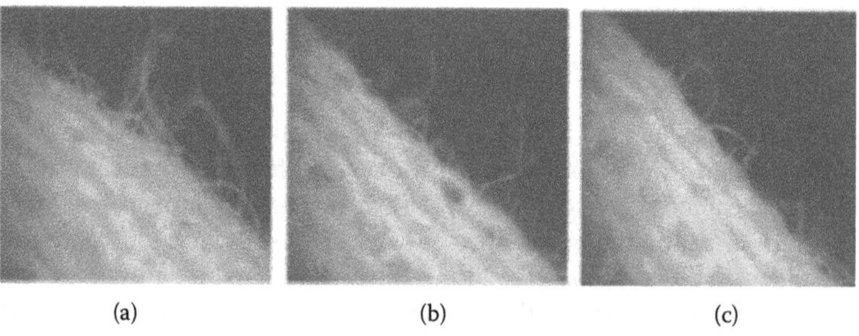

FIGURE 7.4 (a) Untreated; (b) 60-s air and (c) 60-s argon plasma-treated wool fabrics. (From Özdogan et al. 2009.)

caused by untangled fibre ends protruding from the surface. After atmospheric plasma treatment, the height and density of protruding fibre ends decreased and the surface became more compact, which is desirable for fabrics that are to be printed. This is probably caused by the etching action of the atmospheric plasma treatment, which makes surface fibres more fragile and increases the surface friction coefficient from 0.3056 to 0.3697 (Lima et al. 2005). These increments also affect the fibre–fibre interaction and make it more difficult for fibres to protrude from the fabric surface. In other words, increased inter-fibre friction reduces fuzziness of the fabric. With the decrease in fabric fuzziness, the multidirectional scattering of light decreases. In other words, the colour of the fabric seems deeper (Figure 7.5) (Özdogan et al. 2009).

Higher K/S values could be obtained even at the lowest plasma exposure time with the lowest steaming time, regardless of the plasma type. The results showed that K/S values increased gradually with an increase in steaming time. On the other hand, both air and argon plasma caused higher K/S values with the intensifying exposure time. Atmospheric plasma activated the surface by forming new hydrophilic groups (Demir et al. 2008). When the exposure time was increased, the plasma etching effect and partial degradation of the fatty layer became more dominant, which increased the dye penetration during printing (Özdogan et al. 2009).

Wetting agent had a positive effect on the printability of untreated and treated wool fabrics. Although wetting agent increased the penetration of printing paste, the results showed that atmospheric plasma treatment was more effective. In many cases, wetting agent would not be required if plasma treatment were carried out. If we take into consideration the plasma type, argon plasma seems more effective than air plasma. This is probably caused by the higher etching effect of noble gases (Poletti et al. 2003).

It is well known that adequate preparation of wool prior to printing provides high-quality prints (Bell 1998). Corona and low-pressure plasma treatments led to significant improvement of wool printability (Radetic, Jocic, et al. 2000; Ryu, Wakida, and Takagishi 1991). K/S values were used as a measure of the colour yield at the surface of the print (Radetic, Jocic, et al. 2000; Ryu, Wakida, and Takagishi 1991; Puac et al. 2005). The influence of treatment time (2.5, 5, 8 and 10 min) and pressure (0.25, 0.50 and 0.75 mbar) on K/S values of argon plasma treated wool were examined by Radetic, Jovancic, et al. (2007). Only 2.5 min of plasma treatment led

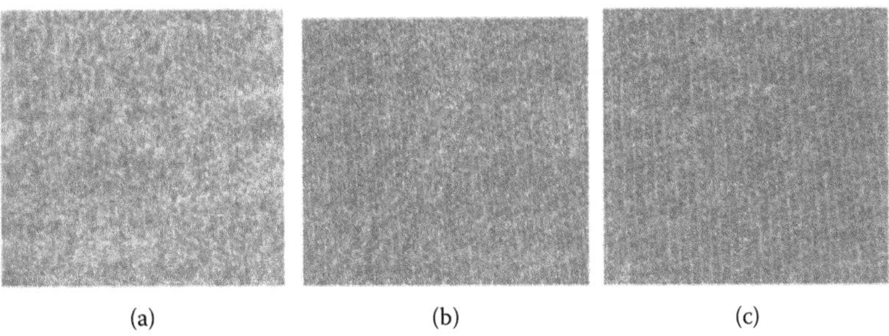

FIGURE 7.5 (a) Untreated; (b) 60-s air and (c) 60-s argon plasma-treated and printed fabrics. (From Özdogan et al. 2009.)

to a remarkable increase in colour yield. The prolongation of treatment time resulted in improved printability. Similar results were obtained in the case of air and oxygen plasma-treated wool (Radetic, Jocic, et al. 2000).

Improved printability can be attributed to enhanced wettability and swelling, since the fibre became more accessible to water and dye molecules, which could penetrate much more easily to the fibre interior (Radetic, Jocic, et al. 2000). The effect of pressure on colour yield increased in the following order: 0.50 mbar > 0.75 mbar > 0.25 mbar, which is explained by the pressure-dependent efficiency of the production of active plasma particles (Radetic, Jocic, et al. 2000). The results also revealed that colour yield of plasma-treated samples did not reach the efficiency of a conventionally chlorinated sample, but even short plasma treatments ensured sufficient preparation of wool knitted fabric prior to printing. This is very important from an environmental point of view, since conventional preparation of wool for dyeing and printing as well as for imparting of felting shrinkage resistance to wool products is carried out by the Chlorine-Hercosett process (Lewis 1977). Despite its high efficiency, high water consumption and generation of adsorbable organic halides (AOX) make this treatment environmentally unfriendly. The adoption of advanced chemicals may improve process efficiency, but these compounds or their by-products can deteriorate already complex wastewater composition. Therefore, the implementation of preferentially dry and clean technologies is required (Özdogan et al. 2009).

A comparative cost analysis of conventional chlorination and plasma processing of wool (Lewis 1977; Rakowski 1989) demonstrated that energy costs for chlorination were 7 kWh/kg wool, whereas the energy for low-pressure plasma treatment was only 0.3–0.6 kWh/kg wool. The application of low-pressure plasma for the modification of 120 t/year of wool can save 27,000 m^3 of water, 44 t of sodium hypochlorite, 16 t of sodium bisulphite, 11 t of sulphuric acid and 685 MWh of electrical energy (Rakowski 1989; Tsai, Wadsworth, and Roth 1997).

In the case of colourfastness to washing, plasma-treated samples showed staining values between 4 and 5. Although washing fastnesses were similar, especially for durations of 40 and 60 s, air and argon plasma-treated fabrics had higher lightfastness regardless of the wetting agent (Özdogan 2009). In the case of light and rubbing fastness values, there were no significant changes in terms of rubbing fastnesses. However, lightfastness showed higher values when compared with the untreated fabric, which could be due to the higher K/S values of plasma-treated fabrics (Özdogan et al. 2009).

7.2.2 Printing Properties of Plasma-Treated Silk Fibre: Ink-Jet Printing

In order to study the influence of treating time, plasma treatments on silk fibre were carried out at 3, 5, 10, 15 and 20 min durations with gas pressure and working power fixed at 50 Pa and 80 W, respectively (Fang et al. 2008). Experimental results showed that K/S values increased with an increase of plasma exposure time. This could be attributed to the increasing number of plasma-created polar groups and roughness on the surface due to etching and other chemical changes of the surface (Temmerman and Leys 2005). It was interesting that when the fabric was treated for more than 10 min, the K/S value remained unchanged. It is generally agreed that a large number of active particles will be generated during plasma treatment, such as electrons, ions, free radicals, photons, and excited atoms/molecules, but the high-speed electrons were especially contributing. At fixed input power, the dissociative reaction rate increased and led to further increasing the amount of high-speed electrons before the exposure time reached 10 min, and it may be that the number of electrons did not increase further when the treating time exceeded 10 min, which led to saturation of the plasma effect on the surface of silk fabric (Fang et al. 2008).

By keeping pressure at 50 Pa and treatment time at 10 min, several power conditions (50, 60, 70, 80 and 90 W) were used to find out the effect of input electrical power to the plasma (Fang et al. 2008). The results showed that K/S values increased with an increase of the input power to the plasma. Higher input power increased the number of high-speed electrons in the plasma and improved the plasma treatment effect (Wang and Qiu 2007), which led to the increased K/S value of the fabrics. This work was aimed at improving the antibleeding performance of silk fabric. We found that when the fabric was treated with 80 W, it could satisfy the antibleeding performance of ink-jet printing. For energy savings and practical considerations, 80 W was the optimum treatment power (Fang et al. 2008).

The relationship between pressure and K/S value was studied under constant power (80 W) and time (10 min) by Fang et al. (2008). K/S value reached its maximum when working pressure was set at 50 Pa. This could be explained by the fact that the number of active particles was very low at low working pressure, and increasing the pressure increased the plasma effect. However, the total energy of plasma was constant at fixed input power, and the mean energy of every active particle was reduced when the pressure was very high, which led to shrinkage of the electron mean free path. Thus electrons could not accumulate enough energy to dissociate when they collided with each other. The amount of real dissociative electrons decreased, resulting in a decrease of the plasma effect (Fang et al. 2008).

Wettability measurement results were obtained by Fang et al. (2008). Samples treated with plasma had a substantial improvement in their wettability when compared with the untreated control sample. Both advanced contact angle and retrograde contact angle decreased after treatment for 1 min. The advanced contact angle was further decreased with an increase of treatment time. This suggested that the hydrophilicity of the silk fibre had been enhanced remarkably. The effect must be attributed to the fact that plasma treatment not only brought etching effects to the surface of silk fibre, but also introduced polar groups (–OH, –COOH, –C=O, $-NH_2$) into the surface layer. Both of these actions could improve the hydrophilicity of the silk fibre (Fang et al. 2008).

Figure 7.6 shows the antibleeding performance of untreated and plasma-treated silk fabrics with ink-jet printing (Fang et al. 2008). As shown in Figure 7.6a, the bleeding performance of untreated silk fabric was severe along the edge of ink-jet printing. The antibleeding performance of the treated sample was dramatically improved, with excellent sharpness after plasma treatment. This was due to the etching and the introduction of polar groups onto the surface layer of the fabric, which improved the hydrophilicity of the fabric and consequently expedited the absorption speed of the ink. In addition, by studying the AFM (atomic force microscopy) image of the fibre, it is apparent that plasma treatment produced more grooves on the surface of the fibre (Fang et al. 2008). These grooves had the ability

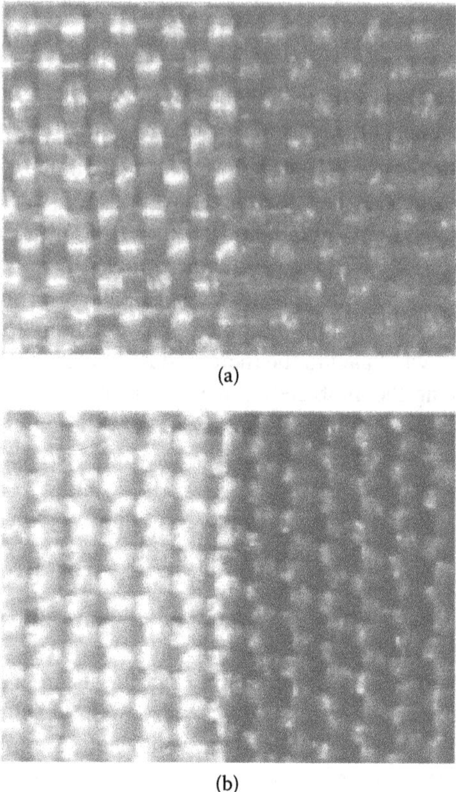

(a)

(b)

FIGURE 7.6 Antibleeding images of silk fabric which were taken after ink-jet printing by DZ3-video focus-exchanged microscope at 375: (a) untreated, (b) treated. The sample was treated at a pressure 50 Pa and power 80 W for 10 min. (From Fang et al. 2008.)

to hold more ink on the surface of the fabric, and this led to the excellent sharpness of ink-jet printing (Fang et al. 2008).

The colour measurement results of plasma-treated and untreated fabrics after ink-jet printing were investigated by Fang et al. (2008). Their study showed that K/S values increased after plasma treatment, the L (lightness) value decreased a little and the C (chroma) value increased. On one hand, the etching and the polar groups on the surface of the fabrics induced by plasma improved the antibleeding performance of the silk fabric, which increased the amount of ink colourant retained on the fabric surface. On the other hand, the etching action of plasma increased the surface roughness of the fabrics. It also contributed to the increase of K/S values of ink-jet-printed specimens by decreasing the fraction of light reflected from the treated rough surfaces compared with untreated smooth surfaces (Fang et al. 2008).

Fang et al. (2008) produced ink-jet-printing images of untreated and plasma treated silk fabrics. From these images, the plasma-treated sample had excellent sharpness along the edge of ink-jet printing, while the untreated one was degraded. In comparison with the untreated fabric, the colour of the treated sample turned deeper and more vivid. It also should be pointed out that some hydroxyl groups can be introduced onto the fabric surfaces following oxygen plasma treatment. The pigment ink that we used was composed of several ingredients, including pigment particles, dispersants, polyols, and deionised water. The dispersant was adsorbed onto the surface of pigment particles, which contain some carboxyl groups, and these carboxyl groups could then interact with the hydroxyl groups through hydrogen bonding. This effect also contributed to the deeper and more vivid colour performance.

7.3 EFFECT OF PLASMA TREATMENT ON PRINTING SYNTHETIC FIBRES

Plasma chemical surface treatment of different synthetic fibres has been reported (Deshmukh and Bhat 2003; Uchida, Uyama, and Ikada 1989; Inagaki, Tasaka, and Suzuki 1994; Joshi et al. 1997). The changes occur mainly on account of bombardment of electrons, ions and neutrals and affect the material only over a range of a hundred to several thousand angstroms in depth. When polymer substrates are used, the main effects are in terms of etching (loss of weight), cross-linking, oxidation and other chemical reactions, depending on the type of gas present. The changes that occur affect several physical and chemical properties of polymers, such as permeability, solubility, melting point and roughness at the surface (Mittal 1983). Adhesion, printability, colouration (dyeing) and adsorption, which depend on wettability, are naturally affected by the plasma treatment.

A standard ESCA (electron spectroscopy for chemical analysis) spectrum for polyester (Klopffer 1984) was examined by Deshmukh and Bhat (2003) in which a C1s spectrum corresponding to three inequivalent carbon atoms was observed. The relative intensities of the structure at 289, 286.5 and 285 eV correspond to C atoms in O–C=O, C–C–O and C–C–C bonding positions in the ratio of 2:2:6, respectively. In addition, there is an O1s spectrum which consists of two peaks corresponding to C–O and C=O at 534.7 and 533 eV, respectively. After plasma treatment with air, it shows that the intensity of aromatic peak at 285 eV has decreased (Deshmukh and Bhat 2003). The relative intensity of peak ($-O-CH_2$) at 286.2 eV has considerably increased. The peak (–COO–) at 289.1 eV has slightly increased in relative intensity. This shows that a large number of oxygen polar functional groups are introduced into the polyester surface when treated with air plasma. Plasma treatments are widely used to improve the wettability of polymeric films. Polar groups are always incorporated onto plasma-treated polymer surfaces regardless of gas type (He, Ar, H_2, N_2, O_2, air, etc.) (J. Chen, Wang, and Wakida 1999). These polar groups are readily detected by ESCA, but are often missed by ATR-FTIR (Attenuated Total Reflectance Fourier Transform Infrared) spectroscopy (Wu 1982). Surface energy (SE) is an indirect measure of polar groups. It has been reported that the treatment carried out in inert gases like Ar introduces oxygen moieties onto the polymer surface because of post-plasma exposure of samples to atmosphere (Gupta et al. 2000). The atomic concentration of the C1s element decreases from 77.17% to 60.65% (Deshmukh and Bhat 2003).

The O1s spectrum of control polyester and polyester treated in air plasma indicated that peaks at 533 eV is due to C=O, while the peak at 534.5 eV is due to C–O. When the polyester film is treated with air plasma, it is observed that the absolute intensities of both the peaks have increased. However, the relative intensity of the peak at 534.5 eV with respect to the first one has increased slightly and, as a result, the resolution of the peaks has disappeared. It thus appears as a single peak, which is quite broad. This observation has been supported through our previous discussion of C1s spectra, wherein it is indicted that contribution due to –O–C has increased. The atomic concentration of the O1s element has increased from 22.65% to 36.48% (Deshmukh and Bhat 2003).

Since polyester film is treated in air plasma, it is obvious that some nitrogen-containing groups will be incorporated onto the surface. For the control polyester, there is no peak due to N1s spectrum, but for air-treated polyester, there appears to be a small peak. (However, the spectra are not produced here.) This means that nitrogen is being incorporated onto the polyester surface. The atomic concentration of N1s element has increased from 0.18% to 2.87% (Deshmukh and Bhat 2003).

Deshmukh and Bhat (2003) show the contact angle (CA) with respect to water (W), glycerol (G) and formamide (F), the polar component of solid surface free energy (γ_s^p), the disperse component of solid surface free energy (γ_s^d) and SE (γ_s). It can be seen that the abundance of polar components (γ_s^p) onto the surface sharply increases initially and then steadily with the plasma treatment time. It is mainly due to the incorporation of polar groups like CO, COO, OH, etc. (Briggs et al. 1980). However there is no appreciable change in γ_s^d. Hence, the increase in SE (γ_s) is due to the incorporation of polar groups onto the polyester surface. One can observe a pronounced rise in SE over a short period of plasma treatment. The wettability and hence SE of polyester is increased because of the interaction between the hydrogen bond and dipoles in the vertical direction of the interface (Westerdani et al. 1974). The properties like wettability, adhesion, printability, etc., strongly depend upon the SE polyester becoming hydrophilic after the plasma treatment; hence, its surface energy increases with the time of treatment. This improves the adhesion and printability properties of polyester (Deshmukh and Bhat 2003).

It is well known that before printing is made, polymers are subjected to a plasma treatment (Brewis 1982). Plasma treatment requires very high power, and the processing cost is also high. In a study by Deshmukh and Bhat (2003), the surface of the polyester was modified using cold plasma to have a good printability and good adhesion to ink. This method offers a simple way to measure the degree of adhesion of the ink (coating) on a substrate, but is rather qualitative. The improvement in printability of air plasma-treated polyester was shown by Deshmukh and Bhat (2003). For control polyester, the ink adhesion is only 6.3%, which improves to more than 80% for a treatment time of only 15 s. This is a very sharp rise in the value of the percent of printability of plasma-processed polyester. After 15 s, there is a slow improvement in the value of the percent of printability, and the increase continues up to 5 min. Improvement in ink adhesion after plasma treatment has also been reported (Jana et al. 2001). Such modifications as the incorporation of polar groups onto the surface and the phenomenon of plasma etching are observed as possible explanations for the improvement in wettability. This in turn increases an effective area of contact for the spreading of ink material. Both processes contribute to the improvement in ink adhesion (Deshmukh and Bhat 2003).

The incorporation of polar groups onto the plasma-processed surface is seen in ESCA. The AFM morphology (Deshmukh and Bhat 2003) shows that the surface becomes rough after plasma treatment. Anchoring of ink takes place at the rough surface, causing better adhesion. The morphology of the polyester was investigated using AFM. In AFM, the surface of the control polyester was very smooth. The mean surface roughness (Ra), measured by AFM, was only 2.328 nm, which changed to a rough surface having Ra = 5.805 nm in a treatment time of 30 min. Deshmukh and Bhat (2003) show mean surface roughness (Ra) of the control polyester and the polyester treated in air plasma for various durations of time. It was found that initially there was a rapid increase in the value of Ra, followed by a slow rise for higher duration of treatment time. The AFM photographs of plasma-treated polyester showed that most of the amorphous portion was etched out in a 30-min treatment time, resulting in rough surface morphology. It was noted that the surface became rougher, and occasionally some re-deposited fragments could be seen. This resulted in an apparent increase in the surface area, resulting in improvements in wettability, bonding strength and printability.

In order to study the influence of treating power, plasma treatment was carried out at several power conditions (40, 60, 80, 100 and 120 W) with gas pressure and treatment time fixed at 40 Pa and 9 min, respectively (Wang and Wang 2010). The results showed that the K/S value of ink-jet-printed polyester fabric increased with an increase of the input power to the plasma. Higher input power increased the number of high-speed electrons in plasma, which would improve the effect of plasma treatment (Wang and Qiu 2007), leading to an increase in the K/S value of the fabric. Because the etching effect on structure played a leading role, the ink molecules could rapidly penetrate into the fabric. Therefore, the K/S value increased with increasing treatment power. It was interesting that with the treatment power increasing, after 80 W and there were no significant changes in the K/S value. Maybe there was no further etching effect.

Wang and Wang (2010) showed that the K/S value of ink-jet-printed polyester fabric reached its maximum value when the working pressure was set at 40 Pa·min. This indicated that there was a direct correlation between the K/S value and plasma treatment. The K/S value increased as the gas pressure increased due to the predominant etching effect on the fabric. The K/S value decreased with increasing gas pressure because of the consumption of the plasma's active particles. This could be explained by the fact that the number of active particles was very low at low working pressure, and increasing the pressure increased the plasma effect. In addition, during the oxygen plasma treatment, more polar groups could be introduced to the fibre, so more ink molecules could diffuse into the polyester fabric.

Wang and Wang (2010) also found that, in the case of ink-jet printing, the K/S values of the plasma-treated polyester fabric with different treatment times were higher than that of the untreated fabric. The K/S value increased gradually with plasma treatment time and reached the maximum at 9 min of treatment time. Further increase of plasma treatment time did not give further enhancement of the K/S value, producing instead a gradual reduction in the K/S value. This could be attributed to the increasing number of plasma-created polar groups and roughness on the surface, due to etching and other chemical changes of the surface

(Temmerman and Leys 2005). The plasma treatment at the levels of power and pressure applied in this experiment could complete its actions fully and with better etching effect on the fabric. However, when the fabric was treated for more than 9 min, the K/S value remained unchanged. This could be attributed to the saturation of plasma effect on the surface of polyester fabric.

The relationship between gap distance and K/S value of ink-jet-printed fabric under the effect of plasma treatment was studied by Zhang and Fang (2009). The highest K/S value of samples was obtained when distance between electrodes was set at 3 mm for both air and air/Ar atmospheres. The most effective gap distance in this experiment was 3 mm, at which the best effect of plasma treatment was obtained. This is due to the fact that the amount of gas between the gaps diminished when the gap distance was too small, leading to a decrease in the number of particles which would be excited during the discharging process. Therefore, the effect of plasma discharging became weaker. On the other hand, the field intensity became weaker when the gap distance was too large, which could also weaken the effect of plasma discharging (Akishev et al. 2002).

The sharpness of the ink-jet-printed pattern was measured by the optical analysis method (Wang and Wang 2010). Obviously, the ink-jet-printed patterns in the weft direction were thicker than those in the warp direction for both the untreated and plasma-treated fabric. This might be due to the differential wicking effect caused by the warp and weft yarns (Yuen and Kan 2007). When comparing the width of the printed patterns, the patterns printed on the plasma-treated polyester fabric were narrower than the untreated fabric in both the warp and weft directions. This could be attributed to the reduced spreading of the printed pigment inks as a result of the strong fibre and pigment attraction. Consequently, the radio frequency (RF) plasma treatment on polyester fabric could enhance the sharpness of the ink-jet prints (Zhang and Fang 2009).

The antibleeding performance of untreated and plasma-treated polyester fabrics was measured by Zhang and Fang (2009). The samples were treated at a power of 300 W for 150 s. The bleeding performance of untreated polyester fabric was severe along the weft and warp edge of the ink-jet-printed fabrics. The antibleeding performance of the treated sample was dramatically improved with excellent sharpness after plasma treatment. This was due to the hydrophilic improvement of the fabric, consequently expediting the speed of ink absorption and increasing the holding ability of the inks. The results for colour measurement of treated and untreated polyester fabrics are listed in Table 7.1 (Zhang and Fang 2009). It shows that K/S and C values increased and the L values decreased after air/Ar plasma treatment. This means that the chroma and saturation of the sample increased and the luminance decreased. On one hand, the etching and the polar groups on the surface of the fabrics induced by plasma treatment improved the antibleeding performance of the polyester fabric, which increased the amount of ink colourant retained per area of the fabric. On the other hand, the etching action of plasma increased the surface roughness of fabrics. It also contributed to the increase of K/S values of ink-jet-printed specimens by decreasing the fraction of light reflected from treated rough surfaces compared with untreated smooth surfaces (Zhang and Fang 2009).

TABLE 7.1
Colour Measurement Results of Treated and Untreated Fabrics

Samples	K/S	L	C
Untreated	3.34	54.72	49.06
Air/Ar plasma treated	4.38	51.95	53.56

Source: Zhang and Fang (2009).

The dry and wet rubbing fastnesses of ink-jet-printed polyester fabrics were also investigated by Zhang and Fang (2009). The results showed that there was almost no difference on both rubbing fastnesses between untreated and air/Ar plasma-treated samples. This result indicates that plasma treatment has no effect on the colourfastness of ink-jet-printed polyester fabrics.

7.4 CONCLUSION

This chapter reviewed the effect of plasma treatment on the printing properties of textile fibres. In the case of printing, the plasma treatment is used as a pretreatment to improve the adhesion of printing paste during the printing operation. In recent years, with the development of ink-jet printing, plasma is playing an important role in this new printing technology. Before the fabric is used for ink-jet printing, conventional printing ingredients such as dye or pigment should be first placed in the fabric. The application of printing ingredients on the fabric surface depends greatly on the adhesion properties of the fabric. Thus, the plasma treatment, being a surface treatment, can help to modify the adhesion properties of the fabric surface, which in turn improves the efficiency of ink-jet printing.

REFERENCES

Achwal, W. B. 2002. Textile chemical principles of digital textile printing (DTP). *Colourage* 49(12): 33–34.

Akishev, Y., M. Grushin, A. Napartovich, and N. Trushkin. 2002. Novel AC and DC non-thermal plasma sources for cold surface treatment of polymer films and fabrics at atmospheric pressure. *Plasmas and Polymers* 7(3): 261–89.

Bell, V. 1998. Recent development in wool printing. *Journal of the Society of Dyers and Colourists* 104: 159–72.

Brewis, M. 1982. *Surface analysis and pretreatments of plastics and metals*. London: Applied Science Publisher.

Briggs, D., D. G. Rance, C. R. Kendall, and A. R. Blythe. 1980. Surface modification of poly(ethyleneterephthalate) by electrical discharge treatment. *Polymer* 21: 895–900.

Chen, J. R., X. Y. Wang, and T. Wakida. 1999. Wettability of poly(ethylene terephthalate) film treated with low-temperature plasma and their surface analysis by ESCA. *Journal of Applied Polymer Science* 72: 1327–33.

Chen, W., G. Wang, and Y. Bai. 2002. Best for wool fabric printing: Digital inkjet. *Textile Asia* 33(12): 37–39.

Demir, A., H. A. Karahan, E. Özdogan, T. Oktem, and N. Seventekin. 2008. Synergetic effects of alternative methods in wool finishing. *Fibres & Textiles in Eastern Europe* 16(2): 89–94.

Deshmukh, R. R., and N. V. Bhat. 2003. The mechanism of adhesion and printability of plasma processed PET films. *Materials Research Innovations* 7: 283–90.

Fang, K. J., S. H. Wang, C. X. Wang, and A. Tian. 2008. Inkjet printing effects of pigment inks on silk fabrics surface-modified with O_2 plasma. *Journal of Applied Polymer Science* 107: 2949–55.

Gupta, B., J. Hilborn, C. H. Hollenstein, C. J. G. Plummer, R. Houriet, and N. Xanthopoulos. 2000. Surface modification of polyester films by RF plasma. *Journal of Applied Polymer Science* 78(5): 1083–91.

Inagaki, N., S. Tasaka, and Y. Suzuki. 1994. Surface chlorination of polypropylene film by $CHCI_3$ plasma. *Journal of Applied Polymer Science* 51(13): 2131–37.

Inbakumar, S., R. Morent, N. De Geyter, T. Desmet, A. Anukaliani, P. Dubruel, and C. Leys. 2010. Chemical and physical analysis of cotton fabrics plasma-treated with a low pressure DC glow discharge. *Cellulose* 17: 417–26.

Jana, T., B. C. Roy, R. Ghosh, and S. Maiti. 2001. Biodegradable film, IV: Printability study on biodegradable film. *Journal of Applied Polymer Science* 79(7): 1273–77.

Joshi, A. H., C. Natarajan, S. M. Pawade, and N. V. Bhat. 1997. Grafting of cellophane films using magnetron-enhanced plasma polymerization. *Journal of Applied Polymer Science* 63(6): 737–43.

Klopffer, W. 1984. *Introduction to polymer spectroscopy*. New York: Springer-Verlag.

Lewis, J. 1977. Superwash wool, Part I: A review of the developments of superwash technology. *Wool Science Review* 54: 2–29.

Lima, M., L. Hes, R. Vasconcelos, and J. Martins. 2005. Frictorq, accessing fabric friction with a novel fabric surface tester. *AUTEX Research Journal* 5(4): 194–201.

Miles, L. W. C. 1994. *Textile printing*. Bradford, UK: Society of Dyers and Colourists.

Mittal, K. L. 1983. *Physicochemical aspects of polymer surfaces*. Vol. 2. New York: Plenum Press.

Özdogan, E., A. Demir, H. A. Karahan, H. Ayhan, and N. Seventekin. 2009. Effects of atmospheric plasma on the printability of wool fabrics. *Tekstil ve Konfeksiyon* 19: 123–27.

Poletti, G., F. Orsini, A. Raffaele-Addamo, C. Riccardi, and E. Selli. 2003. Cold plasma treatment of PET fabrics: AFM surface morphology characterization. *Applied Surface Science* 219: 311–16.

Puac, N., Z. L. Petrovic, M. Radetic, and A. Djordjevic. 2005. Low pressure RF capacitively coupled plasma reactor for modification of seeds, polymers and textile fabrics. *Materials Science Forum* 494: 291–96.

Radetic, M., D. Jocic, R. Trakovic, and Z. L. Petrovic. 2000. The effect of low-temperature plasma treatment on wool printing. *Textile Chemist and Colorist & American Dyestuff Reporter* 32(4): 55–60.

Radetic, M., P. Jovancic, N. Puac, and Z. L. Petrovic. 2007. Environmental impact of plasma application to textiles. *Journal of Physics: Conference Series* 71: 012017.

Rakowski, W. 1989. Plasma modification of wool under industrial conditions. *Melliand Textilberichte/International Textile Report* 70: E334–37.

Ryu, J., T. Wakida, and T. Takagishi. 1991. Effect of corona discharge on the surface of wool and its application to printing. *Textile Research Journal* 61(10): 595–601.

Temmerman, E., and C. Leys. 2005. Surface modification of cotton yarn with a DC glow discharge in ambient air. *Surface and Coatings Technology* 200(1–4): 686–89.

Tsai, P. P., L. C. Wadsworth, and J. R. Roth. 1997. Surface modification of fabrics using a one-atmosphere glow discharge plasma to improve fabric wettability. *Textile Research Journal* 67: 359–69.

Uchida, E., Y. Uyama, and Y. Ikada. 1989. Surface graft polymerization of acrylamide onto poly(ethylene terephthalate) film by UV irradiation. *Journal of Polymer Science Part A: Polymer Chemistry* 27(2): 527–37.

Wang, C. X., and Y. P. Qiu. 2007. Two sided modification of wool fabrics by atmospheric pressure plasma jet: Influence of processing parameters on plasma penetration. *Surface and Coatings Technology* 201: 6273–77.

Wang, C., and C. Wang. 2010. Surface pretreatment of polyester fabric for ink jet printing with radio frequency O_2 plasma. *Fibers and Polymers* 11(2): 223–28.

Westerdani, C. A. L., J. R. Hall, E. C. Schramm, and D. W. Levi. 1974. Gas plasma effects on polymer surface. *Journal of Colloid and Interface Science* 47(3): 610–20.

Wong, K. K., X. M. Tao, C. W. M. Yuen, and K. W. Yeung. 1999. Low temperature plasma treatment of linen. *Textile Research Journal* 69: 846–55.

Wu, S. 1982. *Polymer interface and adhesion*. New York: Marcel Dekker.

Yasuda, T., M. Gazicki, and H. Yasuda. 1984. Effects of glow discharges on fibres and fabric. *Journal of Applied Polymer Science: Applied Polymer Symposium* 38: 201–14.

Yuen, C. W. M., and C. W. Kan. 2007. Influence of low temperature plasma on the ink-jet printed cotton fabric. *Journal of Applied Polymer Science* 104(5): 3214–19.

Yuen, C. W. M., S. K. A. Ku, P. S. Choi, and C. W. Kan. 2004. The effect of the pretreatment print paste contents on colour yield of an ink-jet printed cotton fabric. *Fibers and Polymers* 5(2): 117–21.

Zhang, C., and K. Fang. 2009. Surface modification of polyester fabrics for inkjet printing with atmospheric-pressure air/Ar plasma. *Surface and Coatings Technology* 203: 2058–63.

8 Application of Plasma Treatment in Finishing of Textiles

8.1 EFFECT OF PLASMA TREATMENT ON FINISHING CELLULOSIC FIBRES

8.1.1 Denim Finishing

The popularity of "worn look" denim products has lasted more than two decades. Originally, the abrasive action of pumice stones on the garment surface was used to achieve this effect. Although efficient, traditional stonewashing often damaged the garments and the machines. This process had additional problems, such as the generation of huge amounts of pumice dust in the laundry environment and the reduction of machine capacity due to the high proportion of stones (up to 1 kg of stones per kg of jeans). The development of enzymatic stonewashing (biostoning) partially or completely replaced stonewashing. Biostoning with cellulase enzymes provides a soft handle and the desired look (Buschle-Diller et al. 1994; Tyndall 1992). However, biostoning requires a high quantity of water and chemicals that are released into effluents, making the process less eco-friendly. Recently, low-pressure plasma and corona treatments for obtaining a "worn look" effect on denim fabrics were proposed (Ghoranneviss et al. 2006). In a study by Radetic et al. (2007), the CIE L*a*b* colorimetric system was used to determine the colour difference between untreated and plasma-treated denim fabrics. Apparently, the higher the power and the greater the number of passes in the case of corona treatment, the higher was the fabric lightness. Similarly, in the case of argon plasma treatment, the prolongation of treatment time and increase in power brought about an increase in fabric lightness. In both cases, under severe treatment conditions, samples became more yellow (higher values of b*), but yellowness disappeared after washing. Mechanical properties of the material were not changed after plasma treatments. These results can encourage further research on the possibility of plasma implementation in denim finishing because of water- and chemical-free processing, lower energy costs and shorter treatment time compared to conventional biostoning (approximately 90 min). However, plasma treatment induces a harsher handle of denim fabric, indicating the need for some additional after treatment (Radetic 2007).

8.1.2 Flame Retardancy of Cotton by Plasma-Induced Graft-Polymerization (PIGP)

The persistent fire resistance of textiles is an unsolved problem, and that will remain so until a durable treatment that is resistant to many cycles of washing is found. This is even more problematic for textiles made from fibres of natural origin, such as cellulose fibres (cotton for instance). Indeed, for this type of fibre, the fire-retardant property can only be conferred by means of a surface treatment of the material. Among all textile fibres, cotton is the one most commonly used in domestic applications (clothes, bedding, furniture, wall hangings, etc.). However, it is also one of the most flammable materials, with a limiting oxygen index (LOI) of 18.4%. It is thus of primary importance for public safety to find ways to render this material less flammable and, of course, in a most economically and environmentally friendly manner (Tsafack and Levalois-Grützmacher 2006b).

Halogen-free phosphorus-based compounds are the flame retardants most frequently used on cotton textiles. Numerous studies (Kandola et al. 1996; Franklin and Rowland 1979) have shown that they reduce the formation of flammable volatiles and catalyse char formation. They fulfil these two functions because, upon heating, they release polyphosphoric acid, which phosphorylates the C(6) hydroxyl group in the anhydroglucopyranose moiety that serves as an acidic catalyst for the dehydration. The first reaction prevents formation of levoglucosan, the precursor of flammable volatile compounds. This ensures that the competing char-forming reaction becomes the favoured pyrolysis route. The acidic catalytic effect of the released polyacid further increases the rate of this reaction (Tsafack and Levalois-Grützmacher 2006b).

Many of the phosphorus-based compounds which evoke these effects also contain nitrogen. This observation has led to the proposal of a synergistic effect in compounds where both elements are present. Several studies (Hendrix, Drake, and Barker 1972; Langley, Drews, and Barker 1980; Lawler, Drews, and Barker 1985; Lewin 1999) have demonstrated that phosphorus compounds containing nitrogen are more efficient flame retardants because the char formation increases. However, not all nitrogen-containing compounds are effective. Only those with nucleophilic nitrogen atoms such as amides and amines show this behaviour (Tsafack and Levalois-Grützmacher 2006b).

Flame-retardant cottons are usually produced by a chemical treatment in a textile finishing process that, depending on the specific chemical process, generates flame-retardant properties having varying degrees of durability to various laundering processes. The fire retardants may be (a) simple salts (e.g., ammonium phosphates, polyphosphate and bromide borate–boric acid mixtures) that give water-soluble, non-durable finishes; (b) functional finishes (e.g., organophosphorous and nitrogen-containing monomers such as alkylphosphonamide derivatives, or polycondensates such as tetrakis(hydroxylmethyl)phosphonium salt), which contain reactive groups which may react chemically with surface functionalities to confer more-durable flame retardancy; or (c) back-coating, which usually comprises a resin-bonded antimony-bromine flame-retardant system (Tsafack and Levalois-Grützmacher 2006b).

Durable flame-retardant properties are more difficult to achieve than the non-durables. Many techniques for imparting durable fire-retardant properties to cotton fabrics, such as pad/dry cure, exhaust, spray/dry cure or coat/dry cure, have been described in the literature (Kandola et al. 1996; Wakelyn, Rearick, and Turner 1998). However, relatively few of them are practiced today, either because of safety concerns or process-control issues.

Researchers have recently developed a process to reduce the flammability of PAN (polyacrylonitrile) textiles (Hochart, De Jaeger, and Levalois-Grützmacher 2003; Tsafack, Hochart, and Levalois-Grützmacher 2004; Tsafack and Levalois-Grützmacher 2006a). The protocol consisted of the simultaneous grafting and polymerization of acrylate monomers bearing phosphate and phosphonate groups on the surface of the material. The polymerization of the monomers was induced by means of an argon plasma (Tsafack and Levalois-Grützmacher 2006b).

Levels of fire retardant to be applied depend upon the degree of flame retardancy required and the area density and structure of the fabric. Generally speaking, a loading of 1.5% to 4% (w/w) of phosphorus on the fabric is required (Kandola et al. 1996; Bajaj et al. 2000). For this, a minimum of 10% to 15% (w/w) of grafted phosphorus polymer is necessary, depending on the monomer. Previously (Hochart, De Jaeger, and Levalois-Grützmacher 2003; Tsafack, Hochart, and Levalois-Grützmacher 2004; Tsafack and Levalois-Grützmacher 2006a), it was demonstrated that the presence of a cross-linking agent is necessary for an efficient grafting of acrylate phosphate and phosphonate monomers on PAN fabrics. Therefore, the influence of the cross-linking agent and the monomer concentration was investigated in order to find the optimum conditions for plasma-induced graft-polymerization of the acrylic monomers containing phosphorus onto cotton fabrics (Tsafack and Levalois-Grützmacher 2006b).

The different experiments were performed with (10% and 20% [w/w]) and without the cross-linking agent ethyleneglycoldiacrylate (EGDA) at monomer concentration of 200 g/L using the PIGP procedure. The grafting yields obtained after washing (Soxhlet methanol and water) and air drying are shown in Figure 8.1 (Tsafack and Levalois-Grützmacher 2006b).

It was observed that without EGDA, the percentage of grafting is undetectable with diethyl-2-(methacryloyloxyethyl) phosphate (DEMEP) and below 7% for the other monomers except diethyl(acryloyloxyethyl)phosphoramidate (DEAEPN), which exhibits 20.9% of grafted polymer. This high percentage of grafting obtained without cross-linking agent can be attributed to the formation of hydrogen bonds between the –OH functional group of the cotton with the nitrogen of the phosphoramidate unit (Tsafack and Levalois-Grützmacher 2006b). However, compared to DEAEPN, the amount of grafted polymer without EGDA on fabrics treated with BisDEAEPN is low. This fact can be explained partly by the difference of the structure between DEAEPN and BisDEAEPN and their different conversion rates. When 10% (w/w) of EGDA is added in the monomer solution, the amount of grafted polymer is less than 10% for DEMEP, between 22% and 28% for diethyl(acryloyloxyethyl) phosphate (DEAEP), diethyl(acryloyloxymethyl)phosphonate (DEAMP) and dimethyl(acryloyloxymethyl)phosphonate (DMAMP), and more than 30% for DEAEPN and acryloyloxy-1,3-bis(diethylphosphoramidate)-propan (BisDEAEPN).

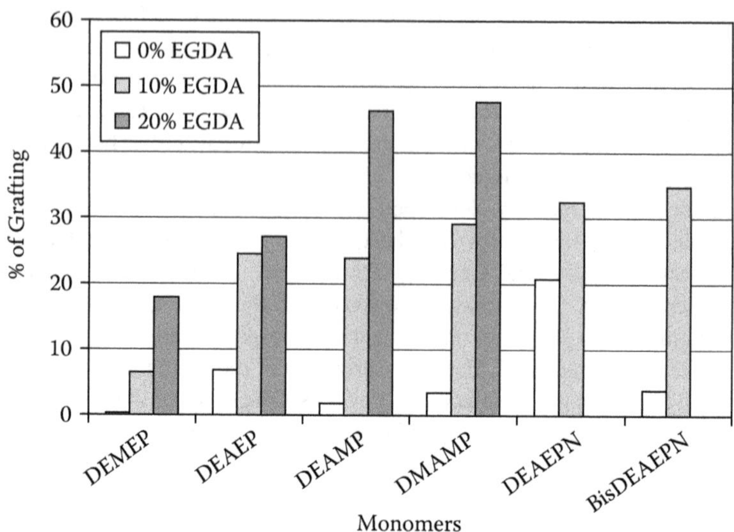

FIGURE 8.1 Effect of the amount of cross-linking agent (EGDA) on the percentage of grafting of the monomers DEMEP, DEAEP, DEAMP, DMAMP, DEAEPN and BisDEAEPN on cotton fabrics treated in microwave (MW) argon plasma after washing and air drying. (From Tsafack and Levalois-Grützmacher 2006b.)

The highest grafting yield of 34.7% is attained with BisDEAEPN. Because of the high grafting rate of acrylate phosphoramidate monomers with 10% of EGDA, no further tests were performed with higher concentrations of EGDA. For almost all tested monomers, the amount of grafted polymer increased considerably (from 6.3% to 17.8% for DEMEP, from 23.8% to 46.2% with DEAMP and from 29% to 47.6% for DMAMP) when the concentration of the cross-linking agent was augmented from 10% to 20%. Exceptions are experiments with DEAEP, where the grafting yields increased only slightly (from 24% to 27%). Clearly, the presence of cross-linking agent is necessary to achieve a good grafting of all the monomers on cotton fabrics excepted for DEAEPN. On the other hand, with the high amount of grafting achieved with DEAMP (46.2%) and DMAMP (47.6%), the stiffness of the fabrics is visibly affected. In order to increase the grafting yield without affecting the stiffness of the fabrics, the effect of varying the concentration of the monomer was investigated (Tsafack and Levalois-Grützmacher 2006b).

A number of experiments were carried out with the cross-linking agent EGDA (10% [w/w]), at monomer concentration of 200 g/L and 300 g/L, using the PIGP procedure. The grafting yields obtained after washing (Soxhlet methanol and then water) and air-drying are shown in Figure 8.2 (Tsafack and Levalois-Grützmacher 2006b).

The results clearly indicated that the amount of grafted polymer increases with the monomer concentration. The lowest grafting yield was obtained with DEMEP. This observation has already been made when DEMEP was grafted and polymerised onto PAN fabrics using the PIGP procedure. The highest grafting was

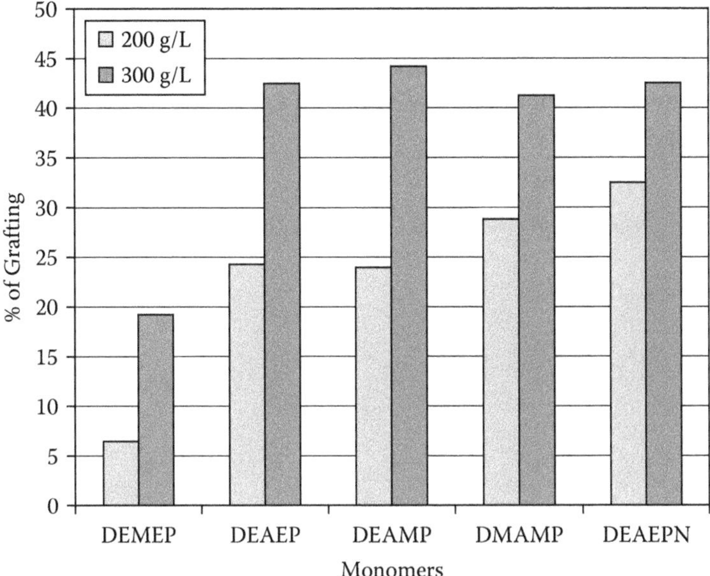

FIGURE 8.2 Effect of the concentration of DEMEP, DEAEP, DEAMP, DMAMP and DEAEP on the percentage of grafting of cotton fabrics (120 g/m^2) treated in MW argon plasma after washing and air drying. (From Tsafack and Levalois-Grützmacher 2006b.)

attained at about 42% with DEAEP, DEAMP, and DMAMP. These results indicate that with a small amount of cross-linking agent, it can be possible to reach a sufficient amount of grafted polymer to impart adequate flame retardancy to cotton fabrics. Therefore, treatments were performed with 5% (w/w) of EGDA for DEAEP, DEAMP, DMAMP and DEAEPN at monomer concentration of 200 g/L or 300 g/L. It can be concluded that by using the PIGP procedure, it is possible to graft quantitatively acrylic monomers containing phosphorus onto the surface of cotton fabrics (120 g/m^2) in the presence of EGDA as a cross-linking agent (Tsafack and Levalois-Grützmacher 2006b).

The flame retardancy of the untreated and plasma-treated cotton fabrics was assessed by limiting oxygen index (LOI) measurements. The results are given in Table 8.1. They clearly indicate that the initial LOI value of the untreated cotton (19) increases by 4 units when treated with DEAMP, up to 7 units when treated with DEMEP, DEAEP, and DMAMP, and up to 9 and 10 units when treated with DEAEPN and BisDEAEPN, respectively (Tsafack and Levalois-Grützmacher 2006b).

When the monomer concentration of DEAEP, DEAMP, DMAMP, and DEAEPN was increased from 200 g/L to 300 g/L under otherwise constant conditions, the amount of grafted polymer increased significantly. However, the LOI was augmented only slightly. This finding is in accord with the law of diminishing effectiveness. Indeed, when applied at higher add-on levels, most flame retardants exhibit this law; there is no linear improvement in flame retardancy, and a ceiling value of LOI is thus approached as the add-on is increased (Tsafack and Levalois-Grützmacher 2006b).

TABLE 8.1
LOI Values of Treated Cotton Fabrics as a Function of Monomer Type, of the Treatment Conditions and Percentage of Grafting after Washing and Air Drying

	Treatment Conditions				
Monomer	**Monomer (g/L)**	**% EGDA**	**%G**	**%P (w/w) Measured on Cotton**	**LOI (%)**
Untreated cotton	...	...	...	...	19
DEMEP	200	...	...	...	19
		10	6.3	...	21
	300	10	19.3	1.96	24.5
		20	29.2	...	25.5–26
DEAEP	200	...	6.9	...	21.5
		10	24.4	...	25.5
	300	5	28.6	2.75	26.0
		10	42.7	...	26.5
DEAMP	200	0	1.7	...	19
		10	23.8	...	22.5–23
	300	5	34.7	3.28	23.0
		10	44.3	...	23–23.5
DMAMP	200	...	3.3	...	20.0
		10	29.0	3.27	25.5–26
	300	5	37.8	4.10	26.0
		10	41.2	...	26.0
DEAEPN	200	...	20.9	...	26.5
		5	24.0	...	26.5
		10	32.4	2.77	27.5
	300	...	36.2	3.16	28.5
		5	38.6	3.36	28.5
		10	42.5	...	28.0
BisDEAEPN	100	10	13.0	1.48	25.0
	200	5	29.7	3.29	29.5
		10	34.7	...	29.5

Source: Tsafack and Levalois-Grützmacher (2006b).

Importantly, the results also show that the fabrics treated with the acrylate phosphoramidate monomers showed a better flame-retardant effect than the fabrics treated with acrylate phosphate and phosphonate monomers. While a LOI value of 26 was achieved with DEAEP at a phosphorus content of 2.75%, a LOI value of 27.5 was obtained with DEAEPN at approximately the same P content (2.77%). Comparing the LOI values obtained with DEAEPN and the one obtained with the acrylate phosphonate monomers, it is seen that whereas a LOI value of 28.5 is obtained with DEAEPN at a phosphorus content of 3.16%, LOI values of 25.5 and 23 are obtained for DMAMP and DEAMP, respectively, at approximately the same P content of 3.27%. The LOI values indicate that the effectiveness of the acrylic monomers containing phosphorus that were studied follow the order BisDEAEPN > DEAEPN > DEAEP > DEMEP > DMAMP > DEAMP. This difference in efficiency between the acrylate phosphoramidate monomers and the acrylate phosphate and phosphonate monomers can be attributed to the fact that phosphorus–nitrogen-containing compounds are better phosphorylating agents than are the related compounds without nitrogen (Hendrix, Drake, and Barker 1972; Langley, Drews, and Barker 1980; Lawler, Drews, and Barker 1985; Lewin 1999). And, it has been reported that an increasing rate of phosphorylation of cellulose hydroxyl functional groups produces a corresponding increase in the flame-retardant efficiency (Lawler, Drews, and Barker 1985).

Clearly and expected, the flame-retardant effect depends greatly on the chemical structure of the compound. The phosphorus content alone is no guarantee for flame-retardant properties, and very similar phosphorus contents can lead to very different LOI values. With the exception of DEAMP, all monomers studied here gave LOI values greater than 25. The highest LOI values of 28.5 and 29.5 were found for fabrics treated with DEAEPN and BisDEAEPN, respectively (Tsafack and Levalois-Grützmacher 2006b).

In order to evaluate the durability of the coating to washing, the plasma-treated samples were submitted to the accelerated laundering method proposed by McSherry et al. (1974). The results obtained for the fabrics treated with DEAEP, DMAMP, DEAEPN and BisDEAEPN are given in Table 8.2 (Tsafack and Levalois-Grützmacher 2006b).

The results also indicate that the percentage of grafting and therefore the LOI values after the washing procedure depend on the amount of cross-linking agent added. Without EGDA, the LOI value of DEAEPN (monomer concentration of 200 g/L) drops from 26.5 to 22.5, whereas with 10% of EGDA, the LOI value drops only from 27.5 to 24.5. In the case of DEAEP, the LOI value decreases from 26 to 23.5

TABLE 8.2
LOI Values of Treated Fabrics before LOI_1 and after LOI_2, the Accelerated Laundering Method

	Treatment Conditions					
Monomer	Monomer (g/L)	% EGDA	$\%G_1$	LOI_1	$\%G_2$	LOI_2
DEAEP	300	5	30.6	26	17	23.5
		10	43.2	26.5	35.5	25.5
DMAMP	200	10	31.1	26.0	24.2	22.5
DEAEPN	200	…	20.9	26.5	8.7	22.5
		10	34.0	27.5	23.6	24.5
	300	…	38.1	28.5	17.8	25.0
		5	38.6	28.5	25.4	25.5
BisDEAEPN	200	5	29.8	29.5	26.7	25.0

Source: Tsafack and Levalois-Grützmacher (2006b).

with 5% of EGDA and only from 26.5 to 25.5 with 10% of EGDA (monomer concentration of 300 g/L). The LOI values of about 25 obtained with DEAEP, DEAEPN and BisDEAEPN demonstrate that the polymers are well grafted onto the cotton fabrics (Tsafack and Levalois-Grützmacher 2006b).

8.1.3 Plasma-Enhanced Synthesis of Flame-Retardant Cellulosic Materials

Natural polymers are a crucial part of today's life; they can be found nearly everywhere. Nowadays, natural polymer materials are having a comeback due to their specific properties (i.e., excellent mechanical properties, biodegradable, sustainable, etc.). However, one weak aspect of natural polymer materials compared with other materials is that they are combustible. Among them, cellulosic textiles are probably the most flammable (Totolin et al. 2010), with cotton being the most commonly used of all textile fibres. Thus, the majority of polymer-containing end products (e.g., cables, carpets, furniture cabinets, etc.) must have a satisfactory degree of fire resistance to ensure public safety from fire. Silicone materials have been produced commercially since the beginning of the 1940s. Over the past 60 years, silicone materials have grown into a billion-dollar industry, and they are used in many industrial applications in civil engineering, construction building, electrical, transportation, aerospace, defence, textiles and cosmetics (Totolin et al. 2010). The main structural elements of siloxanes have a direct or indirect influence on their stability at elevated temperatures, including inherent strength of the Si–O bond and pronounced flexibility of the siloxane $-[Si-O]_x-$ chain segments (Totolin et al. 2010).

Siloxanes have comparatively low heat-release rates (HRR), minimal sensitivity to external heat flux, and low yields of carbon monoxide release (Buch 1991). They also show a slow burning rate without a flaming drip and, when pure, no emissions of toxic smoke. Based on these fire properties, silicones offer significant advantages for flame-retardant (FR) applications. Unlike organic polymers, silicones exposed to elevated temperatures under oxygen leave behind a silica residue. Its shielding effects provide some of the fundamentals for the development of silicone-based fire retardants. Silica residue serves as an "insulating blanket," which acts as a mass-transport barrier, delaying the volatilization of decomposition products. Therefore, it reduces the amount of volatiles available for burning in the gas phase and, thus, the amount of heat that feeds back to the polymer surface. The silica residue also serves to insulate the underlying polymer surface from incoming external heat flux (Totolin et al. 2010).

It is believed that silicone-based flame retardants (FRs) used for synthetic or natural polymers act in two basic ways: by thermal quenching (endothermic decomposition of the FR) and by forming a protective coating (liquid or char barrier) (Quede et al. 2004). Because of increased market and societal demand for materials that can be processed by environmentally friendly methods, new and innovative production techniques are needed. These have led to further development of alternative physicochemical processing methods. In this field, plasma technology shows distinct advantages because it is environmentally friendly, and the surface properties of even inert materials can be easily modified (Totolin et al. 2009). The overall scheme of functionalizing cotton surfaces with silicone materials is shown in Figure 8.3.

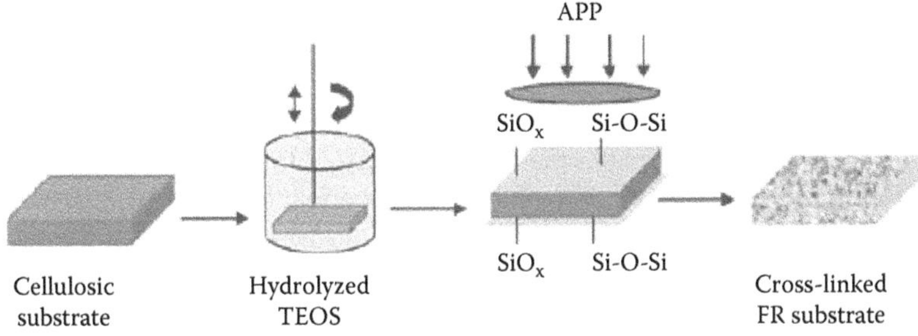

FIGURE 8.3 Scheme of cotton's dip-coating process. (From Totolin et al. 2010.)

The experimental parameters and the results from burning spread time of the plasma-treated samples showed that 30 (wt%) concentrations of the precursor (tetraethyl orthosilicate, TEOS) and medium radio frequency (RF) power (70 W) led to longer burning time spread for the treated samples in comparison to control (non-treated) ones. The difference in burning time spread was about two times longer for concentrations as high as 20 wt% and more than three times longer for 30 and 40 wt% of TEOS (Totolin et al. 2010).

In order to investigate the influence of experimental parameter space on the process and select the optimal conditions, a statistical design of experiments was conducted (Totolin et al. 2010). Four parameters were taken into consideration: TEOS concentration (wt%), RF power (W), plasma exposure time (minute) and temperature of the solution (°C) during the coating process. It can be noted that the TEOS concentration was the most influential parameter in the process, while the temperature of the solution was less effective. The burning time regression is a quadratic equation with 99.99% chance to express the factors' influences; the power versus time term has been removed due to its statistically low significance. Concentrations of 20–40 wt%, RF power of 35–80 W and plasma exposure time between 3 and 7 min yield to longer burning spread times. The regression model predicts burning time of 32.26 s for many sets of parameters around 33.7% concentration, 35.5 W RF power, 3.7 min plasma treatment time and 72.7°C TEOS solution temperature (Totolin et al. 2010).

The presence of the coating can be easily demonstrated by comparing SEM photomicrographs of an extracted control sample with SiO_2-coated (with different concentrations) and plasma-treated samples (Totolin et al. 2010). The surface morphology along the control fibre was wrinkled, which contrasts with the treated samples. The fibres covered with SiO_2 layers showed an absence of the wrinkled morphology due to the coating and plasma treatment. It should be noted that the space around the fibres was free of coating, which guarantees the breathability of the fabric (Totolin et al. 2010).

Cotton fibres have been coated with silicone dioxide–based layers using an atmospheric-pressure plasma (APP) technique. SiO_2 network armour was obtained through hydrolysis and condensation of the precursor TEOS, which was then cross-linked onto the surface of cotton fibres. Because of the protective effects of the SiO_2 network armour, the modified cellulose fibres exhibit improved flame-retardant properties. In the future, SiO_2-APP-coated textiles could have numerous applications in the development of upholstered furniture, clothing and military applications (Totolin et al. 2010).

8.1.4 Metal Absorption and Antibacterial Property on Cotton Modified by Plasma

Absorption of silver particles and subsequent antibacterial activity of cotton fabric modified by low-temperature plasma were investigated by Shahidi et al. (2010). The modification consisted of plasma prefunctionalization followed by one-step wet treatment with silver nitrate solution. Oxygen and nitrogen were used as the working gases in the system, and the results were compared. The results showed that nitrogen plasma-treated samples can absorb more silver particles than oxygen-treated samples, and thus the antibacterial activity of the nitrogen plasma samples was increased considerably, as analysed by the counting bacteria test.

Fourier transform infrared spectroscopy (FTIR) was used to examine the functional groups of the untreated and plasma-treated samples. An increase in absorbance at the 1720-cm^{-1} (C=O) and 3400-cm^{-1} (O–H) bands and the 2900-cm^{-1} (N–H) (positive) and 1080–1300-cm^{-1} (C–O) groups was noticed after plasma treatment (Nastase et al. 2005; Errifai et al. 2004; Carmen et al. 2005). These functional groups were produced on the fabric by the reaction between the active species induced by the plasma in the gas phase and the C-surface atoms. As can be seen, the effect of nitrogen plasma treatment is greater than that of oxygen. Compared with normal cotton, nitrogen plasma-treated cotton has an obvious new peak at 2900 cm^{-1}, which should be attributed to the N-H groups. These results indicate that the nitrogen plasma treatment successfully converted cotton into a cationic material.

The activation of the surfaces can help to absorb more silver particles. SEM micrographs of untreated and plasma-treated fabrics show the development of ripple-like patterns oriented in the fibre axis after plasma treatment. However, the surface morphology of the oxygen plasma-treated sample changed more significantly compared with that of nitrogen. As shown in SEM images (Shahidi et al. 2010), more silver particles were attached to the surface of prefunctionalised samples, and the size of the particles was reduced from 800 nm for untreated cotton to 250 nm after plasma treatment. It was also shown that the number of silver particles on the surface of the N_2 plasma-treated sample was more than on the O_2 plasma-treated samples (Shahidi et al. 2010).

The EDX (energy-dispersive X-ray analysis) results showed that the silver content on the plasma-treated cotton was more than that on normal cotton, and the absorption of silver particles due to nitrogen plasma prefunctionalization of the cotton samples was increased five times. Oxygen plasma pretreatment also increased the amount of silver absorption compared to the untreated cotton, but the difference was negligible (Shahidi et al. 2010).

The results of reflective spectrophotometry showed that the reflection factor of silver-loaded cotton decreased with plasma prefunctionalization, and the reflection factor of the sample pretreated with nitrogen plasma after incubation in silver nitrate had the lowest value. The K/S values of plasma pretreated samples were more than those of the untreated ones because, by using plasma prefunctionalization, more silver from the silver nitrate solution attached and was absorbed by the cotton samples. The absorption of silver particles was higher for the N_2 plasma prefunctionalised sample (Shahidi et al. 2010).

TABLE 8.3
Results of the Bacterial Counting Test

Samples	Percentage Reduction (%)	Percentage Reduction after Washing (%)
Untreated	15	8
Ag-loaded	90	85
Plasma-treated Ag-loaded O_2	95	92
Plasma-treated Ag-loaded N_2	100	99

Source: Shahidi, Rashidi, and Ghoranneviss et al. (2010).

The much higher silver content of the nitrogen plasma-treated cotton can be explained by considering that when cotton fibres are immersed in a silver colloid bath, negative charges resulting from dissociation of the functional groups of cellulose repulse the anions on the surface of the particles. Silver particles dispersed in aqueous solutions usually have a negative surface charge (Shateri Khalil-Abad, Yazdanshenas, and Nateghi 2009) and then create a repulsive force against the cellulose fibres, which inhibits the sedimentation of silver particles on the surface of the fibres. Consequently, silver particle adsorption is much lower on normal cotton than on nitrogen-plasma-treated cotton fibres. In the case of the N_2 plasma-treated cotton, large quantities of positive charges decrease the zeta potential of the fibre surfaces, which in turn increases the sedimentation of silver particles because of the greater attractive forces between the fibres and the silver (Z. Liu et al. 2007).

The silver absorption mechanism can be viewed as the complexation of the Ag with carboxyl groups and amino groups on the nitrogen plasma-treated samples during the absorption. From these preliminary evaluations, it is possible to conclude that the nitrogen plasma-pretreated cotton has great potential for application in water treatment for the removal of heavy metal ions. The antibacterial activity of the samples was tested, and the results are shown in Table 8.3. The results of the counting test showed more reduction of survival of bacteria in the case of pre-plasma modification. This is because the interaction between silver ions with bacteria can change the metabolic activity of bacteria and eventually cause their death (Shahidi et al. 2010).

The antibacterial efficiency of washed samples was also tested. The results showed that after 10 cycles of washing, the antibacterial activity was reduced to 85% for untreated Ag-incubated cotton. But in the case of pretreating with nitrogen plasma, washing did not have any effect on the efficiency of antibacterial activity because of the greater absorption of silver particles by the sample pretreated with nitrogen plasma (Shahidi et al. 2010).

8.1.5 Antimicrobial Efficacy of Plasma- and Neem Extract-Treated Cotton

A thorough investigation on the antimicrobial activity of oxygen plasma and azadirachtin (neem extract)-treated cotton fabric was conducted by Vaideki et al. (2007). The hydrophilicity of cotton fabric was found to improve when treated with RF oxygen plasma. The process parameters such as electrode gap, time of exposure and oxygen pressure were varied to study their effect on improving the hydrophilicity of the cotton fabric. The static immersion test was carried out to assess the hydrophilicity of the oxygen plasma-treated samples, and the process parameters were optimised based on these test results. The formation of carbonyl groups during surface modification in the plasma-treated sample was analysed using FTIR studies. The surface morphology was studied using SEM micrographs. The antimicrobial activity was imparted to the RF oxygen plasma-treated samples using methanolic extract of neem leaves containing azadirachtin. The antimicrobial activity of these samples was analysed and compared with the activity of the cotton fabric treated with neem extract alone. The investigation revealed that the surface modification due to RF oxygen plasma was found to increase the hydrophilicity and hence the antimicrobial activity of the cotton fabric when treated with azadirachtin (Vaideki et al. 2007).

The samples were oxygen plasma treated at low pressures of 0.04, 0.06, 0.08 and 0.1 mbar, for exposure times of 5, 7, 10, 12, 15 and 20 min, having the electrode gap as 2, 3, 4 and 5 cm, for a fixed RF power of 40 W. At this low power, there is no possibility of sputtering of electrode material, since they are made up of stainless steel. The investigation was carried out at low pressure, since the mean free path of the chemically active species in the gas phase is greater than the distance between the fibres that form the thread (Poll, Schladitz, and Schreiter 2001). As a result, the collisions in the gas phase are greatly reduced. The particles in the gas phase react with the cellulosic radicals formed due to plasma exposure on the textile surface, thus modifying it very effectively (Vaideki et al. 2007).

The hydrophilicity of these samples was assessed using a static immersion test, and after receiving the antimicrobial finish, the same samples were subjected to an agar diffusion test and a modified Hohenstein test to check the efficacy of their antimicrobial activity. The maximum absorption percentage of the sample due to oxygen plasma treatment for a time of exposure of 10 min had the maximum antimicrobial activity from the treatment with neem extract. There was variation in the absorption percentage of the samples treated at various oxygen pressures, and the maximum absorption percentage was achieved at a pressure of 0.06 mbar, which provided the maximum efficacy of the antimicrobial treatment. The variation of hydrophilicity (absorption percentage) and antimicrobial efficacy with respect to electrode gap showed that the maximum hydrophilicity—and hence antimicrobial activity—was obtained for an electrode gap of 3 or 4 cm (Vaideki et al. 2007).

The effect of plasma exposure time, oxygen pressure and electrode gap on the hydrophilicity of the oxygen plasma-treated sample confirms the attainment of maximum hydrophilicity for 40-W power, oxygen pressure of 0.06 mbar and for a time of exposure of 10 min while maintaining the electrode gap at either 3 or 4 cm. This result may be attributed to the production of a high concentration of chemically active species—excited oxygen molecules, oxygen radicals, free electrons and oxygen ions—for the optimised process parameters. These chemically active species impinge on the sample surface, which results in the generation of cellulosic radicals on the fabric surface. These radicals are a result of any one of the following mechanisms (Figure 8.4) (McCord et al. 2003; T. Ward et al. 1979):

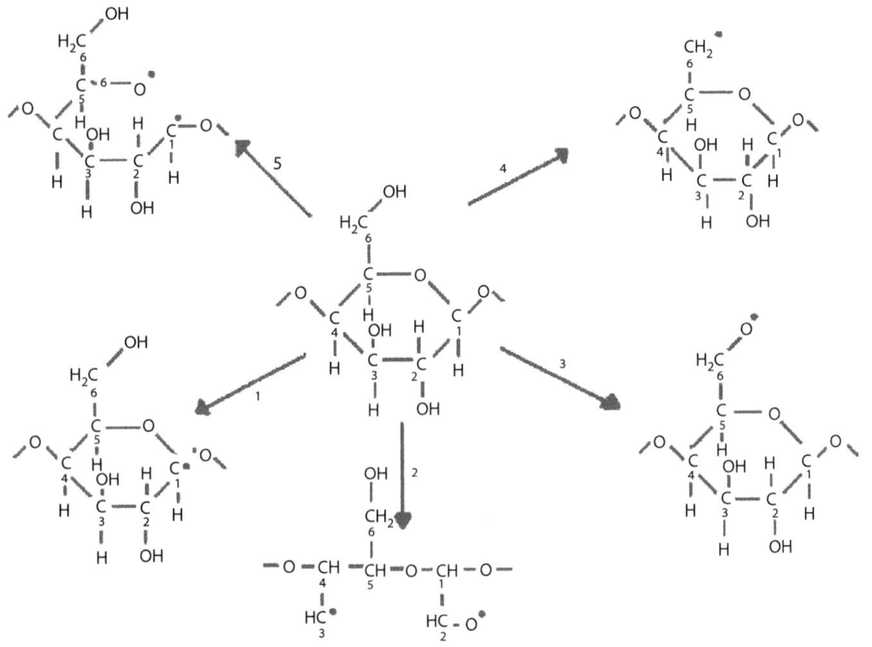

FIGURE 8.4 Cellulosic radical formation (1) bond breakage between C_1 and glycosidic bond oxygen, (2) dehydrogenation and dehydroxylation between C_2 and C_3 after the ring opening, (3) dehydrogenation at C_6, (4) Dehydroxylation at C_6 and (5) bond breakage between C_1 and ring oxygen. (From Vaideki et al. 2007.)

1. Bond breakage between C_1 and glycosidic bond oxygen
2. Dehydrogenation and dehydroxylation between C_2 and C_3 after the ring opening of anhydroglucose
3. Dehydrogenation at C_6
4. Dehydroxylation at C_6
5. Bond breakage between C_1 and ring oxygen

Of the five mechanisms, those that would result in the formation of carbonyl (aldehyde) groups, which have a greater polarity than hydroxyl groups, are bond breakage between C_1 and glycosidic bond oxygen, dehydrogenation at C_6 and bond breakage between C_1 and ring oxygen. The (0 2 0) cellulose peak in the X-ray diffractogram of the untreated and plasma-treated cotton fabric revealed the crystalline nature of the samples, which confirms the presence of a folded chain structure, where the polymer chain folds back and forth to form a parallel alignment (Vaideki et al. 2007). Because of this, the first mechanism, i.e., bond breakage between C_1 and glycosidic bond oxygen, can also be eliminated, as it would result in chain scission. The chemically active species in the plasma phase impinge on the sample surface and modify the surface molecules by means of chemical reactions with the active cellulosic radicals. This reaction leads to oxidation of aldehyde as well as hydroxyl groups, which results in the formation of carboxylic acid (carbonyl) groups which also have an increased polarity. This improves the hydrophilicity of the cotton fabric (Yousefi et al. 2003; Abidi and Hequet 2004; Tsai, Wadsworth, and Roth 1997). The hydrophilic nature of the carbonyl groups present in the oxygen plasma-treated samples increases the antimicrobial activity of the azadirachtin treatment (Vaideki et al. 2007).

8.1.6 Plasma-Induced Durable Water-Repellent Functionality and Antimicrobial Functionality on Cotton/Polyester Blend

Nanolayer thickness can be controlled through process parameters such as the plasma power, the precursor flow rate, and the speed at which the substrate moves through the plasma system (Albaugh, O'Sullivan, and O'Neill 2008). Research was conducted to determine if gas flow also affected the nanolayer thickness. The results of this research found that gas flow did not impact the thickness of the nanolayer produced (Davis, El-Shafei, and Hauser 2011). Previous studies have found that the thickness of the nanolayers formed through plasma treatments is usually in the range of angstroms to nanometers (10^{-10}–10^{-9} m) in thickness (Davis, El-Shafei, and Hauser 2011).

Water repellency on textile surfaces can be achieved through a fluorocarbon repellent finish. These fluorocarbon finishes have the advantage of very high repellency at relatively low percent add-on (usually less than 1% owf). However, they are relatively more expensive compared to silicon-based finishes, and there is a potential environmental concern about fluorocarbon derivatives. Certain fluorocarbon finishes, especially those consisting of eight carbons in the perfluoroalkyl chain, can degrade to form perfluorooctanoic acid (PFOA). PFOA is of environmental concern because it bioaccumulates in the human body (Armitage, Macleod, and Cousins 2009). The Environmental Protection Agency has taken measures to limit the use of PFOAs and PFOA precursors in the industry (Davis, El-Shafei, and Hauser 2011). There has also been research into the potentially harmful effects of such compounds. This research found that mice exhibit adverse neurobehavioral responses to PFOA exposure (N. Johansson, Fredriksson, and Eriksson 2008). Research is currently being conducted in the laboratory at NC State University to find an alternative, such as a fluorocarbon with only six carbons in the perfluoroalkyl chain or aromatic-based fluorocarbons, which has similar or better water-repellent characteristics (Davis, El-Shafei, and Hauser 2011).

Several studies have been done on the use of plasma to impart water-repellent characteristics to fabric (Ceria and Hauser 2010; Leroux et al. 2008; Di Mundo et al. 2009). Most of these studies used fluorocarbon-based chemistry; however, none of them attempted a dual treatment with antimicrobials. One study attempted to combine a water-repellent finish with a flame-retardant finish (Tsafack and Levalois-Grützmacher 2007). Moreover, a review paper entitled "Non-Thermal Plasma Treatment of Textiles" provided a comprehensive review on how non-thermal plasma was used effectively to impart different properties to textiles such as hydrophilic, hydrophobic and oleophobic properties, etc. (Morent et al. 2008).

Microbial activity can be detrimental to textiles. It can cause unpleasant odours, lead to weakening of the substrate, produce discoloration, and even contribute to the spread of disease. Because of this, antimicrobials have been investigated as a finish for textiles. A large class of antimicrobial agents is based on silver chemistry. Organo-silver compounds, silver zeolites, and silver chloride nanoparticles are all possible antimicrobial textile finishes. While these antimicrobials are leaching types and are "used up" after a set amount of time on the fabric, there are several alternatives which are bound types. For example, silicone quaternaries can exhaust to the fabric surface. During a curing step, they self-react to form a coating, and can be used on a wide range of fibres. Being ionic, they can act as a multifunctional finish, imparting both antistatic and hydrophilic properties while also providing antimicrobial activity (Davis, El-Shafei, and Hauser 2011).

Polymers can also be used as antimicrobial finishing agents. For example, polyhexamethylene biguanide is a polymer which is durable on cotton. It can be exhausted or pad applied, without the requirement of a curing step, and yet still has good bound antimicrobial activity. It is durable on cotton through strong hydrogen bonding between the amine groups and the cellulose structures (Wallace 2001).

Studies have been conducted on the use of plasma treatment to impart an antimicrobial finish to textiles. Some of these studies used the plasma to treat the substrate with a precursor with which an antimicrobial agent could be reacted (Gawish, Matthews, et al. 2007; Wafa et al. 2007; Gawish, Ramadan, et al. 2007; Vaideki et al. 2008), while a different study used plasma to graft polymerise an antimicrobial polymer (Thone et al. 2003). The monomer used in that study was a quaternary ammonium-based antimicrobial, diallyldimethylammonium chloride (DADMAC). However, none of these previous studies attempted to impart an antimicrobial treatment along with a separate functional finish to a single substrate (Davis, El-Shafei, and Hauser 2011).

The plasma used was an atmospheric glow-discharge plasma (Davis, El-Shafei, and Hauser 2011). Compared to corona-discharge plasma and dielectric barrier-discharge plasma, plasmas generated by glow discharge have the highest electron density, concentration of reactive species, and power density of all at 10^{12} electrons/cm^3, 10^{16}/cm^3 and >10 W/cm^3, respectively. Moreover, plasmas generated by glow discharge are classified as non-thermal plasma because the temperature of the electrons is much higher than that of ions and neutral species. Because electrons are extremely light, they move much faster than the other species present and have no heat capacity (Selwyn et al. 2001). Atmospheric non-thermal high-density plasma produced by glow discharge is composed of a mixture of highly reactive species, e.g., ions, radicals, photons, electrons, and excited-state species. The ratio of the number of electrons to the number of ions and neutrals mainly depends on the geometry of the radio frequency (RF) electrode, gas flow rate, power input (Watts) and frequency (MHz). Furthermore, surface treatment of materials depends on all the aforementioned parameters and plasma exposure time (Davis, El-Shafei, and Hauser 2011).

Davis, El-Shafei and Hauser (2011) proposed that the fluorocarbon acrylate-based water repellents (1,1,2,2-tetrahydroperfluorodecyl acrylate [THPFDA] and 1,1,2,2-tetrahydroperfluorododecyl acrylate [THPFDDA]), react through a chain-growth free-radical polymerization of the vinyl group ($CH_2{=}CH$) of the monomer. The π bond breaks homolytically via a free-radical initiator, which is supplied by the plasma treatment and exists on the activated fabric surface and is deposited on the monomer as well following the monomer deposition. Once the π bond breaks homolytically, the free-radical graft chain polymerization ensues.

As seen in Figure 8.5, the graft polymerization occurs when the monomer reacts with a free radical on the surface of the fabric. The free radical attacks the double bond of the monomer, opening the bond and forming a new bond between the fabric and the monomer, through Reaction 1. This leaves the free radical on the monomer, which may undergo different reaction pathways. Two such possible reactions are shown where the free radical begins on the second carbon atom of the fluorocarbon monomer. The reaction may continue through pathway A, in which the free radical attacks an unreacted monomer. This continues the free-radical chain polymerization until the propagating species are terminated. Or the polymerization reaction may terminate by abstracting a hydrogen atom, as seen in path B. Path A will lead to polymer or an oligomer grafted to the surface of the substrate, while path B will lead to monomer grafted to the surface of the substrate. As both paths result in a fluorocarbon chain oriented vertically on the substrate, both are expected to provide durable water-repellent properties (Davis, El-Shafei, and Hauser 2011).

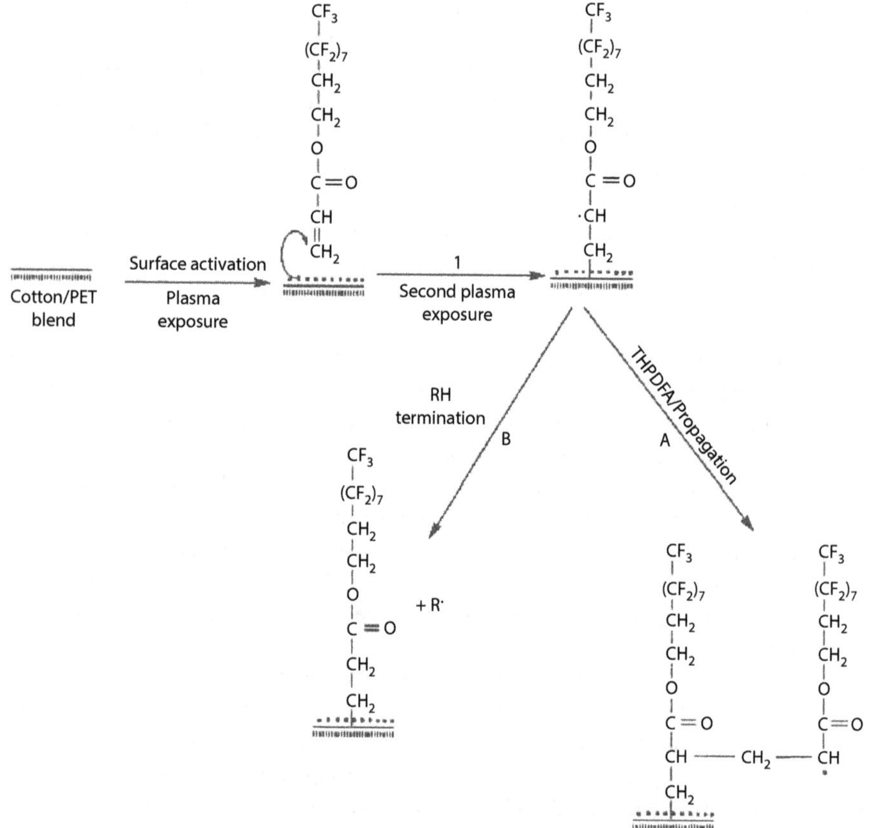

FIGURE 8.5 Proposed reaction mechanisms and polymerization. (From Davis, El-Shafei, and Hauser 2011.)

The contact angles of water droplets on the surface of the treated fabrics were measured using a goniometer. The contact angle was measured on the treated side of the fabric. The high contact angles of water droplets on the surface showed that the treatment resulted in a fabric which was hydrophobic. The treatment was also semi-durable, in that some hydrophobicity remained after washing in most of the samples. Samples treated with 2.13% add-on of THPFDA, and subjected to a 10-s surface activation and a 10-s surface polymerization treatment had the best durability, both at 600 W of RF plasma power. Samples treated with the same percent add-on of monomer and the same surface polymerization plasma treatment, but without a surface-activation step, had the best durability and good initial water repellency. However, after the accelerated wash cycle, the textile absorbed the water drop completely, as evidenced by a contact angle of zero. This shows that the surface-activation step, which aids in the free-radical chain polymerization *from* the surface furnishes a grafted polymer to the surface that is more durable than free-radical chain polymerization that commence from the monomer *to* the surface. This confirmed that graft polymerization *from* the surface is significantly more effective than that *to* the surface of the substrate. It is likely that the presence of free radicals on the substrate increases the degree of graft polymerization, while if the free radicals are generated on the monomer, homopolymerization ensues, which can be washed off easily (Davis, El-Shafei, and Hauser 2011).

By contrast, the samples which were treated with the water-repellent treatment, followed by the DADMAC antimicrobial treatment, failed to maintain a hydrophobic surface. These samples readily absorbed water on contact, making contact-angle measurements impossible to perform (Davis, El-Shafei, and Hauser 2011).

The log reduction of the bacterial activity was calculated, and it was noted that the sample treated with water repellent only decreased the antimicrobial activity of the fabric. A negative log reduction indicates more bacterial growth than an untreated sample. Although the reasons for this are unknown, it makes the difference with the samples that were treated with the DADMAC-based antimicrobial finish more significant. These samples were antimicrobial enough to overcome the starting disadvantage caused by the water-repellent finish and impart substantial antimicrobial functionality to the fabric (Davis, El-Shafei, and Hauser 2011).

8.1.7 Plasma Pretreatment on the Wrinkle-Resistance Properties of Cotton Treated with 1,2,3,4-Butanetetracarboxylic Acid

The cross-linking of cotton with 1,2,3,4-butanetetracarboxylic acid (BTCA) has provided an alternative route for the formaldehyde-free wrinkle-resistance treatment of cotton fabrics (B. Kim, Jang, and Ko 2000; Yoon, Woo, and Seo 2003; Bhattacharyya, Doshi, and Sahasrabudhe 1999; Sricharussin et al. 2004). The cross-linking reaction is enhanced by catalysts, such as sodium hypophosulfite (SHP) and titanium dioxide (TiO_2) (Yuen et al. 2007; C. C. Wang and Chen 2005a, 2005b; C. Chen and Wang 2006). To enhance the BTCA treatment with TiO_2, surface modification of the cotton fibre is required. However, most of the surface modification processes involve the use of chemicals. Because of environmental concerns, a dry treatment provides an alternative for surface modification. Among various dry surface treatments, plasma is a promising surface-modification technique that can modify the material surface properties and composition without altering the bulk properties. In the past, plasma treatment has been conducted in vacuo and, thus, may not be a continuous process. However, in recent years, atmospheric-pressure plasma jets have been developed and widely used in the textile industry to modify the fabric surface in an environmentally friendly way that reduces the consumption of wet chemicals and energy (C. X. Wang and Qiu 2007; De Geyter et al. 2007).

Plasma gas contains activated species that are able to initiate chemical and physical modifications at the fabric surface to cause an etching effect (De Geyter et al. 2007; C. X. Wang et al. 2008; Stephen 2004; Rajpreet et al. 2004). The process itself can improve the wrinkle-recovery ability of fabrics. The nature of the gas plays an important role in plasma treatment. One can vary the characteristics of the plasma by changing the gas used. In addition, the effectiveness of plasma treatment may also be affected by the plasma-treatment conditions, such as treatment duration, gas flow rate, applied power, jet-to-substrate distance, and pore size or structure (C. X. Wang and Qiu 2007; C. X. Wang et al. 2008).

Table 8.4 shows that after BTCA treatments, the wrinkle recovery angle (WRA) results of the plasma-pretreated cotton fabrics continuously increased from 65.7% up to 70.0% compared to the untreated cotton fabric. However, the WRA results of samples without plasma pretreatment were only increased slightly, from 52.2% to 61.3% compared with the untreated cotton fabric. This was because the plasma pretreatment had some unique features that enhanced the wrinkle resistance of the cotton fibres (Lam, Kan, and Yuen 2011).

TABLE 8.4
Optimised Plasma Pretreatment Conditions Subjected to Different BTCA Treatment Systems

BTCA Treatment System	Treatment Speed (mm/s)	Oxygen Flow Rate (L/min)	Jet-to-Substrate Distance (mm)	Dominating Factor	WRA of the Untreated Fabric (°)	WRA of Plasma-Pretreated Fabric (°)
0% BTCA	…	…	…	…	67.8	…
0% BTCA	10	0.1	4	…	…	71.9
5% BTCA and 10% SHP	10	0.1	4	Treatment speed	103.2	112.4
5% BTCA, 10% SHP, and 0.1% TiO_2	10	0.1	4	Treatment speed	107.5	113.7
5% BTCA, 10% SHP, and 0.2% TiO_2	10	0.1	4	Treatment speed	109.4	115.3

Source: Lam, Kan, and Yuen (2011).

The plasma could remove organic contamination from the fibre surface and could, thus, prevent the interference of bonding between the fibres and BTCA. The formation of chemical functional groups by a plasma process led to acid–base interactions and covalent linkages with the cross-linking agent BTCA; this resulted in an improvement of the adhesion between the fibre and the cross-linking agent. The etching effect of plasma reduced weak boundary layers and increased the surface area to allow more chemicals to be attached. Moreover, the microroughness of the fibre surface increased friction forces, and thus the wrinkle recovery increased obviously after the plasma pretreatment, which used helium as an inert gas. The plasma pretreatment induced a higher cohesive strength by the formation of a thin cross-linking layer, which stabilised the surface mechanically and against the diffusion of low-molecular-weight species to the interface. Therefore, plasma pretreatment improved the wrinkle resistance of the cotton fabrics (Lam, Kan, and Yuen 2011).

When only 5% BTCA and 10% SHP were used, WRA increased significantly to 52.2%. This was because the catalyst SHP accelerated the formation of anhydride intermediates, which in turn esterified the cotton cellulose. However, there were several disadvantages to the use of SHP as catalyst. It had a high possibility of causing shade changes in the dyed fabrics. In addition, the use of phosphorus compounds in textile finishing raises environmental concerns.

When phosphorous compounds are discharged into streams and lakes, they may serve as nutrients and accelerate the growth of algae. To reduce the disadvantages of using a phosphorous compound, some researchers have carried out studies to investigate the possibility of decreasing the amount of SHP by using TiO_2 as a cocatalyst.

As shown in Table 8.4, 5% BTCA in the presence of 10% SHP and 0.1% TiO_2 (as a cocatalyst) was used, and the addition of TiO_2 as a cocatalyst further increased WRA by 58.5%. This was because both TiO_2 and SHP accelerated the catalytic reaction through the formation of ester bonds between the cyclic anhydride ring and the hydroxyl group of cellulose. The improvement of WRA by the addition of TiO_2 in the BTCA treatment was probably due to the unique photocatalytic properties of TiO_2, which is a kind of N-type semiconductor. The hydroxyl radical (–OH) and superoxide anion ($-O_2-$) formed may have acted as catalysts to accelerate the formation of anhydrides from poly(carboxylic acid)s. Furthermore, the hydroxyl radical (–OH) and superoxide anion ($-O_2-$) may have had a significant effect on the increase of the charge localization of the solid cellulose medium in which the esterification and cross-linking occurred. Therefore, the WRA results of the cotton fibre treated with 5% BTCA, 10% SHP, and 0.2% TiO_2 further increased by 61.3% compared with those of the untreated cotton fabric. The increment was proportional to the increased amount of TiO_2 from 0.1% to 0.2% in the BTCA treatment bath (Lam, Kan, and Yuen 2011).

As shown in Table 8.4, there was an increasing trend after BTCA treatment in the presence of SHP (catalyst) and TiO_2 (cocatalyst). The WRA results of the plasma-pretreated fabrics further improved compared with those of the untreated fabrics. This confirmed that the plasma treatment with a 10-mm/s speed, a 0.1-L/min oxygen flow rate, and a 4-mm jet-to-substrate distance was the most effective method for improving the wrinkle-resistance properties of the BTCA-treated fabric in the presence of SHP (Lam, Kan, and Yuen 2011).

The treatment speed controlled the extent of material surface modification, and the optimum conditions prevented damage to the materials. The slow treatment speed might have provided sufficient time for the high concentration of active species generated from the plasma jet to accumulate on the fabric surface and cause an etching effect to alter the material surface or interior characteristics. Table 8.4 shows that a speed of 10 mm/s was the best treatment speed and provided enough time for the substrate to be bombarded by the concentrated active species produced in plasma gases, which enhanced WRA and minimised fibre damage. A high oxygen flow rate carried a high concentration of active species in the plasma jet and caused a relatively severe etching effect to alter the material surface characteristics. When the plasma-treated fabrics were subjected to the 5% BTCA treatment in the presence of 10% SHP, the treatment required a 0.1-L/min oxygen flow rate to maximise the potency of the treatment to contribute a better WRA. When the cocatalyst TiO_2 was added in the treatment, a 0.1-L/min oxygen flow rate was sufficient to roughen the fabric surface and allow TiO_2 particles to fill up the etched surface. The particles probably restricted the molecular movement of cellulose. In addition, the optimum jet-to-substrate distances for the different BTCA treatments were dissimilar. When the distance between the plasma jet nozzle and fabric surface was too small, the flow of the gas from the nozzle was almost blocked by the fabric, and the gas could only be bounced off the surface and fly out in a generally parallel direction to the fabric surface, which greatly reduced the effectiveness of the treatment. However, when the distance reached 6 mm, the velocity and activity of the active species in the plasma had greatly decreased by the time it reached the fabric surface, and thus, no effective plasma pretreatment occurred. Therefore, 4 mm was an acceptable range for the plasma gas flow on the fabric surface in this study (Lam, Kan, and Yuen 2011).

8.2 EFFECT OF PLASMA TREATMENT ON FINISHING PROTEIN FIBRES

8.2.1 Finishing Effect Imparted on Wool Fibre by Plasma Treatment

8.2.1.1 Shrinkproofing of Plasma-Treated Wool Fabric

It can be seen in Table 8.5 that there was a significant improvement in shrink resistance after the subsequent plasma treatment. Clearly, the shrinkage will increase as the processing is changed from relaxation shrinkage to felting shrinkage.

The wool fabric shrinkage is correlated with the frictional coefficient of the constituent wool fibres. It is also common knowledge that plasma treatment increases the dry and wet frictional coefficients in the scale and anti-scale directions. However, the effect of the plasma process is attributed to several changes in the wool surface such as (a) the formation of new hydrophilic groups, (b) partial removal of covalently bonded fatty acids belonging to the outermost surface of the fibre, and (c) the etching effect (Rakowski 1997). The first two changes contribute mainly to the increased wettability, while the last basically reduces the differential friction coefficients of the fibres, leading to the decreases in the natural shrinkage tendency (Kan and Yuen 2007).

TABLE 8.5
Results of Dimensional Changes (Lengthwise) of the Samples

	Relaxation Dimensional Change (%)		Consolidation Dimensional Change (%)		Felting Dimensional Change (%)	
Sample	**Warp**	**Weft**	**Warp**	**Weft**	**Warp**	**Weft**
Untreated	5.0	2.0	6.8	2.6	9.6	3.6
Plasma-treated	0.6	0.2	0.8	0.3	1.1	0.4

Source: Kan and Yuen (2007).

8.2.1.2 The Effect of Enzyme, Peroxide, Plasma and Chitosan Processes on Wool Fabrics and Evaluation for Antimicrobial Activity

Pretreated (enzyme and enzyme+hydrogen peroxide) knitted wool fabrics were treated with atmospheric argon and air plasma to improve their adsorption capacity (Demir et al. 2010). After plasma treatments, chitosan solution was applied to impart an antimicrobial effect on wool fabrics. The treated fabrics were evaluated in terms of washing stability as well as antimicrobial activity. The surface morphology was characterised by SEM images and FTIR analysis. From the results, it was observed that atmospheric plasma treatment had an etching effect and increased the functionality of a wool surface. Atmospheric plasma treatment also enhanced the adhesion of chitosan to the surface and improved the antimicrobial activity of the wool sample.

Argon was found to be more effective than air, since argon radicals played an important role in killing and removing bacteria. No significant difference in washing durability was observed in terms of plasma treatments. The samples of combined pretreatment processes had good washing durability even after 10 washing cycles. From the SEM images, it was observed that a combination of plasma and the other pretreatment processes gave less damage than only one process.

Protease enzyme was used to remove the scale cuticles or smooth the edges. Protease enzymes penetrate and degrade the internal structure of wool during processing. An oxidation process applied after enzyme treatment catalyses the fibre for further applications. For example, as an oxidizing agent, hydrogen peroxide (H_2O_2) in an aqueous alkaline medium favours the formation of the unstable perhydroxyl (HO_2-) species that transfers oxygen, and under these conditions, the disulfide bond is attacked, but this action also causes some fibre damage (Shen et al. 2007; Silva et al. 2005; Cardamone, Yao, and Nunez 2004).

Plasma-assisted coating of surfaces is a new and very promising application for effective antimicrobial treatments. These plasma-based antimicrobial treatments have been in the focus of scientific research and development for several years (Weltmann et al. 2008; Nasadil and Benesovsky 2008).

The effect of plasma, which is an eco-friendly process, is thought to be due to oxidation and etching reactions, which enhance the hydrophilicity of the fibre surface (Mori and Inagaki 2006). After plasma treatments, the carbon content of the sample is reduced because of the etching effect of atmospheric air and argon plasma treatment. This fact is clearly observed by X-ray photoelectron spectroscopy (XPS) analysis in a previous study (Demir et al. 2008). On the other hand, as a result of the oxidation of hydrocarbon chains located on the wool surface, an increase in carboxylic acid occurs (Demir et al. 2008; Molina et al. 2005).

In order to evaluate the changes on the surface of the untreated and plasma-treated wool samples, SEM observations were made (Demir et al. 2010). As seen in the SEM images, the enzyme, peroxide and atmospheric plasma treatments cause degradation on the surface of the fibre. The dominant effects of the two processes are different. In the plasma processes, etching is dominant and causes partial degradation, such as rounding scales and microcracks (Chvalinova and Wiener 2008; Demir et al. 2008; Mori and Inagaki 2006). Small amounts of substances are thought to be formed by the decomposition of lipid and protein material that are scattered over the surface of the plasma-treated wool fibre (Mori and Inagaki 2006). On the other hand, enzyme treatment and enzyme+peroxide treatment cause intensive degradation. When chitosan is considered, it can be observed that chitosan covers the scaly surface of the fibre and provides a smoother appearance without any damage (Demir et al. 2008). When the combinations are evaluated, it can be seen that they are more effective and less degradative than one process. Chitosan is known as an antimicrobial biopolymer (Lim and Hudson 2003). Although the antimicrobial mechanism is not clear, it is generally accepted that the primary amine groups provide positive charges which interact with negatively charged residues on the surface of bacteria. Such interaction causes extensive changes in the cell surface and cell permeability, leading to leakage of intracellular substances. In order to enhance the bonding efficiency of chitosan polymer with wool and to increase the uniformity of its distribution on the surfaces, the surface energy and anionic character of a wool surface must be promoted. In this way there will be an increase in the reactivity of the wool surface, and antimicrobial efficiency will be better due to the enhanced chitosan binding (Demir et al. 2008).

In another study, different pretreatment processes (enzyme treatment, enzyme+peroxide treatment, air plasma, argon plasma and combinations of these processes) were applied to modify the wool surface, promote chitosan binding and increase antimicrobial efficiency (Demir et al. 2010). By measuring the diameters of clear zones of inhibition against *Staphylococcus aureus* and *Klebsiella pneumonia* in all the treated and untreated samples, it was concluded that enzyme+peroxide+argon plasma pretreatment was the most efficient process.

Plasma is not only effective in killing bacteria and fungi, but it can also remove dead bacteria and viruses from the surface of the samples (Park et al. 2007; Demir et al. 2010). The electrons and ions in the discharge zone of the plasma have a great etching effect on the surface of bacteria, resulting in the crack of the cell walls or membranes and the leakage of protein and nucleic acid (Demir et al. 2010). Plasma treatment also enhances chitosan binding by promoting new anionic groups on the fibre, and these new functional groups cause some changes in the surface composition.

In the study by Demir et al. (2010), plasma treatment was found to have a sterilizing effect to some extent in the samples pretreated by enzyme+peroxide and not post-treated with chitosan. On the other hand, after the chitosan application, excellent antibacterial effects were obtained. When air and argon plasma were compared, argon plasma treatment was found to be more effective than air plasma treatment in terms of antimicrobial efficiency. So it can be said that argon radicals play an important role in killing and removing bacteria, especially against *Klebsiella pneumonia*, which is one of the Gram-negative bacteria. After the washing processes, the diameters of clear zones had decreased, but under the samples of combined pretreatment processes, an antimicrobial effect was still observed even after 10 washing cycles.

8.2.2 Finishing Effect Imparted onto Silk Fibre by Plasma Treatment

8.2.2.1 Hydrophobicity Improvement of Silk

The SF_6 plasma was applied to improve the hydrophobic property of Thai silk (Chaivan et al. 2005). It was produced by an inductively coupled 13.56-MHz RF discharge in a cylindrical stainless steel chamber that was 31.2 cm in diameter and 42.5 cm long. The plasma was confined by arrays of permanent-magnet buttons. The operating pressure was at 1, 3, 5 and 7 mTorr, while the RF power was varied from 25 to 75 W. The absorption times and contact angles were utilised to analyse the results of the treated samples. The hydrophobicity improvement of silk was achieved. The treated samples reached the limit of the absorption times at 180 min, and the contact angles increased to 130°–140°. These results show a significant increase in the hydrophobic property compared with the untreated sample. The optimum operating conditions were at an RF power of around 50 W and a pressure of 3–5 mTorr.

The plasma-treated samples show significant improvement in hydrophobic properties. Treatment time of 3 min showed a dramatic increase of absorption times compared with a 1-min treatment time. The absorption times of 3-min treatment reached the same limit as that of the 180-min treatment. The optimum treatment conditions were at a pressure of 3–5 mTorr and the RF power was about 50 W. Absorption times began to decrease at a pressure higher than 5 mTorr because of plasma instability (Chaivan et al. 2005).

The contact angles of the plasma-treated samples showed impressive results in improving hydrophobic properties. At all operating conditions, the contact angles increased up to 130°–140°. The contact angles derived from treatment times of 1 and 3 min were indistinguishable. This result also indicated that the surface energy decreased from 95 to 20 dyn/cm with a treatment time of 1 min. It should be noted that after 30 days and longer, the hydrophobic property could still be observed for the 3-min treatment. This hydrophobic life time was shorter for the 1-min treatment. Figure 8.6 displays SEM microscopic images of a single silk fibre. The SF_6 plasma tends to form thin films on the surface and increased microroughness of the sample surface. It is known that microroughness is due to the large van der Waals radius of fluorine attached with carbon composite in silk fibre. Because of the carbon composition in fibroin, which constitutes the main structure of silk, the presence of this microroughness is expected. The small amount of thin film as seen in Figure 8.6b should be the C-F film (Chaivan et al. 2005).

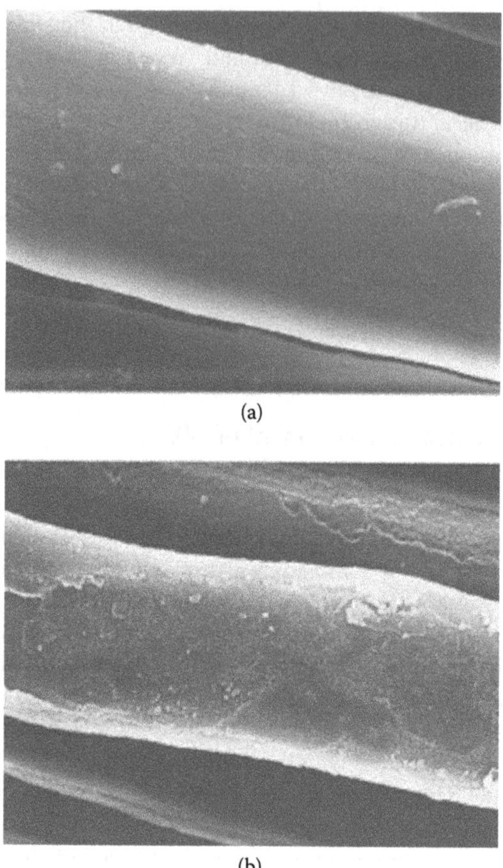

(a)

(b)

FIGURE 8.6 SEM micrographs of silk fibre at 5000´ showing different surface roughness between untreated (a) and treated (b) sample. The treatment condition is at an RF power of 50 W, a pressure of 3 mTorr and a treatment time of 3 min. (From Chaivan et al. 2005.)

8.2.2.2 Plasma-Induced Graft Polymerization of Flame-Retardant Compound onto Silk

Over the past decades, concerns over materials that represent a potential fire hazard have steadily increased. As a consequence, textile applications are subject to material-flammability regulations. Improving the flame-retardant property of textiles has become necessary to minimise the risks of fire hazard under many circumstances (Wichman 2003). Because of its luxurious appearance, silk is one of the most commonly used textiles for interior decoration, such as upholstery, curtains, and bedding. It is therefore of primary significance to improve the flame-retardant property of silk fabrics to comply with relevant safety regulations (Chaiwong et al. 2010).

Flame-retardant fabrics are typically prepared by treating the fabrics chemically with flame-retardant agents. Phosphorous-based compounds are the most extensively used (H. Yang and Yang 2005; Gaan and Sun, 2007; W. Wu and Yang 2007).

For natural-fibre textiles, a number of studies focus on the flame-retardant properties of cotton fabrics (Gaan and Sun, 2007; W. Wu and Yang 2007; Reddy, Agathian, and Kumar 2005; W. Wu and Yang 2006; Tsafack and Levalois-Grützmacher 2006b) and silk fabrics (Achwal, Mahapatrao, and Kaduska 1987; Guan, Yang, and Chen 2009). It was shown that a high level of flame retardancy could be achieved when silk fabric was treated by a reaction mixture of urea and phosphoric acid through a pad/dry process (Achwal, Mahapatrao, and Kaduska 1987). However, the treated silk had limited laundering durability. The flame-retardant agent—under the commercial name Pyrovatex CP, which is N-hydroxymethyl (3-dimethylphosphono) propionamide (HDPP)—was applied to induce flame retardancy on silk (Guan and Chen 2006). This compound needs formaldehyde, which is a human carcinogen, as the bonding agent.

A formaldehyde-free flame-retardant finishing process was recently developed by Guan, Yang, and Chen (2009). The treated silk showed improved flame retardancy with limited laundering durability. Although varying degrees of flame retardancy were obtained, the durability issue was difficult to solve. It is even more problematic when the textiles are from natural origins. The development of a satisfactory, durable, flame-retardant silk is indeed challenging, and alternative eco-friendly processes have to be considered.

Plasma treatment is a potential technique to impart flame-retardant properties to textiles. The reactive species in the plasma interact with the surface atoms or molecules and modify the surface properties without affecting bulk properties. Recently, it was reported that microwave plasma had been employed in a flame-retardant finishing process (Tsafack and Levalois-Grützmacher 2006b). However, low-pressure plasma systems need to operate under vacuum which, in turn, adds to the cost and complexity of the process. An alternative system would be one using an atmospheric-pressure plasma source. A few different designs have been developed and employed to modify the surface of materials (Cheng, Liye, and Zhan 2006; Schafer et al. 2008; Guimin et al. 2009). A system using an atmospheric-pressure plasma source is promising to industrial application, since it eliminates the vacuum system (Chaiwong et al. 2010).

Chaiwong et al. (2010) utilised an atmospheric-pressure plasma jet to graft phosphorus-based flame-retardant agent onto silk. The treated silk fabrics were submitted to a 45° flammability test. The incorporation of phosphorus was studied via energy-dispersive X-ray spectroscopy (EDS). The durability of the treatment was evaluated.

Figure 8.7 shows the burning behaviour of the silk samples prepared with different procedures. In the case of untreated silk (Figure 8.7a), the sample ignited instantly, with a rapid flame spread of 1.43 cm/s. The flame extended to the entire sample without evidence of burning smoke. For the sample directly immersed in PBS (phosphate buffer solution, PBS silk), the ignition character was identical to that of the untreated sample, but the flame spread terminated immediately (Figure 8.7b). The sample did not exhibit an afterglow. Burning smoke, as a consequence of char formation, was observed. The char formation is an indication of phosphorus-containing residue on the surface of the sample (Tsafack and Levalois-Grützmacher 2006b). The compound decomposed to polyphosphoric acid when

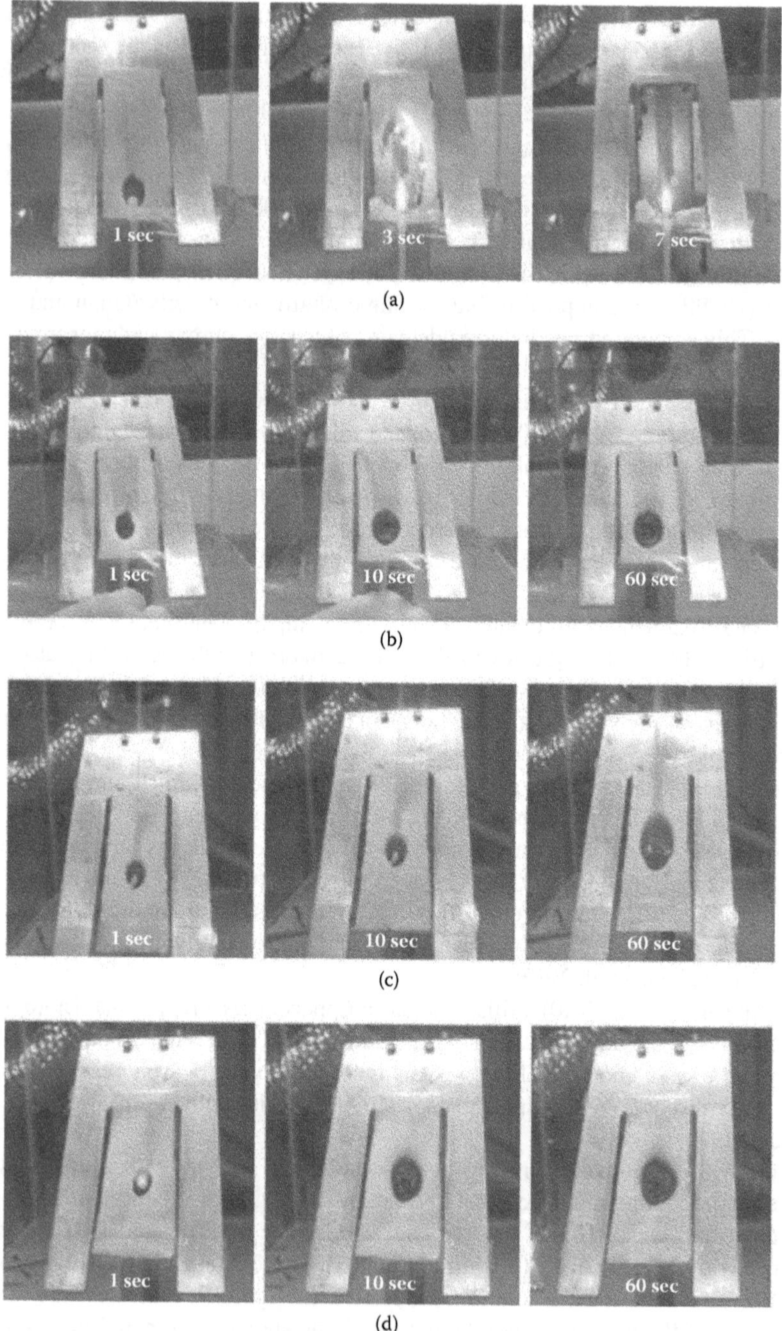

FIGURE 8.7 Burning behaviour of (a) untreated silk, completely burned within 7 s; (b) PBS silk; (c) washed Ar-PBS-Ar silk and (d) ethanol wash only. (From Chaiwong et al. 2010.)

heated and formed a viscous surface layer. This layer prevents oxygen from reaching the silk fibre. As a consequence, the fibre decomposition is inhibited (Chaiwong et al. 2010).

After the washing process, the burning behaviour of the PBS silk was similar to that of the untreated sample. Some burning smoke was observed. This is due to the fact that PBS is water soluble and can thus be removed from the silk during the washing process. The smoke indicated that some PBS remained in the silk. In contrast, the Ar-PBS-Ar silk (Figure 8.7c) behaved differently. Its flame spread rate was higher than the PBS sample. However, the flame vanished immediately without the afterglow. Char formation was observed, but the burning smoke was dramatically reduced to the amount that is close to the untreated sample. Since burning smoke mainly comes from the residual PBS on the surface of the sample, it can be said that most of the PBS molecules were grafted homogeneously into the silk molecular chains by the Ar plasma. A washing process might take away the un-grafted PBS molecules from the silk structure, but the majority remained intact in the silk structure. Hence, with an adequate level of grafted PBS molecules, silk samples can generate char to prevent flame spread without excess burning smoke (Chaiwong et al. 2010).

By comparing the SEM micrographs of the PBS silk (Figure 8.8a) and the washed Ar-PBS-Ar silk (Figure 8.8b), the grafting of PBS can be observed. As shown in Figure 8.8a, PBS particles deposited locally on the knot of the silk yarn. The surface topography along the yarn was relatively smooth. In contrast, the yarn of the washed Ar-PBS-Ar silk was rough and uniformly covered with the PBS particles. It is evident that the durable flame-retardant property of silk can be obtained via Ar plasma grafting (Chaiwong et al. 2010).

The EDS spectrum (Figure 8.9) showed evidence of phosphorus arising from the PBS compound in silk after plasma treatment. Peaks of silk compositions, such as N, C and O, were revealed. Calcium is one of the fingerprints of natural silk. Quantitative analysis of phosphorus content in the samples was done by means of EDS. The phosphorus content in the Ar-PBS-Ar was found to be 11 wt% higher than that in the PBS silk, whose phosphorus content was 7 wt%. This high level of phosphorus content in the Ar-PBS-Ar silk remained constant after the washing process. The results clearly indicate that in order to achieve durable flame-retardant properties, graft polymerization is necessary. The Ar plasma jet used in this work allowed us to bind covalently the flame-retardant compound to the silk fabric. One can say that after the washing process, the Ar-PBS-Ar sample was similar to the ordinary silk with the addition of a flame-retardant property (Chaiwong et al. 2010).

The washed Ar-PBS-Ar silk sample was characterised by ATR-FTIR (Attenuated Total Reflectance Fourier Transform Infrared) spectroscopy in comparison with the untreated control sample (Chaiwong et al. 2010). Graft polymerization via Ar plasma was indicated by the presence of bands at 1196 cm^{-1} (C–O stretching vibration), 1078 cm^{-1} and 919 cm^{-1} (P–O–C stretching vibration). The P=O stretching vibration that indicates the PBS compound overlapped within the C–O band. The IR peak intensity changes seem relatively low, indicating the very thin layer of graft-PBS on the silk surface from plasma treatment.

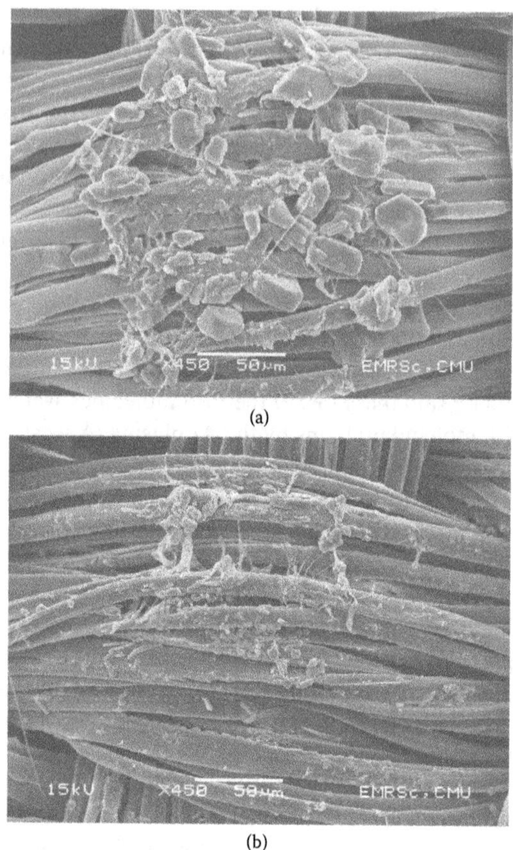

FIGURE 8.8 SEM micrograph of (a) PBS silk and (b) washed Ar-PBS-Ar silk. (From Chaiwong et al. 2010.)

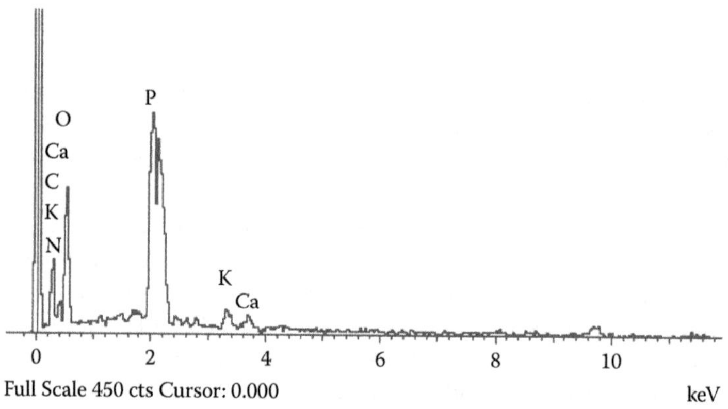

FIGURE 8.9 EDS spectrum of washed Ar-PBS-Ar silk. (From Chaiwong et al. 2010.)

8.3 EFFECT OF PLASMA TREATMENT ON FINISHING OF SYNTHETIC FIBRES

8.3.1 Finishing Effect Imparted on Polyester Fibre by Plasma Treatment

8.3.1.1 Antistatic Effect

The effect of discharge power of low-temperature plasma treatment on the antistatic property of polyester fabric was studied by Kan (2007). The results revealed that when the discharge power increased, the half-life decay time decreased accordingly, which means that a better antistatic property was achieved. When the discharge power increased, the plasma gas could obtain more energy for ionisation and could be ionised more easily. As a result, the concentration of active species would be increased. On the other hand, the speed of the electrons under a strong electric field would be accelerated, resulting in an increment of kinetic energy for the electrons. Both factors added together would greatly increase the action of the plasma on the fibre surface. The surface action would cause the introduction of surface roughness and oxygen polar functional groups in the polyester fabric, which may increase the static dissipation (Wong et al. 2000a, 2000b; Y. Liu, Xiong, and Lu 2006).

The effect of system pressure of low-temperature plasma treatment on the antistatic property of polyester fabric was studied by Kan (2007). Figure 8.10 shows that when the system pressure decreased, the half-life decay time increased accordingly, which means that the antistatic property was adversely affected. When the system pressure is low, the number of collisions between the plasma species and other reactive species would be reduced. As a result, there would be less kinetic energy lost during collisions, and the reactive species could carry a relatively higher kinetic energy when interacting with the surface. This interaction would result in the modification of the polyester fibre surface and hence affect the physical and chemical compositions of the polyester fibre surface.

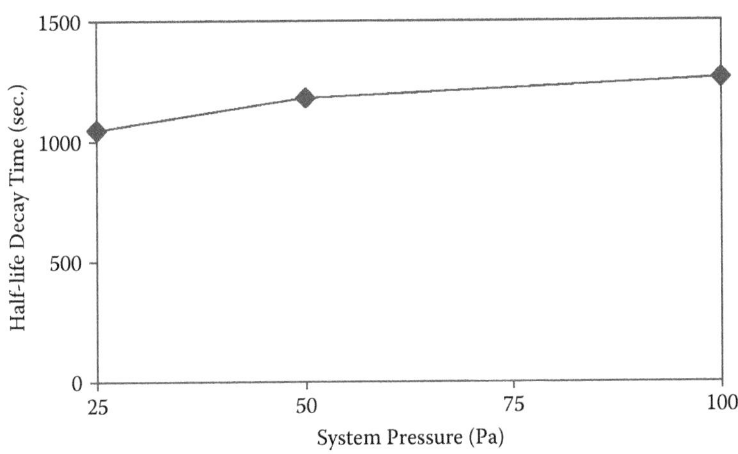

FIGURE 8.10 Effect of system pressure of low-temperature plasma treatment on the antistatic property of polyester fabric. (From Kan 2007.)

The effect of pressure-treatment duration of low-temperature plasma treatment on the antistatic property of polyester fabric was examined by Kan (2007), and the results show that when the treatment duration increased, the half-life decay time decreased accordingly, which means that the antistatic property is improved with a longer treatment duration. When a longer treatment duration is used, more interaction would occur between the fibre surface and the plasma species. As a result, the alteration on the fibre surface would become more significant.

An increase in the moisture content of polyester fabrics would reduce the half-life decay time, i.e., it would improve the antistatic property of polyester fabric with statistical significance (Kan 2007). As moisture contains water which is polar in nature, the conductivity of the water molecules are better than the polyester fabric; therefore the localised static charge on the polyester fabric surface would leak away and the static charges would thus be dissipated more easily. Also, the moisture film formed on the polyester fabric surface may evaporate in air while carrying a sufficient amount of static charges away from the surface. Therefore, if the static charges are dissipated into the air, then the amount of static charges remaining on the polyester fabric would be decreased. As the moisture content was inversely proportional to the half-life decay time of the polyester fabrics, the mechanism of the low-temperature plasma treatment for improving the antistatic property of polyester would be to increase the moisture regain of the polyester fabric, thereby decreasing the half-life decay time of the polyester fabrics.

SEM images were used to comprehend the alteration of surface morphology of the polyester fabrics (Kan 2007). Images of the untreated polyester fibre clearly demonstrate that the fibre surface is smooth and free from roughness, indicating that there is no damage on the fibre surface. This smooth outer surface of polyester can not enhance the absorption of moisture, and hence the moisture regain of the untreated polyester fabric was generally very poor. However, SEM images of fabric samples treated with low-temperature plasma illustrate a change in the fibre surface morphology with voids and pores. The low-temperature plasma treatment clearly imparted a significant alteration on the fibre surface due to its etching action, thereby causing surface roughness (Sharnina 2004; Wong et al. 2000a, 2000b; Y. Liu, Xiong, and Lu 2006; Kan et al. 1998a, 1998b). The rough surface can provide greater capacity for capturing moisture in the air, and this moisture could subsequently penetrate into the polyester fibre, where the increased moisture content would improve the antistatic property of the polyester fabric. In the case of a commercial antistatic finishing agent, it was clearly shown that the agent was deposited on and well covered the fibre surface to perform its function. Therefore, based on the surface morphology of the differently treated polyester fabrics, it could be concluded that the low-temperature plasma treatment and commercial antistatic finishing agent have different mechanisms for improving the antistatic property of polyester fabric (Kan 2007).

Table 8.6 shows a comparison of half-life decay time and moisture content between differently treated polyester fabrics. The half-life decay times of polyester fabric after low-temperature plasma treatment and antistatic finishing were greatly improved from an average 1675.5 s to 286 s and 157.5 s, respectively. Therefore, static problems were reduced for both of the differently treated polyester fabrics. The moisture contents of the untreated polyester and the polyester finished with

TABLE 8.6
Comparison of Half-Life Decay Time and Moisture Content between Differently Treated Polyester Fabrics

Sample	Half-Life Decay Time (s)	Moisture Content (%)
Untreated	1675.5	1.009
Low-temperature plasma treated	286	4.99
Antistatic agent finished	157.5	1.41

Source: Kan (2007).

a commercial antistatic agent were not significantly different. However, this was not the case for the low-temperature plasma-treated polyester fabric, which showed a significant increase in moisture content. This means that the improved antistatic properties obtained on polyester fabrics from two different treatments—one based on low-temperature plasma treatment and the other based on an antistatic finishing agent—were achieved via different antistatic mechanisms (Kan 2007).

The antistatic finishing agent used in the study (Kan 2007) was a hydrophilic polymer with hydroxyl-functional polysiloxane, which will adhere on the polyester fabric surface to increase fibre surface conductivity by forming an intermediate layer on the surface. This layer is typically hygroscopic, with the presence of mobile ions which may increase the moisture content, leading to higher conductivity (Goyal and Deshpande 2006; Xu, Wang, and Chen 2005) and hence increasing the dissipation of static charges.

In the case of low-temperature plasma treatment with oxygen, based on the experimental results, the surface wettability can alter the antistatic property of the polyester fabric. Surface wettability is directly related to surface energy, with a more energetically stable surface resulting in a less wettable surface. It is now established that low-temperature plasma modification of textile fibres results in oxidation and degradation (voids and pore formation) of the fibre surfaces. The oxidation creates oxidised functionalities, which lead to an increase in surface energy, while the degradation mainly changes the surface morphology of the fibres (Luo and van Ooij 2002; Rashidi et al. 2004). SEM images have shown that low-temperature plasma treatment increases the surface roughness. According to the Wenzel equation ($\cos\theta_{rough} = r \cos\theta_0$) (Y. Liu, Xiong, and Lu 2006), the surface roughness influences the contact angle (Sprang, Theirech, and Engermann 1995, 1998), where θ_{rough} is the contact angle on a sample surface, θ_0 is the contact angle on the smooth surface, and r is the roughness (ratio of the actual area of the interface to the geometric surface area). When the surface has a contact angle smaller than 90°, increasing surface roughness probably reduces the contact angle and will contribute to the improved surface wettability. Water is a conductor of electricity, and the improved surface wettability will therefore decrease the accumulation of electrostatic charges (Kan 2007).

The increase of surface roughness also induces an increase in the specific surface area, leading to a more moisture-rich surface, which enhances the conductivity of the fibres. The low-temperature plasma treatment not only increases surface roughness, but also introduces hydrophilic groups onto the fibre surface. XPS analysis has shown that low-temperature plasma treatment can increase oxygen content on the fibre surface, suggesting the possibility of introducing oxygen-containing polar groups such as –OH, –OOH and –COOH (Yip et al. 2003; Wong et al. 2000a, 2000b). There are two possibilities of generating the polar groups: (a) generated by reacting with the ambient gas during the low-temperature plasma treatment and (b) generated when the polyester samples are exposed to air after low-temperature plasma treatment. In either case, low-temperature plasma treatment produces a considerable amount of unsaturated bonds, and then the unsaturated bonds react with atmospheric oxygen to form polar groups on the polyester fabric surface.

As low-temperature plasma treatment increases the amount of oxygen-containing polar groups on the polyester fibre surface, these polar groups will incorporate with moisture through hydrogen bonding and enhance moisture penetration and binding on the fibre surface (Y. Liu, Xiong, and Lu 2006). Under the action of water molecules, these polar groups will generate ionisation and lead to a structural layer of conducting electricity on the fibre surface, which enhances the electrostatic dissipation. Therefore, the half-life decay time of the fibres decreases after low-temperature plasma treatment (Kan 2007).

8.3.1.2 Water-Repellency Property

Polyester fabric was treated with dichlorodimethylsilane (DCDMS) solution by two methods: (a) dipping the fabric directly in DCDMS solution for different intervals and (b) dipping the fabric in DCDMS solution after exposure in a RF plasma chamber for different durations at optimised conditions of exposure power. The physical properties of polyester fabric treated with DCDMS in the presence or absence of air plasma were compared with a control fabric. The results show that exposure of polyester fabric in plasma caused a continuous weight loss, which may be attributed to the etching action and cleaning process (Jahagirdar and Tiwari 2007).

In the case of polyester fabric pretreated in plasma followed by DCDMS treatment, a continuous weight gain was observed with increasing treatment time. For those polyester fabrics treated directly with DCDMS, there was no weight gain. Exposure of polyester fabrics to plasma before DCDMS treatment resulted in greater deposition of silane groups onto the surface compared to the fabrics directly treated with DCDMS solution (Jahagirdar and Tiwari 2007). The weight gain in the plasma-pretreated fabrics can be attributed to the formation of silane layers onto the polyester surface during plasma pretreatment (Jahagirdar and Tiwari 2007).

The effect of DCDMS on the visual appearance of fabric was assessed by measuring the colour parameters of the untreated and treated polyester samples (Jahagirdar and Tiwari 2007). The results showed that DCDMS treatment did not make any difference on the fabric's visual appearance, as the yellowness index (YI) and b* values did not show any remarkable change. This lack of colour change is mainly due to the non-reaction of the polyester fabric with the HCl vapours that are generated during the treatment of DCDMS solution on polyester fabric. The bulk property of the polyester fabric also was not altered much after its treatment with DCDMS solution, which is again mainly due to non-reaction of the polyester fabric with liberated HCl vapours during immersion. On the other hand, the polyester fabrics that were exposed to plasma for a short duration prior to treatment with DCDMS solution did show some improvement in bulk properties (Jahagirdar and Tiwari 2007).

It is now known that the formation of silane groups on any textile material improves its water retardancy. Once this had been established, the next step was to check the water-repellency behaviour of the treated polyester fabrics. The wettabilities (or hydrophilicities) of untreated and modified polyester fabrics were measured using a water-drop test applied according to the AATCC (American Association of Textile Chemists and Colorists) standard without using any soap or detergent. The samples were then dried in air at room temperature (Jahagirdar and Tiwari 2004). It was found that the DCDMS-modified polyester fabric with prior exposure to plasma treatment even for 10 s did not absorb the water droplets for up to 1 hour, whereas those fabrics treated directly with DCDMS solution absorbed the water droplets at around 50–55 min. Both samples were also washed up to 10 cycles, but no changes in the modified property were observed (Jahagirdar and Tiwari 2007).

An ATR-IR study was carried out in order to investigate the effect of plasma treatment with DCDMS on the surface of the polyester fabric. Figure 8.11 demonstrates

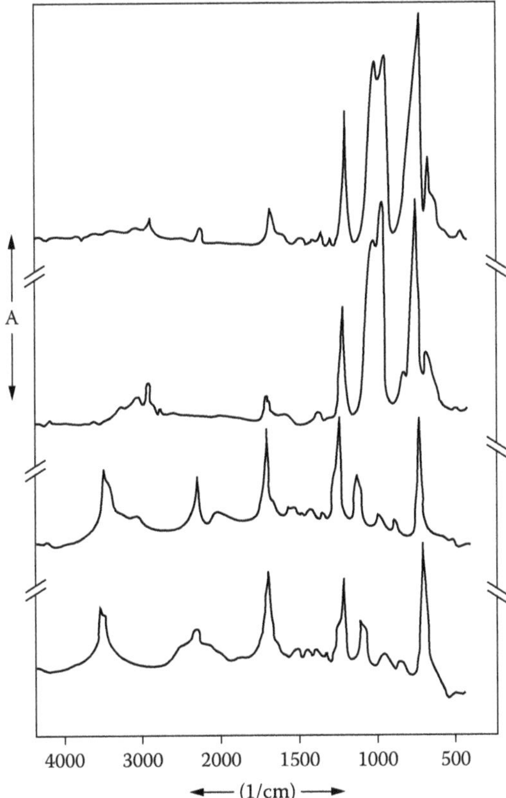

FIGURE 8.11 IR spectra of polyester fabric: (a) untreated, (b) 2-min plasma exposed, (c) 2-min DCDMS treated and (d) 2-min DCDMS treated after exposure to plasma for 2 min. (From Jahagirdar and Tiwari 2007.)

the ATR spectra of untreated and differently processed polyester fabrics. From Figure 8.11, it can be seen that plasma exposure of polyester fabric did not cause any remarkable change in the spectrum as compared to untreated polyester fabric except for changes in the relative intensities of the stretching vibrations of the functional groups, which determine its various properties. But when the fabric was treated with DCDMS solution, some new peaks can be seen when compared with the spectrum of the control fabric. Similar observations can also be made in the case of prior exposure to plasma before treatment (Jahagirdar and Tiwari 2007).

In the sample treated with DCDMS solution, the absorption at 2964.4 cm^{-1} is due to the CH_3 stretching vibrations. The peak at 1713.7 cm^{-1} is due to the stretching vibrations of the C=O group. This can be seen by the reduction in the relative intensity of the C=O group after the treatment with DCDMS solution. The peak at 1258.4 cm^{-1} is due to the Si-CH_3 deformation vibration. The shoulder at 1079.1 cm^{-1} is because of the Si–O linkage. The peaks at 1020.1, 864.7 and 794.5 cm^{-1} are due to Si–O–C, stretching vibration of Si–C and stretching vibration of the Si–Cl group, respectively. So in general, CH_3, Si–CH_3, Si–O, Si–O–C, Si–C and Si–Cl groups are formed onto the polyester fabric after DCDMS treatment. These are also the groups found when the fabrics were treated with plasma followed by immersion in DCDMS solution, the only difference being the change in their relative intensities (Jahagirdar and Tiwari 2007).

8.3.1.3 Formation of Hydrophobic and Water-Repellent Surface

The formation of a water-repellent surface on poly(ethylene terephthalate) (PET) fibres via plasma polymerization at atmospheric pressure was reported by Ji et al. (2008). PET fibres were treated with a middle-frequency (MF) plasma in a mixture of argon gas and liquid-phase hexamethyldisiloxane (HMDSO). The surface morphologies and the chemical functional groups of plasma-treated fibres were characterised by scanning electron microscopy (SEM) and Fourier transform infrared (FTIR) spectrometry (Ji et al. 2008). Water repellency was also characterised by comparing it with the American Association of Textile Chemists and Colorists (AATCC) standard spray test chart. The results of this study are discussed later in this section.

In the textile industry, there has been an ongoing thrust to achieve improved fabric performance properties. The properties of interest range from the quantifiable, such as water repellency, fire resistance, and adhesion performance, to the subjective, such as comfort. Plasma-enhanced chemical vapour deposition (PECVD) is a suitable approach to preparing various types of film by initiating chemical reactions in a gas-liquid phase with an electrical discharge. PECVD using low-temperature plasmas has found important applications in microelectronics, optics, solar cells, mechanical industries, and plastics industries. The use of plasmas for surface modification and polymerization to deposit thin films is also very attractive owing to the control over surface chemistry (Yasuda 1985). The films deposited by plasma polymerization have a good adhesion to the substrate, are pinhole-free, and have mechanical and chemical stabilities. The plasma polymerization technique has also been employed conventionally to fabricate thin functional films that facilitate surface modification of materials (Szeto and Hess 1981; Harada 1981; Grundmeier et al. 2003).

When polymeric materials are exposed to plasma, radicals are generated in the polymeric chain. These radicals can initiate polymerization reaction when they are in contact with monomers in a liquid or gaseous phase. Electrons in plasma near a polymer surface generate radicals at the surface of the polymer through excitation of polymer molecules. As a result, a grafting polymer is formed on the surface of the polymeric material. The produced film has an ample range of prospective applications in anticorrosive surfaces, electrical resistors, scratch-resistant coatings, optical filters, chemical-barrier coatings, and water-repellent coatings (Yasuda and Matsuzawa 2005; Q. Zhang et al. 2001; Cökeliler and Mutlu 2002; G. Yang et al. 2003; Zou et al. 2001). Hence, an additional surface modification is required to achieve the desired properties while maintaining the characteristics of the volume (Li, Ye, and Mai 1997; Favia et al. 2003; Shibuchi et al. 1998). In particular, hydrophobic and water-repellent surface treatments of various substances have become of great interest in recent years (Francesch et al. 2005; Kokkoli and Zukoski 2000; W. Chen et al. 1999). Among many techniques that can be utilised for superhydrophobic and water-repellent treatments, plasma-based processes have been widely studied and developed (Woodward et al. 2003; Evju et al. 2004; Mukhopadhyay et al. 2002; Ricketts et al. 2004). To improve superhydrophobicity and water repellency, a coating material must be added, which can amplify the hydrophobicity. It is well known that the chemical modification of synthetic materials can be used in such applications as fluoropolymeric coating (Cicala et al. 2003; Rinsch et al. 1996). However, the plasma polymerization of silicon compounds is a more environmentally friendly technology for attaining superhydrophobic surfaces (Ji et al. 2008).

In a study by Ji et al. (2008), HMDSO as a monomer was employed for coating of a silicon compound on a polymer. In general, HMDSO is well known as one of the most common materials for hydrophobic coating (Aumaille et al. 2000; Wavhal et al. 2006; Zheludkevich et al. 2006; Bellel et al. 2006). In this study, Ji et al. demonstrated a manufacturable method for formation of hydrophobic and water-repellent coating on PET fibres. The coating was obtained using an in-line atmospheric MF glow-discharge plasma in a mixture of Ar and HMDSO. The plasma system did not require any vacuum line and was operated not in a batch mode, but in an in-line mode. Therefore, it can be easily scaled up for application to large substrate surfaces or continuous processing (Guimond and Wertheimer 2004; Schutze et al. 1998; Hermann et al. 1999; L. Ward et al. 2003; Yamamoto et al. 2004).

Figures 8.12a–d (Ji et al. 2008) show the plot of the number of plasma-treatment passes vs. the ratings corresponding to the AATCC standard spray test. As shown in Figure 8.12a, the PET fibre without plasma treatment completely absorbed sprayed water. As the number of plasma-treatment passes increased, on the other hand, the ratings increased from 50 to 90. This means that hydrophobic species produced from fragmentation of HMDSO by plasma were formed on the PET fibre (Ji et al. 2008).

The chemical composition of the PET fibres before and after plasma treatment was characterised by FTIR spectroscopy. The FTIR spectra of plasma-treated PET fibres in the study by Ji et al. (2008) showed peaks related to Si–O–Si at 1080 cm^{-1}, Si–O–C at around 1030 cm^{-1}, Si–$(CH_3)_2$ at 790 cm^{-1} and Si–C at 851 cm^{-1}, and the assigned peaks were checked from the actual spectra of the films prepared using HMDSO and plasma (Kashiwagi, Yoshida, and Murayama 1991;

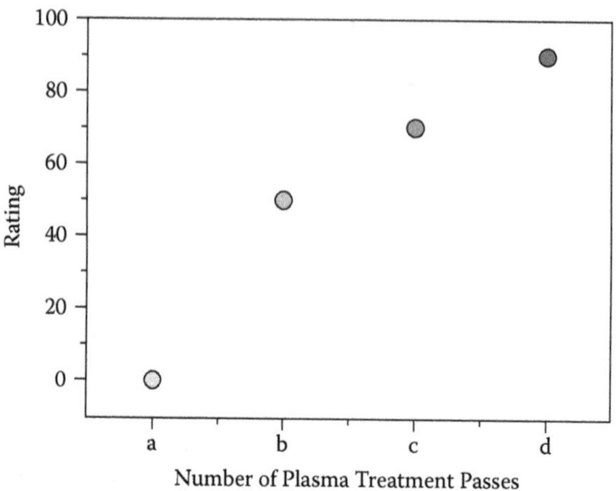

FIGURE 8.12 Water repellency ratings along with the number of plasma treatment passes: (a) 0, (b) 10, (c) 15, and (d) 20. (From Ji et al. 2008.)

Wrobel et al. 1990; Hamada and Morishita 1983). From the FTIR results, as the time of exposure to plasma increases, the intensities of Si–O–Si, Si–$(CH_3)_2$, and Si–C bonds increase, revealing that more hydrophobic surfaces are formed. Also, the inherent peaks of the PET fibre were decreased by HMDSO–plasma treatment. For instance, the peaks of aromatic C–H (730 cm^{-1}), O–C=O (1250 cm^{-1}), and ester C=O (1750 cm^{-1}) slightly decreased, and the peaks at 1350 and 1410 cm^{-1} indicate that the aliphatic C–H substantially decreased relative to untreated PET (Ji et al. 2008).

On the basis of these FTIR results, the fragmentation mechanism of HMDSO and the mechanism of reaction between PET and HMDSO induced by the plasma treatment are described as follows. From electrons generated by the plasma process, monomer fragmentation was suggested by Goujon, Belmonte and Hention (2004). The Si–O bond (8.3 eV) is broken by high-energy electron collisions. Also, Si–C (4.5 eV) and C–H (3.5 eV) bonds are broken by low-energy electrons (Figure 8.13) (Ji et al. 2008). The C–H radical groups were generated by plasma on the fibre surface, because the chain scission of PET took place (Jasso et al. 2006) in the order from a low dissociation energy to a high dissociation energy (C–H [3.5 eV], C–O [3.7 eV], C=O [7.6 eV]). Consequently, the radical groups bonded to the activated species of HMDSO (Figure 8.13) (Ji et al. 2008).

The hydrophobicity of PET fibres strongly depends on the content of silicon compounds or plasma treatment time. However, it was observed that the PET surfaces became covered with white silicon oxide powder when the number of plasma-treatment passes was more than 30. Therefore, plasma treatment of 20 passes is expected to be the best experimental condition that does not produce deposition of visible silicon oxides. The change in PET surface morphology is mainly caused by the bonding of activated species and depends on the plasma condition during the coating processes (Ji et al. 2008).

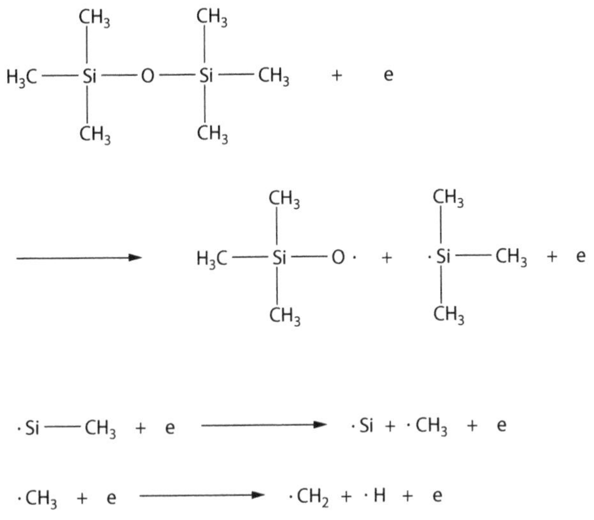

$$(CH_3)_3Si{-}O{-}Si(CH_3)_3 + e \longrightarrow (CH_3)_3Si{-}O\cdot + \cdot Si(CH_3)_3 + e$$

$$\cdot Si{-}CH_3 + e \longrightarrow \cdot Si + \cdot CH_3 + e$$

$$\cdot CH_3 + e \longrightarrow \cdot CH_2 + \cdot H + e$$

FIGURE 8.13 Mechanism of fragmentation HMDSO. (From Ji et al. 2008.)

8.3.1.4 Chitosan Modification of Polyester Fabric and Its Antibacterial Effects

Chitosan is a natural nontoxic biopolymer used widely in various fields due to its antimicrobial activities. Chang et al. (2008) studied the properties of polyester fabrics grafted with chitosan oligomers/polymers after being activated by atmospheric-pressure plasmas. The antibacterial effect was most evident when the surface of fabrics was activated by atmospheric-pressure plasma for 60 to 120 s and grafted with chitosan oligomers. The modified fabrics also exhibited good biocompatibility. This process can be applied to a large area and used to produce antibacterial polymer fibres.

The surface chemical compositions of PET fabrics were characterised by XPS (Table 8.7) (Chang et al. 2008). Treated PET fabrics had higher O/C ratios than untreated samples due to oxygen gas introduced during plasma activation. The surface nitrogen contents showed increases of more than 20-fold in PET fabrics grafted with chitosan oligomer, while those in plasma-activated PET fabrics only increased slightly. The remarkable increases in nitrogen contents on the surface of modified PET fabrics were attributed to the amine groups from grafted chitosan, which was in accordance with previous findings (Huh et al. 2001).

PET fabrics and *Bacillus subtilis* were used for antibacterial experiments following the method of AATCC 90. As shown in Figure 8.14, the growth of *Bacillus subtilis* was inhibited by Group 1 fabrics grafted with chitosan. The inhibitory effects were stronger when fabrics were grafted with chitosan oligomer rather than polymer. No et al. (2002) examined antibacterial activities of six chitosan oligomers and six chitosan polymers with different molecular weights. Chitosan oligomers showed higher antibacterial activities than chitosan polymers and markedly inhibited the growth of most bacteria tested. Chitosan grafted onto PET fabrics

TABLE 8.7
XPS Results for PET Fabrics Treated with Ar-O_2 Plasma

Activation Time [a] (s)	C1s (C,%)	O1s (O,%)	N1s (N,%)	O/C	N/C	(O+N)/C
Untreated	71.9	28.1	<0.1	0.391	0.000	0.391
60A	69.1	30.7	0.2	0.444	0.003	0.447
60G	66.8	30.7	2.5	0.460	0.037	0.497
120A	68.8	30.6	0.6	0.445	0.009	0.454
120G	66.4	31.6	2.0	0.476	0.030	0.506

Source: Chang et al. (2008).

[a] A = fabrics activated by Ar-O_2 plasma treatment (150 W) only; G = fabrics grafted with chitosan oligomer following plasma activation.

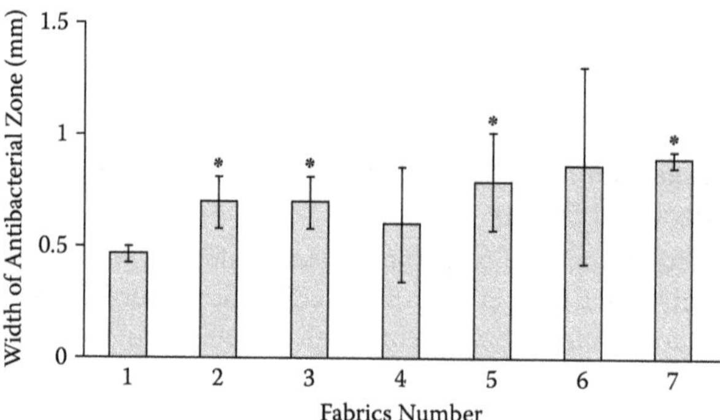

FIGURE 8.14 Antibacterial activities of PET fabrics grafted with chitosan (Chang et al. 2008).

continued to show antibacterial effects in this study. The hydrophilicity of chitosan oligomer facilitated its release from the fabric; thus it could react with bacteria more effectively and showed stronger antibacterial effects than chitosan polymer (Chang et al. 2008).

When PET fabrics were activated by plasma for 120 s, fabrics activated for one side (Fabric 2) showed better antibacterial effects (see Figure 8.14) than for two sides (Fabric 4). While the activation time was 80 s, the antibacterial effects of fabrics (see Figure 8.14) activated for both sides (Fabric 5) were better than for one side only (Fabric 3). Excessive activation time of plasma treatment might result in surface cross-linking or increased grafting of chitosan. The structure of PET fabrics might be destroyed due to overtreatment of plasma and consequently bond to chitosan less efficiently (Chang et al. 2008).

Alternatively, increased reaction of chitosan might leave fewer amino groups ($-NH_2$) on chitosan to be exposed. In either case, the antibacterial effectiveness was affected (Huang and Chen 2002). The fabrics activated for 80 s with both sides (Fabric 5)

TABLE 8.8
Experimental Parameters for PET Fabrics under Different Activation-Time Grafting Polymerization for 5 min and Antibacterial Ratios of Group 3 PET Fabrics

	Fabric No.									
	1	2[a]	3	4	5	6	7	8	9	10
Activation time (s)	0	60	120	120	60	60	30	30	15	15
Culturing time (h)	48	48	48	48	48	48	48	48	48	48
CFU/ml ($\times 10^4$)	2.15 ± 0.16	0.51 ± 0.08	0.25 ± 0.04	0.29 ± 0.03	0.23 ± 0.05	0.24 ± 0.03	0.69 ± 0.12	0.58 ± 0.06	0.72 ± 0.16	0.77 ± 0.11
Antibacterial ratio (%)	...	76.28 ± 3.88	88.60 ± 1.76	86.74 ± 1.40	89.30 ± 2.24	88.84 ± 1.21	67.91 ± 5.45	73.26 ± 2.64	66.51 ± 7.38	64.42 ± 5.27

Source: Chang et al. (2008).
Note: Initial bacterial concentration = 3.2×10^5 CFU/fabric (CFU = colony forming units).
[a] Fabric 2 was not grafted.

might have reached saturation of treatment and showed the best antibacterial effects. Prolonging activation time to 120 s was disadvantageous for antibacterial activities in fabrics activated on two sides (Chang et al. 2008).

PET fabrics (Table 8.8) and *Staphylococcus aureus* were used for antibacterial experiments following the method of JIS Z 2801:2000. The antibacterial ratios (approaching 90%) were the highest when activation times were 60 s (Fabrics 5–6) or 120 s (Fabrics 3–4). The ratio of Fabric 2 that was plasma-activated but not grafted was 75%. Shortening the activation times to 30 and 15 s (Fabrics 7–10) remarkably decreased the ratios to 60%–70%. According to the results presented, the major factor to determine antibacterial effects of the modified fabrics appeared to be the activation time (Chang et al. 2008).

8.3.1.5 Functionalization of Polyester Fabrics with Nanosilver

The antibacterial and other properties of polyester fabrics previously functionalised by corona and/or silver nanoparticles have been studied by Gorensek et al. (2010). Corona air plasma was used as a pretreatment of raw, washed and washed-thermostabilised polyester fabrics to increase the adhesion of nanosilver particles, resulting in an excellent antibacterial effect. X-ray photoelectron spectroscopy (XPS) was applied to analyse the surface composition and chemical bonding of the surface atoms on untreated and treated fabrics. The surface morphological changes of polyester fibres were observed by scanning electron microscopy (SEM). The quantity of silver on the polyester fabrics was determined by the use of the inductively coupled plasma-atomic emission spectrometry method.

The antimicrobial properties of functionalised polyester fabrics were tested according to American Society for Testing and Materials ASTM Designation: E 2149-01.

The surface of raw fabrics was richer in carbon, and the concentrations of the C–O and O–C=O groups were lower than on the other samples. An opposite effect was observed for washed and washed-thermostabilised fabrics. SEM analyses show that the plasma treatment also affects the surface morphology. The chemical surface composition and morphology are highly related to the hydrophobicity and hydrophilicity and, thus, to the achievement of better nanosilver adhesion and enhanced dyeing and antimicrobial properties of differently prepared corona plasma-treated polyester fabrics. Therefore, corona air-treated raw polyester fabrics demonstrated optimum antimicrobial properties due to the excellent adhesion of nanosilver (Gorensek et al. 2010).

SEM analysis of raw polyester sample R (Figure 8.15a) shows a normal smooth surface with coincidental dust particles or abrasions. A tiny amorphous veil resulting from a layer of finishing agent is characteristic of an untreated raw sample. A disrupted surface structure with a high surface area and enhanced porosity were observed on the surface of the plasma-treated raw sample R1 (Figure 8.15b). Etched porous areas containing 100-nm hollows are present among larger islands of about 1 μm with more or less smooth surfaces (Gorensek et al. 2010).

The washed plasma-treated sample W1 (Figure 8.16a) shows a uniform grainy surface with various particle sizes and a prevailing particle size of 100 nm. The plasma-etched surface of W1 fibres is not as porous as that of R1, although the specific

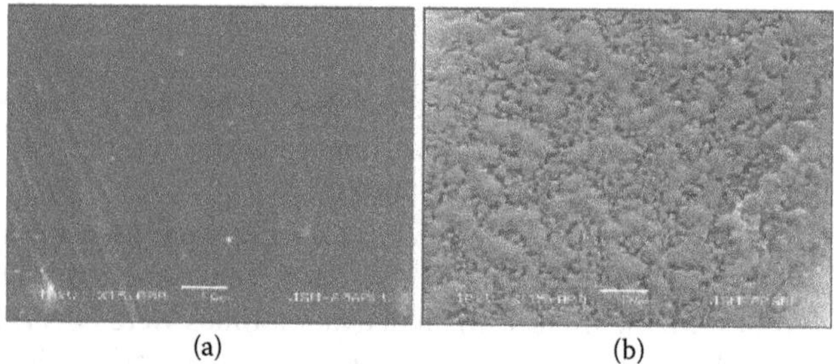

(a) (b)

FIGURE 8.15 SEM micrographs: (a) raw untreated sample R, (b) plasma-treated sample R1. (From Gorensek et al. 2010.)

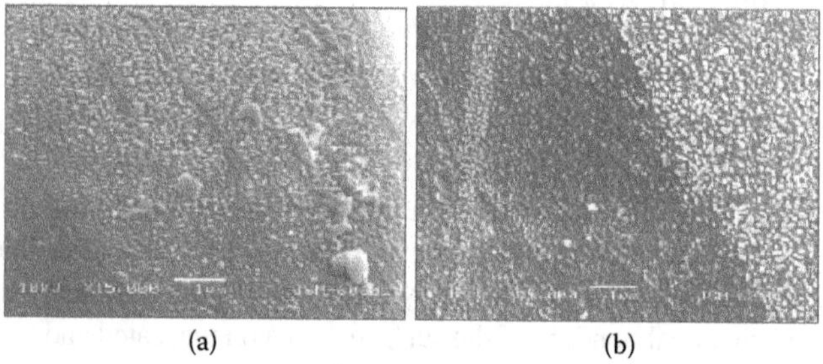

(a) (b)

FIGURE 8.16 SEM micrograph of plasma-treated samples: (a) W1, (b) T1. (From Gorensek et al. 2010.)

surface area is certainly enlarged as well. The shallow relief depth is comparable to the height of the particles caused by etching (Gorensek et al. 2010).

The surface of the plasma-treated washed-thermostabilised sample T1 is similar to the plasma-treated washed sample W1 with a well-marked, uniform, and slightly more roughly grained structure. Particles from 100–200 nm are observed as well and, as expected, smaller specific surface area in the T1 sample (Gorensek et al. 2010)

The increased adsorption of nanosilver on the samples can be explained by the increase in surface area and the change in surface composition and chemical bonding resulting from the corona air-plasma treatment. The raw fabrics became more porous, which seems to be preferable for Ag absorption. Fabrics also show increased Ag adsorption when treated with plasma, although not to the same extent as that with raw plasma-treated fabrics (Gorensek et al. 2010)

The results of microbial reduction were obtained according to the ASTM Designation: E 2149-01 method (Gorensek et al. 2010). The results of microbial reduction were in accordance with the concentrations of nanosilver. The raw plasma-treated fabrics showed the highest adsorption of Ag and, as a consequence, exhibited the best microbial reduction. *Staphylococcus aureus* and *Streptococcus faecalis* were adequately reduced (>60%) by the treated polyester fabrics (Gorensek et al. 2010)

8.3.2 Finishing Effect Imparted on Nylon Fibre by Plasma Treatment

8.3.2.1 Water-Repellency and Antibacterial Activities of Plasma-Treated Cleavable Silicone Surfactants on Nylon Fabrics

In a paper by L. Lin et al. (2006), the authors described how cleavable surfactants decompose into water-insoluble silanols and two water-soluble products when subjected to vacuum plasma treatment. They used Raman spectroscopic analysis to confirm these structural changes, and we performed contact angle measurements and employed scanning electron microscopy to observe the surface morphologies of these compounds. Contact angle measurements confirm that the products had degraded on nylon fabrics during argon gas plasma treatment. All of the PEG-silicone polyesters displayed excellent water repellency; PEG6000-silicone exhibited the largest contact angle (130°) and, hence, the greatest water repellency. Their results indicate that the silanols that form upon plasma treatment may be useful in coatings applications. They also found that the PEG-silicone polyester surfactants impart a new function, i.e., water-repellency, to nylon fabrics. In addition, they investigated the antibacterial properties of a nylon fabric containing two biocides. After vacuum plasma treatment, the hydrophobic surface of the nylon fabric retained its antibacterial activity and was water repellent (L. Lin et al. 2006).

The linkage bonds—those between silicon and oxygen atoms (Si–OCH_2) and within ester units [C(O)–O]—in the structures of the cleavable polymeric surfactants are susceptible to degradation during plasma treatment (Figure 8.17). Figure 8.18 displays the Raman spectra of the cleavable surfactant and the degradation products obtained after plasma treatment. We observe that the adsorption band at 3625 cm^{-1} disappeared after plasma treatment; presumably this signal was the interference peak of the universe of the beam. In addition, the intensities of the signals at 3610 and 3525 cm^{-1} increased after cleavage; this finding indicates that the ester group

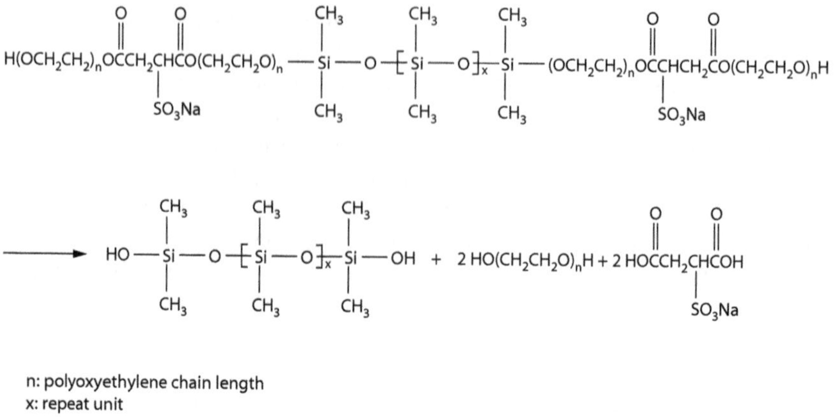

$$H(OCH_2CH_2)_nO\overset{O}{\overset{\|}{C}}CH_2\underset{SO_3Na}{\underset{|}{C}}H\overset{O}{\overset{\|}{C}}O(CH_2CH_2O)_n-\underset{CH_3}{\underset{|}{\overset{CH_3}{\overset{|}{Si}}}}-O-\left[\underset{CH_3}{\underset{|}{\overset{CH_3}{\overset{|}{Si}}}}-O\right]_x-\underset{CH_3}{\underset{|}{\overset{CH_3}{\overset{|}{Si}}}}-(OCH_2CH_2)_nO\overset{O}{\overset{\|}{C}}\underset{SO_3Na}{\underset{|}{C}}HCH_2\overset{O}{\overset{\|}{C}}O(CH_2CH_2O)_nH$$

$$\longrightarrow HO-\underset{CH_3}{\underset{|}{\overset{CH_3}{\overset{|}{Si}}}}-O-\left[\underset{CH_3}{\underset{|}{\overset{CH_3}{\overset{|}{Si}}}}-O\right]_x-\underset{CH_3}{\underset{|}{\overset{CH_3}{\overset{|}{Si}}}}-OH + 2\,HO(CH_2CH_2O)_nH + 2\,HO\overset{O}{\overset{\|}{C}}CH_2\underset{SO_3Na}{\underset{|}{C}}H\overset{O}{\overset{\|}{C}}OH$$

n: polyoxyethylene chain length
x: repeat unit

FIGURE 8.17 Cleavable modified silicone surfactants. (From L. Lin et al. 2006.)

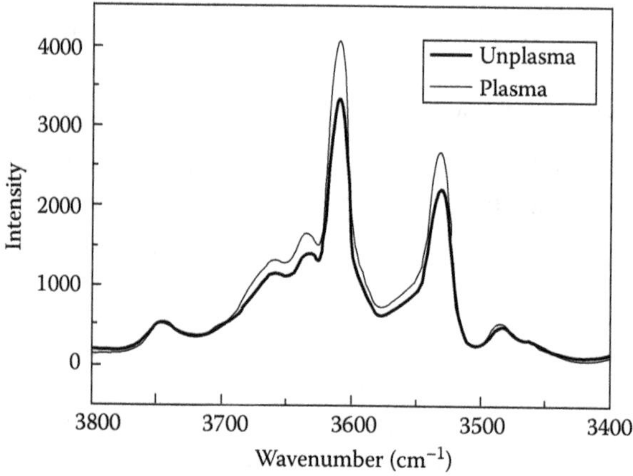

FIGURE 8.18 Raman spectra of PEG6000-silicone before and after plasma treatment. (From L. Lin et al. 2006.)

of the surfactant decomposed into carboxylic acid (RCOOH) and hydroxyl (ROH) groups (L. Lin et al. 2006).

A number of textile finishing agents and finishing processes have been developed to overcome some of the defects of natural and man-made textile fabrics and to impart any desirable properties that they lack or possess to only a limited extent. Water-repellent finishing—the process of imparting water repellency—is one of the most important finishing procedures for many different fabrics (Shiozawa 1991; Noll 1968; Schmidt 1988; Shenai 1976). A number of organopolysiloxanes, e.g., the hydroxyl-terminated polydimethylsiloxane, are used widely to produce water-repellent fabrics. Unfortunately, the presence of emulsifiers in formulated commercial silicone water-repellent agents decreases the hydrophobicity of the agent and decreases its water repellency. In a study done by L. Lin et al. (2006), the authors revealed that the cleavage of the prepared surfactants would lead to the formation of a water-insoluble silanol moiety and two water-soluble products after plasma treatment at different pressures and powers and for different lengths of time. The hydrophobic silanol moiety readily undergoes polymerization through silanol condensation and deposits a hydrophilic film onto the fibre. This film, which consists of hydrocarbon chains attached to and oriented on the fibre surface through polar groups, imparts water repellency and softness to the fabrics (Shenai 1976).

The contact-angle measurements evaluated the effects that the cleaved surfactants have upon the water repellencies of the nylon fabrics (Yamauchi et al. 1996). L. Lin et al. (2006) evaluated the contact angles of aqueous solutions upon the cleavable silicone surfactant after plasma treatment at various powers. The results indicate that the PEG6000-silicone surfactant imparted a higher water repellency than did the other cleavable surfactant; i.e., the use of the PEG6000-silicone surfactant led to the highest contact angle. More specifically, increasing the length of the polyoxyethylene chain improves the ability of the silicone surfactant to increase the contact angle. Water repellency was achieved more rapidly by the longer-chain surfactants upon plasma treatment at high power. We believe that these findings arise from an increase in the hydrophobic portion of the cleavable surfactant, which results in an increase in its water repellency at the surface.

SEM images show that silicone has filled and sealed the inter-monofilament pores and voids and indicates that the PEG-silicone surfactants coating the fabrics decomposed after plasma treatment. The cleaved products coating the nylon fabric appear to be distributed very uniformly. This phenomenon is consistent with our results of the contact-angle measurements: Cleavage of the prepared surfactants leads to water-insoluble silanol moieties and two water-soluble products; this process imparts excellent water repellency to the nylon fabric. Z. Zhang et al. (2003) determined the antibacterial properties qualitatively by measuring the areas from which *S. aureus* and *K. pneumoniae* had been eradicated. The silicone coat on the nylon fabric of the eradicated area was not transparent. Biocidal treatments with both the silicone and PEG2000-silicone surfactants improved the antibacterial properties of the nylon fabrics. The results we obtained before and after plasma treatment were similar, especially for *S. aureus*. Thus, the sample treated with PEG2000-silicone retained its antibacterial activity and water repellency after plasma treatment (L. Lin et al. 2006).

8.3.2.2 Nylon Grafted with Chitosan by Plasma and Its Antimicrobial Effect

Global demand has stimulated research into the use of nylon textile as fabrics. McCord et al. (2002) treated nylon fabrics with atmospheric-pressure plasma, and Zhu, Wang, and Qiu (2007) treated nylon fabrics with radio frequency (RF) plasma. Both research groups found that oxygen was bound to the surfaces of treated nylon fabrics. After the treatment, the absorption capacity for moisture and dyeability of the nylon fabrics were improved. Yip et al. (2002) treated nylon fabrics with low-temperature plasma with oxygen, argon, or tetrafluoromethane. Such treatment did not affect the mechanical properties of nylon fabrics. The antimicrobial treatment of nylon fabrics has been mentioned in the literature (J. Lin et al. 2001; Son and Sun 2003). However, the antimicrobial treatment of nylon fabrics by chitosan modification has not been reported so far. To make nylon fabrics antimicrobial in an environmentally friendly way, nylon fabrics were first activated in this study by open-air plasma, which did not need a chamber. They were then grafted with chitosan oligomer or polymer, and the antimicrobial effect was evaluated (Tseng et al. 2009).

The hydrophilicity of nylon fabrics under different plasma treatments were studied by Tseng et al. (2009). The wicking time of water on the untreated nylon textile was 0.82 s. The procedures such as solution, washing, and drying did not change the wicking time of nylon fabrics. After the surfaces of nylon fabrics were activated by air plasma, the wicking time was slightly raised. Paynter (1998) treated polystyrene (PS) and poly(ethyleneterephthalate) (PET) with RF plasma of He, He-O_2 or N_2-H_2 and observed the formation of new C–N and C–O bonds on the surfaces of plasma-treated PS and PET by XPS. In addition, the water contact angles were positively correlated with the surface N/O ratio, i.e., smaller with more oxygen and larger with more nitrogen (Paynter 1998). In contrast to the previous finding, the wicking time of nylon fabrics in this study changed only slightly after air plasma treatment. This may be related to the fact that nylon fabrics are intrinsically more hydrophilic than PET fabrics. Comparing the effect of plasma scanning speeds on wicking time, the plasma scanning speed of 2 m/min resulted in coking the fabric surface. This indicated an overexposure of plasma to nylon textiles. Scanning at 26 m/min led to a smaller value in wicking time than scanning at 6 m/min. Scanning for three times further decreased the wicking time. When the activated surface was further grafted with chitosan oligomer or polymer, the wicking time was significantly increased. Moreover, grafting of chitosan oligomer instead of polymer resulted in a smaller increase in wicking time. The increase in wicking time after grafting was due to the presence of chitosan on the surface. M. Yang et al. (2002) made PET fabrics more hydrophilic by grafting them with acrylamide and itaconic acid. The hydrophilicity remained similar regardless of the amount of chitosan subsequently immobilised on the surface of PET. Post-grafting scan(s) (PGS) of air plasma further reduced the wicking time of the grafted nylon fabrics (Tseng et al. 2009).

Tseng et al. (2009) showed SEM images of nylon fabrics and the modified nylon fabrics. The surface of the original nylon fabrics was smooth. After plasma activation, a few particulates showed up. This observation was consistent with that reported in the literature (McCord et al. 2002). Nylon fabrics grafted with chitosan oligomer appeared to be covered by a layer of substance. On nylon fabrics grafted with chitosan polymer, the covering layer showed the features of rippling and waviness. When the scanning speed of plasma activation was lower, the surface demonstrated more rippling and waviness. After PGS, the surface change was even more pronounced.

Once the surface was grafted with chitosan oligomer or polymer after plasma activation, the surface chemistry of nylon textiles was changed. When nylon was activated by air plasma at 6 m/minute or 26 m/min, the O/C ratio increased and N/C ratio decreased. Therefore, bond cleavage may have occurred as a result of plasma activation. The remaining chain ends may have then become cross-linked through the formation of C–C, C–O, O–H, or other new covalent bonds (S. Wu et al. 1997; Tsai, Wadsworth, and Roth 1997; Aouinti, Bertrand, and Poncin-Epaillard 2003; Wagner, Fairbrother, and Reniers 2003). The XPS results on air-plasma-treated nylon agreed with those of McCord et al. (2002) and Zhu, Wang, and Qiu (2007), who found that oxygen was bound to the surfaces of atmospheric pressure or RF plasma-treated nylon fabrics (Tseng et al. 2009).

In nylon fabrics grafted with chitosan oligomer or polymer after plasma activation, the O/C ratio further increased. Nylon textiles grafted with chitosan polymer had a higher O/C ratio than those grafted with chitosan oligomer. When the fabrics were activated by plasma scanning once and grafted with chitosan, the O/C ratio was similar for different scanning velocities. When the fabrics were scanned by air plasma for three times and grafted with chitosan oligomer or polymer, the O/C ratio was greater for those scanned at lower velocities. The greater O/C ratio was probably associated with the higher amount of chitosan grafted. Overall, the O/C and (O+N)/C ratios were highest for nylon fabrics activated at 26 m/min, grafted with chitosan polymer and treated by PGS (Tseng et al. 2009).

The antibacterial properties of modified nylon fabrics were evaluated following the method of JIS L 1902-1998. The results obtained by Tseng et al. (2009) showed that under different scanning speeds of plasma activation, the microbiostasis ratios for fabrics scanned at 26 m/min were, in general, higher than those scanned at 6 m/min, except that the microbiostasis ratio for fabrics activated at 26 m/min for one time was similar to that at 6 m/min for one time. Scanning of plasma for three times resulted in better antibacterial activities than scanning for one time. Comparing the effects of chitosan oligomer and chitosan polymer on antibacterial activities, nylon fabrics grafted with chitosan polymer after plasma treatment showed significantly better antibacterial effects than those grafted with chitosan oligomer in general (Tseng et al. 2009).

8.3.3 Finishing Effect Imparted on Acrylic Fibre by Plasma Treatment

8.3.3.1 Improved Surface Wettability and Antistatic Ability

In a study by Y. Liu, Xiong, and Lu (2006), contact angle and moisture regain were used to evaluate surface wettability. The results are shown in Table 8.9. The original untreated sample was quite hydrophobic (moisture regain was only 0.85 and contact angle was 77.80). However, surface wettability was significantly increased following

TABLE 8.9
Wettability and Antistatic Ability of the Untreated and Plasma-Treated Samples

Samples	Moisture Regain (%)	Contact Angle (°)	Half-Decay Time (s)
Untreated	0.85	77.80	9.57
	Nitrogen Plasma-Treated		
1 min	1.35	65.64	3.98
3 min	2.24	34.92	1.35
5 min	1.65	56.41	2.62
7 min	1.59	54.73	3.16

Source: Y. Liu, Xiong, and Lu (2006).

plasma treatment. Under the given plasma conditions, a time of 3 min was long enough for good wettability (the highest moisture regain and the smallest contact angle). With increasing treatment time, wettability was decreased slightly and then remained at a relatively constant level, which seems to suggest equilibrium between introducing and removing functional groups during plasma treatment.

The half-decay time of the treated samples is also shown in Table 8.9. Here, we see that plasma treatment has caused a drastic reduction in the half-decay time of the fibres. The half-decay time is found to decrease from 9.57 s for the untreated sample to 1.35 s for the 3-min plasma-treated sample. This result shows that the antistatic ability of the fibres is drastically improved by plasma treatment (Y. Liu, Xiong, and Lu 2006).

Surface wettability is directly related to surface energy, with a more energetically stable surface resulting in a less wettable surface. It is now established that plasma modification of the fibres results in oxidation and degradation of the fibre surfaces. The oxidation creates oxidised functionalities, which lead to an increase in surface energy, while the degradation mainly changes surface morphology of the fibres. SEM pictures have shown that plasma treatment increases the surface roughness. According to the Wenzel equation, the roughness of the surface influences the contact angle (Sprang, Theirech, and Engermann 1995).

$$\cos\theta_{rough} = r\cos\theta_0 \tag{8.1}$$

where θ_{rough} is the contact angle on a surface of sample, θ_0 is the thermodynamic contact angle on the smooth surface, and r is the roughness (ratio of the actual area of the interface to the geometric surface area). Equation (8.1) indicates that for the surface having a lower contact angle than 90°, increasing surface roughness probably decreases the contact angle, which will contribute to the improved surface wettability. Water is a conductor of electricity. Therefore, the improved surface wettability will decrease the accumulation of electrostatic charges. The increase of surface roughness also induces the increase in the specific surface area. The increased specific surface area will lead to a more moisture-rich surface, which enhances the conductivity of the fibres.

Plasma treatment not only causes the increase in surface roughness, but also introduces the hydrophilic groups onto the fibre surface. XPS analyses have shown that amide and carboxyl groups are created on the fibre surface after plasma treatment. There are two possibilities of generating the polar groups. The first one is that they are generated by reacting with the ambient gas during the processing. The second is that they are generated when the samples are exposed to air after plasma processing, that is, plasma treatment produces a considerable amount of unsaturated bonds, and then the unsaturated bonds are reacted with atmospheric oxygen to form polar groups on the sample surface. A comparison of the data listed in Table 8.10 (Y. Liu, Xiong, and Lu 2006) shows that cyanogen and ester groups are less stable than methylene and methenyl groups. Thus, in the plasma treatment, oxidation mainly occurs at the cyanogen groups and ester groups on the fibre backbone (Y. Liu, Xiong, and Lu 2006).

Plasma treatment introduces an amount of amide and carboxyl groups on the fibre surface. These polar groups will incorporate with moisture through hydrogen bonding and help moisture penetration and binding on the fibre surface. Under the

TABLE 8.10
XPS Data of Untreated and 3-min Plasma-Treated Samples

Group	Peak Position (eV)	% Total Area	
		Untreated	Plasma-Treated
	C1s		
(–CH_2-, -CH=)	284.71	52.83	48.75
(–CN)	285.67	33.43	17.96
(–COOR)	288.53	13.74	10.42
(–$CONH_2$)	286.54	…	13.09
(–COOH)	287.41	…	9.78
	O1s		
(–COOR)	533.01	80.56	40.25
(–SO_3)	534.47	19.44	15.83
(–$CONH_2$)	530.83	…	13.13
(–COOH)	531.87	…	30.79
	N1s		
(–CN)	401.07	100	53.49
(–$CONH_2$)	400.02	…	46.51

Source: Y. Liu, Xiong, and Lu (2006).

action of water molecules, these polar groups will also generate ionization and lead to a structural layer to conduct electricity on the fibre surface, which enhances the electrostatic dissipation. Therefore, the half-decay time of the fibres decreases after plasma treatment. A 3-min plasma treatment is enough for effective surface modification. As the processing time increases further, some weakly attached polar groups will be removed from the fibre surface, which causes a slight increase in the half-decay time (Y. Liu, Xiong, and Lu 2006).

8.3.3.2 Water Repellent, Stain-Resistant or Flame-Retardant Acrylic (PAN) Fibres

Water-repellent, stain-resistant or flame-retardant characteristics can be conferred to polyacrylonitrile (PAN) textiles by using the plasma-induced graft polymerization technique. The monomers used are perfluoroalkylacrylate, (meth)acrylate phosphates, and phosphonates which are well known to be effective for the waterproofing and the fireproofing of polymeric substrates (Tsafack, Hochart, and Levalois-Grützmacher 2004). In a recent paper, Hochart, De Jaeger and Levalois-Grutzmacher (2003) were able to demonstrate that it is possible to achieve the desired properties by treating PAN fabric samples according to the following procedure: i.e., with 50 mL petroleum ether solution containing 20 g/L of AC8, 2.5 wt% of Darocur and 3.5 wt% of ethylene glycol dimethacrylate (EGDMA); Ar plasma exposed (argon flow rate $F_{Ar} = 500$ sccm, time $t = 10$ min, pressure $p = 40$ Pa); washed in a Soxhlet freon and dried 1 h at 100°C. The textile, which initially absorbed water and oil, became water repellent and stain resistant after treatment (Figure 8.19) (Tsafack, Hochart, and Levalois-Grützmacher 2004).

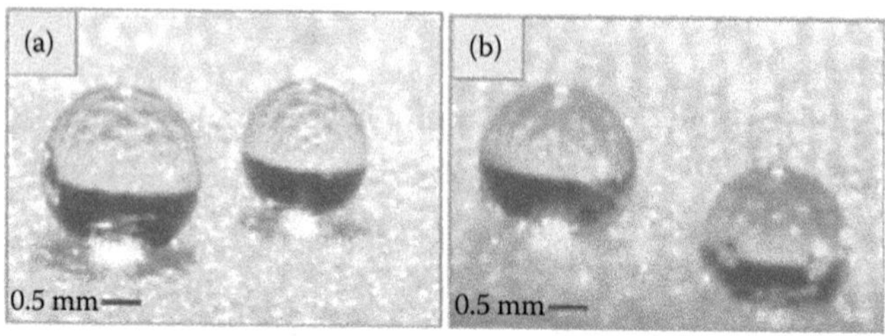

FIGURE 8.19 Repellent properties of polyAC8-grafted PAN tissue samples toward (a) water and (b) olive oil. (From Tsafack, Hochart, and Levalois-Grützmacher 2004.)

TABLE 8.11
Bragg Spacing (nm) Obtained at Room Temperature for PolyAC8 Films

	Diffraction Maxima				
d_1	d_2	d_3	d_4	d_5	References
3.32	1.61	1.08	0.65	0.50	Shimizu et al. (1994)
3.33	1.62	1.07	...	0.50	Volkov et al. (1992)
3.20	1.60	...	...	...	Katano, Tomono, and Nakajima (1994)
3.20	1.60	...	...	0.50	Tsafack, Hochart, and Levalois-Grützmacher (2004)

Source: Tsafack, Hochart, and Levalois-Grützmacher (2004).

The droplets of water roll onto the surface with an apparent contact angle of 132°. In order to better understand these results which are directly correlated to a low surface tension of the treated PAN fabric, X-ray diffraction measurements at room temperature were taken on a polyAC8 film (the monomer was previously coated on a glass support) obtained by Ar-plasma-induced polymerization (F_{Ar} = 500 sccm; Power P = 100 W; t = 10 min; thickness = 190 μm). The results obtained (Table 8.11) (Tsafack, Hochart, and Levalois-Grützmacher 2004) are compared to those extracted from the literature (Shimizu et al. 1994; Volkov et al. 1992; Katano, Tomono, and Nakajima 1994).

Bragg spacings of a polyAC8 film resulting from Ar plasma polymerization of AC8 are comparable to those reported for polyAC8 films obtained by other techniques. The values indicate a bi-layer formation either with a double-comb organization or alternating orientation of the fluorocarbon chains (Tsafack, Hochart, and Levalois-Grützmacher 2004).

8.3.3.3 Fire-Retardant Property

The reduction of the flammability of PAN textiles can be achieved by using (meth) acrylates containing phosphorus monomers that are known to be effective as flame-retardant compounds (Price et al. 2002). A plasma-grafted thin layer of phosphorus polymer can be obtained by plasma-induced graft-polymerization (Tsafack, Hochart, and Levalois-Grützmacher 2004). The formation of a characteristic protective char layer during the burning test was observed for the treated compounds,

(a) (b)

FIGURE 8.20 Burning behaviour of (a) untreated and (b) treated PAN fabric. (From Tsafack, Hochart, and Levalois-Grützmacher 2004.)

while the untreated ones burned without residuals (Figure 8.20) (Tsafack, Hochart, and Levalois-Grützmacher 2004).

The fire retardancy of the treated fabrics using the same procedure as given in the previous subsection was evaluated by LOI (limiting oxygen index) measurements according to ISO 4589-2 (Tsafack, Hochart, and Levalois-Grützmacher 2004). The results obtained for the acrylate phosphate and phosphonate monomers showed that LOI values increases with the relative phosphorus content in the monomer and therefore in the grafted polymer. This can be explained by the fact that phosphate monomers compared to phosphonate monomers are more subjected to chain transfers that lead to highly cross-linked polymers and decrease the flammability (Tsafack, Hochart, and Levalois-Grützmacher 2004).

The durability of the treatment on PAN fabrics to washing was tested according to the accelerated laundering method: The samples are boiled for 4 h in a solution of 0.5% $Na_3PO_4 \cdot 12H_2O$ and 0.1% Triton X-100 at approximate liquor ratio of 40:1 (McSherry et al. 1974). The LOI values of the fabrics drop slightly for the phosphate monomers and relatively more for the phosphonate monomers. This may be due to a slight ablation of the polymer layer. On the other hand, the burning rates decrease considerably, which indicates a better quality of the resulting protective layer. This demonstrates that the monomers were tightly grafted on the PAN fabrics and act effectively as flame retardants (Tsafack, Hochart, and Levalois-Grützmacher 2004).

8.4 CONCLUSION

This chapter reviewed some important finishing effects induced by plasma treatment on the textile materials. Plasma treatment can provide physical and chemical surface modifications that facilitate reactions between the textile materials and the finishing agents. Depending on the plasma treatment, some finishing processes can be carried out as plasma polymerization and plasma grafting. However, in some finishing processes, the role of plasma is to activate the material surface to accept the finishing agent.

REFERENCES

Abidi, N., and E. Hequet. 2004. Cotton fabric graft copolymerization using microwave plasma, I: Universal attenuated total reflectance–FTIR study. *Journal of Applied Polymer Science* 93: 145–54.

Achwal, W. B., C. R. Mahapatrao, and P. S. Kaduska. 1987. Flame retardant finishing of cotton and silk fabrics. *Colourage* 34(6): 16–30.

Albaugh, J., C. O'Sullivan, and L. O'Neill. 2008. Controlling deposition rates in an atmospheric pressure plasma system. *Surface and Coatings Technology* 203: 844–47.

Aouinti, M., P. Bertrand, and F. Poncin-Epaillard. 2003. Characterization of polypropylene surface treated in a CO_2 plasma. *Plasmas and Polymers* 8: 225–36.

Armitage, J. M., M. Macleod, and I. T. Cousins. 2009. Comparative assessment of the global fate and transport pathways of long-chain perfluorocarboxylic acids (PFCAs) and perfluorocarboxylates (PFCs) emitted from direct sources. *Environmental Science and Technology* 43: 5830–36.

Aumaille, K., C. Vallee, A. Granier, A. Goullet, F. Gaboriau, and G. Turban. 2000. A comparative study of oxygen/organosilicon plasmas and thin Si$O_x$$C_y$$H_z$ films deposited in a helicon reactor. *Thin Solid Films* 359: 188–96.

Bajaj, P., A. K. Agrawal, A. Dhand, and N. Kasturia Hansraj. 2000. Flame retardation of acrylic fibers: An overview. *Journal of Macromolecular Science: Reviews in Macromolecular Chemistry & Physics* C40: 309–37.

Bellel, A., S. Sahli, Z. Ziari, P. Raynaud, Y. Segui, and D. Escaich. 2006. Wettability of polypropylene films coated with SiOx plasma deposited layers. *Surface and Coatings Technology* 201: 129–35.

Bhattacharyya, N., B. A. Doshi, and A. S. Sahasrabudhe. 1999. Cost effective catalyst for polycarboxylic acid finishing. *Textile Chemist and Colorist* 31: 33–37.

Buch, R. R. 1991. Rates of heat release and related fire parameters for silicones. *Fire Safety* 17: 1–12.

Buschle-Diller, G., S. H. Zeronian, N. Pan, and M. Y. Yoon. 1994. Enzymatic hydrolysis of cotton, linen, ramie, and viscose rayon fabrics. *Textile Research Journal* 64: 270–79.

Cardamone, J. M., J. Yao, and A. Nunez. 2004. Controlling shrinkage in wool fabrics: Effective hydrogen peroxide systems. *Textile Research Journal* 74: 887–98.

Carmen, M., A. Almazan, J. Paredes, M. Perez-Mendoza, M. Domingo-Garcia, F. Lopez-Garzon, A. Martinez-Alonso, and J. Tascon. 2005. Effects of oxygen and carbon dioxide plasmas on the surface of poly(ethylene terephthalate). *Journal of Colloid and Interface Science* 287: 57–66.

Ceria, A., and P. Hauser. 2010. Atmospheric plasma treatment to improve durability of a water and oil repellent finishing for acrylic fabrics. *Surface and Coatings Technology* 204: 1535–41.

Chaivan, P., N. Pasaja, D. Boonyawan, P. Suanpoot and T. Vilaithong. 2005. Low-temperature plasma treatment for hydrophobicity improvement of silk. *Surface and Coatings Technology* 193: 356–60.

Chaiwong, C., S. Tunma, W. Sangprasert, P. Nimmanpipug, and D. Boonyawan. 2010. Graft polymerization of flame-retardant compound onto silk via plasma jet. *Surface and Coatings Technology* 204: 2991–95.

Chang, Y. B., P. C. Tu, M. W. Wu, T. H. Hsueh, and S. H. Hsu. 2008. A study on chitosan modification of polyester fabrics by atmospheric pressure plasma and its antibacterial effects. *Fibers and Polymers* 9: 307–11.

Chen, C. C., and C. C. Wang. 2006. Crosslinking of cotton cellulose with succinic acid in the presence of titanium dioxide nano-catalyst under UV irradiation. *Journal of Sol-Gel Science and Technology* 40: 31–38.

Chen, W., A. Y. Fadeev, M. C. Hsieh, D. Öner, J. Youngblood, and T. J. McCarthy. 1999. Ultrahydrophobic and ultralyophobic surfaces: Some comments and examples. *Langmuir* 15: 3395–99.

Cheng, C., Z. Liye, and R. J. Zhan. 2006. Surface modification of polymer fibre by the new atmospheric pressure cold plasma jet. *Surface and Coatings Technology* 200: 6659–65.

Chvalinova, R., and J. Wiener. 2008. Sorption properties of wool fibres after plasma treatment. *Chemické Listy* 102: s1473–77.

Cicala, G., A. Milella, F. Palumbo, P. Favia, and R. d'Agostino. 2003. Morphological and structural study of plasma deposited fluorocarbon films at different thicknesses. *Diamond and Related Materials* 12: 2020–25.

Cökeliler, D., and M. Mutlu. 2002. Performance of amperometric alcohol electrodes prepared by plasma polymerization technique. *Analytica Chimica Acta* 469: 217–23.

Davis, R., A. El-Shafei, and P. Hauser. 2011. Use of atmospheric pressure plasma to confer durable water repellent functionality and antimicrobial functionality on cotton/polyester blend. *Surface & Coatings Technology* 205: 4791–97.

De Geyter, N., R. Morent, C. Leys, L. Gengembre, and E. Payen. 2007. Treatment of polymer films with a dielectric barrier discharge in air, helium and argon at medium pressure. *Surface and Coatings Technology* 201: 7066–75.

Demir, A., B. Arik, E. Ozdogan, and N. Seventekin. 2010. The comparison of the effect of enzyme, peroxide, plasma and chitosan processes on wool fabrics and evaluation for antimicrobial activity. *Fibers and Polymers* 11: 989–95.

Demir, A., H. A. Karahan, E. Ozdogan, T. Oktem, and N. Seventekin. 2008. The synergetic effects of alternative methods in wool finishing. *Fibres and Textiles in Eastern Europe* 16(2): 89–94.

Di Mundo, R., V. De Benedictis, F. Palumbo, and R. d'Agostino. 2009. Fluorocarbon plasmas for nanotexturing of polymers: A route to water-repellent antireflective surfaces. *Applied Surface Science* 255: 5461–65.

Errifai, I., C. Jama, M. Le Bras, R. Delobel, L. Gengember, A. Mazzah, and R. De Jaeger. 2004. Elaboration of a fire retardant coating for polyamide-6 using cold plasma polymerization of a fluorinated acrylate. *Surface and Coatings Technology* 180–181: 297–301.

Evju, J. K., P. B. Howell, L. E. Locasicio, M. J. Tarlov, and J. Hickman. 2004. Atmospheric pressure microplasmas for modifying sealed microfluidic devices. *Applied Physics Letters* 84: 1668–70.

Favia, P., E. Sardella, R. Gristina, and A. Mile. 2003. Novel plasma processes for biomaterials: Micro-scale patterning of biomedical polymers. *Surface and Coatings Technology* 169: 707–11.

Francesch, L., E. Garreta, M. Balcells, E. R. Edelman, and S. Borros. 2005. Fabrication of bioactive surfaces by plasma polymerization techniques using a novel acrylate-derived monomer. *Plasma Processes and Polymers* 2: 605–11.

Franklin, W. E., and S. P. Rowland. 1979. Effects of phosphorus-containing flame retardants on pyrolysis of cotton cellulose. *Journal of Applied Polymer Science* 24: 1281–94.

Gaan, S., and G. Sun. 2007. Effect of phosphorus and nitrogen on flame retardant cellulose: A study of phosphorus compounds. *Journal of Analytical and Applied Pyrolysis* 78: 371–77.

Gawish, S. M., S. R. Matthews, D. M. Wafa, F. Breidt, and M. A. Bourham. 2007. Atmospheric plasma-aided biocidal finishes for nonwoven polypropylene fabrics, I: Synthesis and characterization. *Journal of Applied Polymer Science* 103: 1900–10.

Gawish, S. M., A. M. Ramadan, C. E. Cornelius, and M. A. Courham. 2007. New functionalities of PA6,6 fabric modified by atmospheric pressure plasma and grafted glycidyl methacrylate derivatives. *Textile Research Journal* 77: 92–104.

Ghoranneviss, M., M. Bahareh, S. Shahidi, A. Anvari, and A. Rashidi. 2006. Decolorization of denim fabrics with cold plasmas in the presence of magnetic fields. *Plasma Processes and Polymers* 3: 316–21.

Gorensek, M., M. Gorjanc, V. Bukosek, J. Kovac, P. Jovancic, and D. Mihailovic. 2010. Functionalization of PET fabrics by corona and nano silver. *Textile Research Journal* 80: 253–62.

Goujon, M., T. Belmonte, and G. Hention. 2004. OES and FTIR diagnostics of HMDSO/O_2 gas mixtures for SiOx deposition assisted by RF plasma. *Surface and Coatings Technology* 188–189: 756–61.

Goyal, R., and R. V. Deshpande. 2006. Reactive softeners. *Colourage* 53(8): 113–16.

Grundmeier, G., P. Thiemann, J. Carpentier, and V. Barranco. 2003. Tailored thin plasma polymers for the corrosion protection of metals. *Surface and Coatings Technology* 174–175: 996–1001.

Guan, J., and G. Q. Chen. 2006. Flame retardancy finish with an organophosphorus retardant on silk fabrics. *Fire and Materials* 30: 415–24.

Guan, J., C. Q. Yang, and G. Chen. 2009. Formaldehyde-free flame retardant finishing of silk using a hydroxyl-functional organophosphorus oligomer. *Polymer Degradation and Stability* 94: 450–55.

Guimin, X., Z. Guanjun, S. Xingmin, M. Yue, W. Ning, and L. Yuan. 2009. Bacteria inactivation using DBD plasma jet in atmospheric pressure argon. *Plasma Science and Technology* 11: 83–88.

Guimond, S., and M. R. Wertheimer. 2004. Surface degradation and hydrophobic recovery of polyolefins treated by air corona and nitrogen atmospheric pressure glow discharge. *Journal of Applied Polymer Science* 94: 1291–1303.

Hamada, K., and H. Morishita. 1983. Raman, infrared and H^1-NMR spectra of hexamethyldisiloxane and hexamethyldisilazane. *Spectroscopy Letters: An International Journal for Rapid Communication* 16: 717–29.

Harada, K. 1981. Plasma polymerized protective films for plated magnetic disks. *Journal of Applied Polymer Science* 26: 3707–18.

Hendrix, J. E., G. L. Drake, and R. H. Barker. 1972. Pyrolysis and combustion of cellulose, III: Mechanistic basis for the synergism involving organic phosphates and nitrogenous bases. *Journal of Applied Polymer Science* 16: 257–74.

Hermann, H. W., I. Henins, J. Park, and G. S. Selwyn. 1999. Decontamination of chemical and biological warfare (CBW) agents using an atmospheric pressure plasma jet (APPJ). *Physics of Plasmas* 6: 2284–89.

Hochart, F., R. De Jaeger, and J. Levalois-Grützmacher. 2003. Graft-polymerization of a hydrophobic monomer onto PAN textile by low-pressure plasma treatments. *Surface and Coatings Technology* 165: 201–10.

Huang, C. Y., and C. L. Chen. 2002. The effect of plasma surface modification from a rotary plasma reactor on the styrene grafting onto a polypropylene surface. *Surface and Coatings Technology* 153: 194–202.

Huh, M. W., I. K. Kang, D. H. Lee, W. S. Kim, D. H. Lee, L. S. Park, K. E. Min, and K. H. Seo. 2001. Surface characterization and antibacterial activity of chitosan-grafted poly(ethylene terephthalate) prepared by plasma glow discharge. *Journal of Applied Polymer Science* 81: 2769–78.

Jahagirdar, C. J., and L. B. Tiwari. 2004. Effect of dichlorodimethylsilane on plasma treated cotton fabric. *PRAMANA: Journal of Physics* 62: 1099–1109.

Jahagirdar, C. J., and L. B. Tiwari. 2007. Plasma treatment of polyester fabric to impart the water repellency property. *PRAMANA: Journal of Physics* 68: 623–30.

Jasso, M., H. Krump, I. Hudec, P. Stahel, D. Kovacik, and M. Sira. 2006. Coating of PET cords at atmospheric pressure plasma discharge in the presence of butadiene/nitrogen gas mixtures. *Surface and Coatings Technology* 201: 57–62.

Ji, Y. Y., H. K. Chang, Y. C. Hong, and S. H. Lee. 2008. Formation of hydrophobic and water-repellent surface on polyester fibers using Ar/hexamethyldisiloxane plasma at atmospheric pressure. *Japanese Journal of Applied Physics* 47: 4687–91.

Johansson, N., A. Fredriksson, and P. Eriksson. 2008. Neonatal exposure to perfluorooctane sulfonate (PFOS) and perfluorooctanoic acid (PFOA) causes neurobehavioural defects in adult mice. *Neurotoxicology* 29: 160–69.

Kan, C. W. 2007. Evaluating antistatic performance of plasma-treated polyester. *Fibers and Polymers* 8: 629–34.

Kan, C. W., K. Chan, C. W. M. Yuen, and M. H. Miao. 1998a. Surface properties of low temperature plasma treated wool fabrics. *Journal of Material Processing Technology* 83: 180–84.

Kan, C. W., K. Chan, C. W. M. Yuen, and M. H. Miao. 1998b. A study of the shrinkage properties of wool substrates following plasma and chemical treatments. *Journal of Federation of Asian Professional Association* 4: 23–32.

Kan, C. W., and C. W. M. Yuen. 2007. Plasma technology in wool. *Textile Progress* 39: 121–87.

Kandola, B. K., A. R. Horrocks, D. Price, and G. V. Coleman. 1996. Flame-retardant treatments of cellulose and their influence on the mechanism of cellulose pyrolysis. *Journal of Macromolecular Science: Reviews in Macromolecular Chemistry & Physics* C36(4): 721–94.

Kashiwagi, K., Y. Yoshida, and Y. Murayama. 1991. Hybrid films formed from hexamethyldisiloxane and SiO by plasma process. *Japanese Journal of Applied Physics* 30: 1803–7.

Katano, Y., H. Tomono, and T. Nakajima. 1994. Surface property of polymer films with fluoroalkyl side chains. *Macromolecules* 27: 2342–44.

Kim, B. H., J. Jang, and S. W. Ko. 2000. Durable press finish of cotton fabric using malic acid as a crosslinker. *Fibers and Polymers* 1: 116–21.

Kokkoli, E., and C. F. Zukoski. 2000. Interaction forces between hydrophobic and hydrophilic self-assembled monolayers. *Journal of Colloid and Interface Science* 230: 176–80.

Lam, Y. L., C. W. Kan, and C. W. M. Yuen. 2011. Effect of plasma pretreatment on the wrinkle-resistance properties of cotton fibers treated with a 1,2,3,4-butanetetracarboxylic acid–sodium hypophosulfite system with titanium dioxide as a cocatalyst. *Journal of Applied Polymer Science* 120: 1403–10.

Langley, J. T., M. J. Drews, and R. H. Barker. 1980. Pyrolysis and combustion of cellulose, VII: Thermal analysis of the phosphorylation of cellulose and model carbohydrates during pyrolysis in the presence of aromatic phosphates and phosphoramides. *Journal of Applied Polymer Science* 25: 243–62.

Lawler, T. E., M. J. Drews, and R. H. Barker. 1985. Pyrolysis and combustion of cellulose, VIII: Thermally initiated reactions of phosphonomethyl amide flame retardants. *Journal of Applied Polymer Science* 30: 2263–77.

Lewin, M. 1999. Synergistic and catalytic effects in flame retardancy of polymeric materials: An overview. *Journal of Fire Sciences* 17(1): 3–19.

Leroux, F., C. Campagne, A. Perwuelz, and L. Gengembre. 2008. Fluorocarbon nano-coating of polyester fabrics by atmospheric air plasma with aerosol. *Applied Surface Science* 254: 3902–8.

Li, R., L. Ye, and Y. W. Mai. 1997. Application of plasma technologies in fibre-reinforced polymer composites: A review of recent developments. *Composites Part A: Applied Science and Manufacturing* 28: 73–86.

Lim, S. H., and S. M. Hudson. 2003. Review of chitosan and its derivatives as antimicrobial agents and their uses as textile chemicals. *Journal of Macromolecular Science, Part C: Polymer Reviews* 43: 223–69.

Lin, J., C. Winkelman, S. D. Worley, R. M. Broughton, and J. F. Williams. 2001. Antimicrobial treatment of nylon. *Journal of Applied Polymer Science* 81: 943–47.

Lin, L. H., C. C. Wang, C. W. Chen, and K. M. Chen. 2006. Water-repellency and antibacterial activities of plasma-treated cleavable silicone surfactants on nylon fabrics. *Surface and Coatings Technology* 201: 674–78.

Liu, Y. C., Y. Xiong, and D. N. Lu. 2006. Surface characteristics and antistatic mechanism of plasma-treated acrylic fibers. *Applied Surface Science* 252: 2960–66.

Liu, Z. T., Y. Yang, L. Zhang, Z. W. Liu, and H. Xiong. 2007. Study on the cationic modification and dyeing of ramie fiber. *Cellulose* 14: 337–45.

Luo, S., and W. J. van Ooij. 2002. Surface modification of textile fibers for improvement of adhesion to polymeric matrices: A review. *Journal of Adhesion Science and Technology* 16(13): 1715–35.

McCord, M. G., Y. J. Hwang, P. J. Hauser, Y. Qiu, J. J. Cuomo, O. E. Hankins, M. A. Bourham, and L. K. Canup. 2002. Modifying nylon and polypropylene fabrics with atmospheric pressure plasmas. *Textile Research Journal* 72: 491–98.

McCord, M. G., Y. J. Hwang, Y. Qiu, L. K. Hughes, and M. A. Bourham. 2003. Surface analysis of cotton fabrics fluorinated in radio-frequency plasma. *Journal of Applied Polymer Science* 88: 2038–47.

McSherry, W. F., G. L. Drake, A. B. Cooper, and A. R. Markezich. 1974. Accelerated laundry method for flame-retardant fabrics. *American Dyestuff Reporter* 63: 52.

Molina, R., J. P. Espinos, F. Yubero, P. Erra, and A. R. Gonzalez-Elipe. 2005. XPS analysis of downstream plasma treated wool: Influence of the nature of the gas on the surface modification of wool. *Applied Surface Science* 252: 1417–29.

Morent, R., N. De Geyter, J. Verschuren, K. De Clerck, P. Kiekens, and C. Leys. 2008. Non-thermal plasma treatment of textiles. *Surface and Coatings Technology* 202: 3427–49.

Mori, M., and N. Inagaki. 2006. Relationship between anti-felting properties and physicochemical properties of wool fibers treated with Ar-plasma. *Textile Research Journal* 79: 687–94.

Mukhopadhyay, S. M., P. Joshi, S. Datta, and J. Macdaniel. 2002. Plasma assisted surface coating of porous solids. *Applied Surface Science* 201: 219–26.

Nasadil, P., and P. Benesovsky. 2008. Plasma in textile treatment. *Chemické Listy* 102:s1486–89.

Nastase, C., F. Nastase, A. Dumitru, M. Ionescu, and I. Stamatin. 2005. Thin film composites of nanocarbons-polyaniline obtained by plasma polymerization technique. *Composites Part A: Applied Science and Manufacturing* 36: 481–85.

No, H. K., N. Y. Park, S. H. Lee, and S. P. Meyers. 2002. Antibacterial activity of chitosans and chitosan oligomers with different molecular weights. *International Journal of Food Microbiology* 74: 65–72.

Noll, W. 1968. *Chemistry and technology of silicones*. Orlando, FL: Academic Press.

Park, B. J., K. Takatori, M. H. Lee, D. W. Han, Y. I. Woo, H. J. Son, J. K. Kim, K. H. Chung, S. O. Hyun, and J. C. Park. 2007. *Escherichia coli* sterilization and lipopolysaccharide inactivation using microwave-induced argon plasma at atmospheric pressure. *Surface and Coatings Technology* 201: 5738–41.

Paynter, R. W. 1998. XPS studies of the modification of polystyrene and polyethylene terephthalate surfaces by oxygen and nitrogen plasmas. *Surface and Interface Analysis* 26: 674–81.

Poll, H. U., U. Schladitz, and Schreiter. 2001. Penetration of plasma effects into textile structures. *Surface and Coatings Technology* 142: 489–93.

Price, D., K. Pyrah, T. R. Hull, G. J. Milnes, J. R. Ebdon, B. J. Hunt, and P. Joseph. 2002. Flame retardance of poly(methylmethacrylate) modified with phosphorus-containing compounds. *Polymer Degradation and Stability* 77: 227–33.

Quede, A., B. Mutel, P. Supiot, C. Jama, O. Dessaux, and R. Delobel. 2004. Characterization of organosilicon films synthesized by N2-PACVD: Application to fire retardant properties of coated polymers. *Surface and Coatings Technology* 181: 265–70.

Radetic, M., P. Jovancic, N. Puac, and Z. Lj Petrovic. 2007. Environmental impact of plasma application to textiles. *Journal of Physics: Conference Series* 71: 012017.

Rajpreet, K. V., N. R. Gita, M. Bourham, and B. L. Bures. 2004. Plasma and antimicrobial treatment of nonwoven fabrics for surgical gowns. *Textile Research Journal* 74: 1073–79.

Rakowski, W. 1997. Plasma treatment of wool today, Part I: Fibre properties, spinning and shrinkproofing. *Journal of Society of Dyers and Colourists* 113: 250–55.

Rashidi, A., H. Moussavipourgharbi, M. Mirjalili, and M. Ghoranneviss. 2004. Effect of low-temperature plasma treatment on surface modification of cotton and polyester fabrics. *Indian Journal of Fibre and Textile Research* 29(1): 74–78.

Reddy, P. R. S., G. Agathian, and A. Kumar. 2005. Ionizing radiation graft polymerized and modified flame retardant cotton fabric. *Radiation Physics and Chemistry* 72: 511–16.

Ricketts, C. L., A. E. Wallis, J. C. Whitehead, and K. Zhang. 2004. A mechanism for the destruction of CFC-12 in a non-thermal, atmospheric pressure plasma. *Journal of Physical Chemistry A* 108: 8341–45.

Rinsch, C. L., C. Chen, V. Panchaligam, R. C. Eberhart, J. H. Wang, and R. B. Timmons. 1996. Pulsed radio frequency plasma polymerization of allyl alcohol: Controlled deposition of surface hydroxyl groups. *Langmuir* 12: 2995–3002.

Schafer, J., R. Foest, A. Quade, A. Ohl, and K. D. Weltmann. 2008. Local deposition of SiO_x plasma polymer films by a miniaturized atmospheric pressure plasma jet (APPJ). *Journal of Physics D: Applied Physics* 41: 194010.

Schmidt, G. L. F. 1988. *Industrial application of surfactants*. London: RSC.

Schutze, A., J. Y. Jeong, S. E. Babayan, J. Park, G. S. Selwyn, and R. F. Hicks. 1998. The atmospheric-pressure plasma jet: A review and comparison to other plasma sources. *IEEE Transactions on Plasma Science* 26: 1685–94.

Selwyn, G. S., H. W. Herrmann, J. Park, and I. Henins. 2001. Materials processing using an atmospheric pressure, RF-generated plasma source. *Contributions to Plasma Physics* 6: 610–19.

Shahidi, S., A. Rashidi, M. Ghoranneviss, A. Anvari, M. K. Rahimi, M. Bameni Moghaddam, and J. Wiener. 2010. Investigation of metal absorption and antibacterial activity on cotton fabric modified by low temperature plasma. *Cellulose* 17: 627–34.

Sharnina, L. V. 2004. Low-temperature plasma as the basis for creation of modern textile chemical technologies. *Fibre Chemistry* 36: 431–36.

Shateri Khalil-Abad, M., Y. Yazdanshenas, and M. Nateghi. 2009. Effect of cationization on adsorption of silver nanoparticles on cotton surfaces and its antibacterial activity. *Cellulose* 16: 1147–57.

Shen, J., M. Rushforth, A. Cavaco-Paulo, G. Guebitz, and H. Lenting. 2007. Development and industrialisation of enzymatic shrink-resist process based on modified proteases for wool machine washability. *Enzyme and Microbial Technology* 40: 1656–61.

Shenai, V. A. 1976. *Chemistry of textile auxiliaries*. Mumbai, India: Sevak Publications.

Shibuchi, S., T. Yamamoto, T. Onda, and K. Tsujii. 1998. Super water- and oil-repellent surfaces resulting from fractal structure. *Journal of Colloid and Interface Science* 208: 287–94.

Shimizu, T., Y. Tanaka, S. Kutsumizu, and S. Yano. 1994. Ordered structures of poly(1H,1H,2H,2H-perfluorodecyl α-substituted acrylate)s. *Macromolecular Symposia* 82: 173–84.

Shiozawa, K. 1991. *Textile wet-processing technology*. Tokyo: Chilinshokan.

Silva, C. J. S. M., M. Prabaharan, G. Gubitz, and A. Cavaco-Paulo. 2005. Treatment of wool fibres with subtilisin and subtilisin-PEG. *Enzyme and Microbial Technology* 36: 917–22.

Son, Y. A., and G. Sun. 2003. Durable antimicrobial nylon 66 fabrics: Ionic interactions with quaternary ammonium salts. *Journal of Applied Polymer Science* 90: 2194–99.

Sprang, N., D. Theirech, and J. Engermann. 1995. Plasma and ion beam surface treatment of polyethylene. *Surface and Coatings Technology* 74–75: 689–95.

Sprang, N., D. Theirech, and J. Engermann. 1998. Surface modification of fluoropolymers by microwave plasmas: FTIR investigations. *Surface and Coatings Technology* 98: 865–71.

Sricharussin, W., W. Ryo-Aree, W. Intasen, and S. Poungraksakirt. 2004. Effect of boric acid and BTCA on tensile strength loss of finished cotton fabrics. *Textile Research Journal* 74: 475–80.

Stephen, K. 2004. Plasma processes for wide fabric, film and non-wovens. *Surface and Coatings Technology* 186: 214–17.

Szeto, R., and D. W. Hess. 1981. Correlation of chemical and electrical properties of plasma-deposited tetramethylsilane films. *Journal of Applied Physics* 52: 903–8.

Thone, J., A. Holländer, W. Jaeger, I. Trick, and C. Oehr. 2003. Ultrathin antibacterial polyammonium coatings on polymer surfaces. *Surface and Coatings Technology* 174–175: 584–87.

Totolin, V., M. Sarmadi, S. O. Manolache, and F. S. Denes. 2009. Low pressure, non-equilibrium plasma-assisted generation of flame retardant cotton. *AATCC Review* 9: 32–37.

Totolin, V., M. Sarmadi, S. O. Manolache, and F. S. Denes. 2010. Atmospheric pressure plasma enhanced synthesis of flame retardant cellulosic materials. *Journal of Applied Polymer Science* 117: 281–89.

Tsafack, M. J., F. Hochart, and J. Levalois-Grützmacher. 2004. Polymerization and surface modification by low pressure plasma technique. *European Physical Journal: Applied Physics* 26: 215–19.

Tsafack, M. J., and J. Levalois-Grützmacher. 2006a. Plasma-induced graft-polymerization of flame retardant monomers onto PAN fabrics. *Surface and Coatings Technology* 200: 3503–10.

Tsafack, M. J., and J. Levalois-Grützmacher. 2006b. Flame retardancy of cotton textiles by plasma-induced graft-polymerization (PIGP). *Surface and Coatings Technology* 201: 2599–610.

Tsafack, M. J., and J. Levalois-Grützmacher. 2007. Towards multifunctional surfaces using the plasma-induced graft-polymerization (PIGP) process: Flame and waterproof cotton textiles. *Surface and Coatings Technology* 201: 5789–95.

Tsai, P. P., L. C. Wadsworth, and J. R. Roth. 1997. Surface modification of fabrics using a one-atmosphere glow discharge plasma to improve fabric wettability. *Textile Research Journal* 67: 359–69.

Tseng, H. J., S. H. Hsu, M. W. Wu, T. H. Hsueh, and P. C. Tu. 2009. Nylon textiles grafted with chitosan by open air plasma and their antimicrobial effect. *Fibers and Polymers* 10: 53–59.

Tyndall, R. M. 1992. Application of cellulase enzymes to cotton fabrics and garments. *Textile Chemist and Colorist* 24: 23–26.

Vaideki, K., S. Jayakumar, R. Rajendran, and G. Thilagavathi. 2008. Investigation on the effect of RF air plasma and neemleaf extract treatment on the surface modification and antimicrobial activity of cotton fabric. *Applied Surface Science* 254: 2472–78.

Vaideki, K., S. Jayakumar, G. Thilagavathi, and R. Rajendran. 2007. A study on the antimicrobial efficacy of RF oxygen plasma and neem extract treated cotton fabrics. *Applied Surface Science* 253: 7323–29.

Volkov, V. V., N. A. Plate, A. Takahara, N. Amaya, Y. Murata, and T. Kajiyama. 1992. Aggregation state and mesophase structure of comb-shaped polymers with fluorocarbon side groups. *Polymer* 33: 1316–20.

Wafa, D. M., F. Breidt, F. S. M. Gawish, S. R. Matthews, K. V. Donohue, and R. M. Roe. 2007. Atmospheric plasma-aided biocidal finishes for nonwoven polypropylene fabrics, II: Functionality of synthesized fabrics. *Journal of Applied Polymer Science* 103: 1911–17.

Wagner, A. J., D. H. Fairbrother, and F. Reniers. 2003. A comparison of PE surfaces modified by plasma generated neutral nitrogen species and nitrogen ions. *Plasmas and Polymers* 8: 119–34.

Wakelyn, P. J., W. Rearick, and J. Turner. 1998. Cotton and flammability: Overview of new developments. *American Dyestuff Reporter* 87: 13–21, 32.

Wallace, M. L. 2001. Testing the efficacy of polyhexamethylene biguanide as an antimicrobial treatment for cotton fabric. *AATCC Review* 1: 18–20.

Wang, C. C., and C. C. Chen. 2005a. Physical properties of crosslinked cellulose catalyzed with nano titanium dioxide. *Journal of Applied Polymer Science* 97: 2450–56.

Wang, C. C., and C. C. Chen. 2005b. Physical properties of the crosslinked cellulose catalyzed with nanotitanium dioxide under UV irradiation and electronic field. *Applied Catalysis A: General* 293: 171–79.

Wang, C. X., Y. Liu, H. L. Xu, Y. Ren, and Y. P. Qiu. 2008. Influence of atmospheric pressure plasma treatment time on penetration depth of surface modification into fabric. *Applied Surface Science* 254: 2499–505.

Wang, C. X., and Y. P. Qiu. 2007. Two sided modification of wool fabrics by atmospheric pressure plasma jet: Influence of processing parameters on plasma penetration. *Surface and Coatings Technology* 201: 6273–77.

Ward, L. J., W. C. E. Schofield, A. J. Goodwin, P. J. Merlin, and J. P. S. Badyal. 2003. Atmospheric pressure glow discharge deposition of polysiloxane and SiO_x films. *Langmuir* 19: 2110–14.

Ward, T. L., H. Z. Jung, O. R. Hinojosa, and R. J. Benerito. 1979. Characterization and use of radio frequency plasma-activated natural polymers. *Journal of Applied Polymer Science* 23: 1987–2003.

Wavhal, D. S., J. Zhang, M. L. Steen, and E. R. Fisher. 2006. Investigation of gas phase species and deposition of SiO_2 films from HMDSO/O_2 plasmas. *Plasma Processes and Polymers* 3: 276–87.

Weltmann, K. D., Th. Von Woedtke, R. Brandenburg, and J. Ehlbeck. 2008. Biomedical applications of atmospheric pressure plasma. *Chemické Listy* 102: s1450–51.

Wichman, I. S. 2003. A review of the literature of material flammability, combustion, and toxicity related to transportation. *Progress in Energy and Combustion Science* 29: 247–99.

Wong, W., K. Chan, K. W. Yeung, and K. S. Lau. 2000a. Morphological modification of poly(ethylene terephthalate) surface after excimer laser irradiation. *Journal of Textile Engineering* 46(2): 25–31.

Wong, W., K. Chan, K. W. Yeung, Y. M. Tsang, and K. S. Lau. 2000b. Surface structure of poly(ethylene terephthalate) fibres with a UV excimer laser and low temperature plasma. *Journal of Materials Processing Technology* 103: 225–29.

Woodward, I., W. C. E. Schofield, V. Roucoules, and J. P. S. Badyal. 2003. Super-hydrophobic surfaces produced by plasma fluorination of polybutadiene films. *Langmuir* 19: 3432–38.

Wrobel, A. M., G. Czeremuszkin, H. Szymanowski, and J. Kowalski. 1990. Plasma polymerization of carbosilanes: Tetramethylsilane as a model monomer for reactivity study of silylmethyl groups. *Plasma Chemistry and Plasma Processing* 10: 277–89.

Wu, S., J. Xing, C. Zheng, G. Xu, G. Zheng, and J. Xu. 1997. Plasma modification of aromatic polyamide reverse osmosis composite membrane surface. *Journal of Applied Polymer Science* 64: 1923–26.

Wu, W., and C. Q. Yang. 2006. Comparison of different reactive organophosphorus flame retardant agents for cotton, Part I: The bonding of the flame retardant agents to cotton. *Polymer Degradation and Stability* 91: 2541–48.

Wu, W., and C. Q. Yang. 2007. Comparison of different reactive organophosphorus flame retardant agents for cotton, Part II: Fabric flame resistant performance and physical properties. *Polymer Degradation and Stability* 92: 363–69.

Xu, P., W. Wang, and S. L. Chen. 2005. Improving the antistatic properties of polyester fabrics. *Textile Asia* 36(9): 45–47.

Yamamoto, T., M. Okubo, N. Imai, and Y. Mori. 2004. Improvement on hydrophilic and hydrophobic properties of glass surface treated by nonthermal plasma induced by silent corona discharge. *Plasma Chemistry and Plasma Processing* 24: 1–12.

Yamauchi, G., J. D. Miller, H. Saito, K. Takai, T. Ueda, H. Takazawa, H. Yamamoto, and S. Nishi. 1996. Wetting characteristics of newly developed water-repellent material. *Colloids and Surfaces A: Physicochemical and Engineering Aspects* 116: 125–34.

Yang, G. H., Y. Zhang, E. T. Kang, K. G. Neoh, W. Huang, and J. H. Teng. 2003. Surface passivation of (100)-oriented GaAs via plasma deposition of an ultrathin S-containing polymer film and its effect on photoluminescence. *Journal of Physical Chemistry B* 107: 8592–98.

Yang, H., and C. Q. Yang. 2005. Durable flame retardant finishing of the nylon/cotton blend fabric using a hydroxyl-functional organophosphorus oligomer. *Polymer Degradation and Stability* 88: 363–70.

Yang, M. R., K. S. Chen, J. C. Tsai, C. C. Tseng, and S. F. Lin. 2002. The antibacterial activities of hydrophilic-modified nonwoven PET. *Materials Science and Engineering: C* 20: 167–73.

Yasuda, H. 1985. *Plasma polymerization*. London: Academic Press.

Yasuda, H., and Y. Matsuzawa. 2005. Economical advantages of low-pressure plasma polymerization coating. *Plasma Processes and Polymers* 2: 507–12.

Yip, J., K. Chan, K. M. Sin, and K. S. Lau. 2002. Low temperature plasma-treated nylon fabrics. *Journal of Materials Processing Technology* 123: 5–12.

Yip, J., K. Chan, K. M. Sin, and K. S. Lau. 2003. Study on the surface chemical properties of UV excimer laser irradiated polyamide by XPS, ToF-SIMS and CFM. *Applied Surface Science* 205: 151–59.

Yoon, K. J., J. H. Woo, and Y. S. Seo. 2003. Formaldehyde free cross-linking agents based on maleic anhydride copolymers. *Fibers and Polymers* 4: 182–87.

Yousefi, H. R., M. Ghoranneviss, A. R. Tehrani, and S. Khamseh. 2003. Investigation of glow discharge plasma for surface modification of polypropylene. *Surface and Interface Analysis* 35: 1015–17.

Yuen, C. W. M., S. K. A. Ku, C. W. Kan, Y. F. Cheng, P. S. R. Choi, and Y. L. Lam. 2007. Using nano TiO_2 as co-catalyst for improving wrinkle-resistance of cotton fabric. *Surface Review and Letters* 14: 571–75.

Zhang, Q., J. Wanga, G. Wua, J. Shena, and S. Buddhudu. 2001. Interference coating by hydrophobic aerogel-like SiO_2 thin films. *Materials Chemistry and Physics* 72: 56–59.

Zhang, Z., L. Chen, J. Ji, Y. Huang, and D. Chen. 2003. Antibacterial properties of cotton fabrics treated with chitosan. *Textile Research Journal* 73: 1103–6.

Zheludkevich, M. L., R. Serra, G. Grundmeier, L. H. Yang, and M. G. S. Ferreira. 2006. Barrier properties of polyurethane coil coatings treated by microwave plasma polymerization. *Surface and Coatings Technology* 200: 4040–49.

Zhu, L., C. Wang, and Y. Qiu. 2007. Influence of the amount of absorbed moisture in nylon fibers on atmospheric pressure plasma processing. *Surface and Coatings Technology* 201: 7453–61.

Zou, X. P., E. T. Kang, K. G. Neoh, Y. Zang, K. L. Tan, C. Q. Cui, and Y. B. Lim. 2001. Plasma polymerization and deposition of glycidyl methacrylate on Si(100) surface for adhesion improvement with polyimide. *Polymers for Advanced Technologies* 12: 583–95.

9 Sustainability and Development of Plasma Treatment in Textile Wet Processing

9.1 ECONOMIC ASPECT

Conventional wet processing of fabrics, such as desizing, scouring, bleaching, dyeing and finishing, consume large amounts of water and energy and eventually cause water and air pollution (Bhat et al. 2011). Therefore, the pretreatment (preparation) and finishing of fabrics by plasma technologies have become popular because such techniques offer numerous advantages over the traditional wet chemical processes (Bhat et al. 2011; Morent et al. 2008; Samanta, Jassal, and Agrawal 2010). In general, the interaction of plasma with substrate occurs when the reactive species are generated by ionization, fragmentation, and excitation. These species lead to chemical and physical interactions between the plasma and the substrate surface, depending on plasma conditions such as gas type, power, pressure, frequency and exposure time. The choice of discharge parameters is determined by the requirements of each particular application.

Generally speaking, plasma treatment is a dry surface modification process allowing modification of surface properties without changing the bulk properties of the substrate (Bhat et al. 2011; Pane and Tedesco 2001). The low power necessary to maintain plasma-gas activation in the chamber saves large quantities of electrical energy (Pane and Tedesco 2001). In addition, from an economic point of view, the small consumption of chemicals makes the process superior to other applications in the textile industry (Bhat et al. 2011; Pane and Tedesco 2001). Because it avoids the use of water as a medium of treatment, the plasma process results in a more economical process with drastic reductions in pollutants and a corresponding cost reduction for effluent treatment (Morent et al. 2008). Moreover, versatile plasma treatment may remove the hydrophobic fibre's outermost layer and/or introduce a large variety of chemically active functional groups, such as $-COOH$, $-OH$ and $-NH_2$, into the textile surface, aimed at improving fabric wettability (Morent et al. 2008). The effect is beneficial because plasma treatment uses less or no wetting agent as well as using less dyeing or finishing product for identical treatment effectiveness (Morent et al. 2008). The extensive work in the last two decades on the possible application of plasma to different textile materials indicates that this technology can be efficiently exploited to obtain many advantageous effects, i.e., high efficiency, economic feasibility, environmental acceptability and production flexibility (Mihailovic et al. 2011; Gotoh and Yasukawa 2011).

9.2 ENVIRONMENTAL ASPECTS

In general, textile wet-processing methods, especially those involving the use of toxic, hazardous and poorly biodegradable compounds and auxiliaries, are the main sources of effluent pollution (Radetic et al. 2007). However, maintenance of a green environment has become a critical global concern. The introduction of rigorous ecological legislation is forcing companies to consider not only the issue of industrial waste disposal, but also a possible replacement of conventional processes with treatments that can provide equal or even higher efficiency and lower environmental impact (Radetic et al. 2007). The pretreatment (preparation) and finishing of textile fabrics by plasma technologies is increasingly replacing conventional wet-chemical applications not only for economical considerations, but also for ecological considerations. The dry and environmentally friendly plasma treatment does not employ harmful chemical solutions and can be carried out without producing contaminated water or creating mechanical hazards for treated fabrics (Kan, Yuen, and Tsoi 2011; De Geyter, Morent, and Leys 2006; Hossain, Herrmann, and Hegemann 2006; Keller et al. 2005; Ceria et al. 2010). The technology significantly reduces pollution caused by residual chemicals. In addition, plasma is able to modify the substrate surface properties, such as generating microroughness and inducing chemically active functional groups, without affecting the bulk properties of the substrate (De Geyter, Morent, and Leys 2006; Ceria et al. 2010). Efficient plasma pretreatments improve the water absorption of fabrics, thereby reducing amount of chemicals needed in conventional processing with better exhaustion of chemicals from the bath (Radetic et al. 2007). Moreover, shortening of the wet-processing time and decreasing the required wet processing temperature may save limited energy resources (Radetic et al. 2007).

In addition, as an example of the usage of plasma treatment in the textile industry, plasma treatment can create the "worn" look of denim products which was originally produced by the abrasive action of pumice stones on the garment surface or by enzymatic stonewashing (biostoning) (Radetic et al. 2007; Ghoranneviss et al. 2006). The traditional stonewashing often caused damage to the garments and machines while also generating a huge amount of pumice dust in the laundry environment as well as reducing machine capacity due to high proportion of stones required during washing. On the other hand, biostoning with enzymes requires high quantities of water and chemicals that are released into effluents, making the process less eco-friendly (Radetic et al. 2007; Buschle-Diller et al. 1994). Plasma treatment may solve the problems of dust and effluent production. Moreover, the mechanical properties of the treated material may not change after plasma treatments, which further encourages the use of plasma technology in denim finishing (Radetic et al. 2007).

9.3 SOCIAL ACCOUNTABILITY ASPECT

Social accountability has become an increasingly important issue to motivate suppliers to maintain a high level of social and environmental performance and to ensure that business partners comply with the company's commitment to social responsibility as well as the regional laws. SA 8000 (SA8000 2008) is a global social accountability standard for decent working conditions based on the UN Universal Declaration of Human Rights, Convention on the Rights of the Child and the conventions of various international labour organization. The principles of SA 8000 includes (a) no child labour, (b) no forced labour, (c) health and safety of workmen to be ensured, (d) freedom of association and right to collective bargaining to be respected, (e) no discrimination on the basis of sex, religion, caste, etc., (f) no unjust disciplinary practices, (g) working hours not more than eight hours, (h) fair compensation (wages) and (i) management systems to be in place to monitor the implementation of the standard (SA8000 2008).

As discussed in the previous section, environmental problems associated with the textile industry are typically those associated with water pollution. It is usually caused by the discharge of untreated effluents, containing a substantial organic and suspended pollution load, that may threaten aquatic life and damage general water quality downstream (Yperen 2006). Other environmental issues relevant to the textile industry include emissions of volatile compounds and excessive noise or emissions that directly affect workspace safety (Yperen 2006). Based on the concept of social responsibility, companies should reduce inputs of raw materials and energy, eliminate the use of toxic materials, and reduce the quantity and toxicity of emissions and waste outputs. These measures are critical to improve working conditions as well as workers' health and safety. For this reason, dry plasma treatment may replace the conventional wet treatment while still providing the same or even better treatment effects (Ghoranneviss et al. 2006; Poll, Schladitz, and Schreiter 2001).

In addition, excessive overtime is a common problem in many textile manufacturing companies due to tight lead times, late sample approval and last-minute alterations to product specifications, and late delivery of raw materials from the suppliers. Excessive overtime is closely linked with low pay. Long hours also lead to tiredness, illness and accidents that contribute to low productivity and high levels of reworking. With reasonable social accountability, companies could provide better working conditions and increase workers' motivation with decreased overtime and improved health and safety in the workplace (Yperen 2006). Plasma treatment is a highly efficient treatment that eventually enhances productivity. Moreover, with automatic and mechanised plasma apparatus, plasma treatment provides better product quality and reduces the costs in input materials, energy and manpower.

9.4 FUTURE DEVELOPMENT

Due to its obvious superior and environmentally friendly effects, plasma treatment is claimed as a potentially promising treatment in the textile industry. In general, physical modifications—including surface etching, chain scission and chemical modifications such as cross-linking, grafting and polymerization—lead to beneficial changes in the properties of textiles (Sun et al. 2011). The recent application of plasma treatment on polymeric substrates includes (a) wettability enhancement by plasma treatment in oxygen-containing gas, (b) hydrophobicity increment by the use of fluorocarbon, hydrocarbon and mixtures of fluorocarbon and hydrocarbon gases, (c) antishrinkage finishing by oxygen gas, (d) desizing in oxygen or nitrogen plasma gas, (e) flame-retardant treatment in nitrogen or argon plasma and (f) adhesion enhancement by the use of ammonia, air, oxygen or mixtures of argon and ammonium plasma, etc. (Inagaki, Tasaka, and Takami 1990; Sarmadi and Kwon 1993; Kassenbeck 1964; Benerito et al. 1981; Shishoo and Herbert 1999; Gancarz et al. 1999). The main shortcoming of plasma treatment of textile fabrics is that it cannot replace all traditional wet processes; nevertheless, it can be a viable pretreatment providing environmental and economic benefits. Therefore, some textile manufacturers may consider investing in expensive plasma equipment that will quickly repay the initial investment through environment-related savings and profits on the sale of higher-value-added products (Radetic et al. 2007).

Low-pressure plasma is generally viewed as a promising application for this purpose, but the process has not yet been successfully commercialised. In general, low-temperature plasma systems are operated in vacuum conditions, with yellowness being induced in the plasma-treated fabric (Radetic et al. 2007; Samanta, Jassal, and Agrawal 2010; Kan, Yuen, and Tsoi 2011). The low-pressure treatment environment also entails high operational and maintenance costs, while the treatment is not flexible enough to be integrated into the continuous and open systems normally used for fabric processing (Radetic et al. 2007; Samanta, Jassal, and Agrawal 2010; Kan, Yuen, and Tsoi 2011; Cai et al. 2003; Gotoh and Yasukawa 2011).

Atmospheric-pressure plasma devices that do not need a vacuum system have been developed to overcome the drawbacks of low-pressure plasma treatments, e.g., the one atmosphere uniform glow-discharge plasma (OAUGDP), the atmospheric pressure surface discharge, the cold-plasma torch and the atmospheric-pressure plasma jet (APPJ). However, the discharge regions of these plasmas are so limited that they are not suitable to treat bulky or complex substrates, with the possible exceptions of APPJ and the cold-plasma torch (Cheng, Liye, and Zhan 2006). For this reason, the APPJ, which is generated in ambient air with significantly milder effect than the plasma torch, but still highly effective, provides the requisite flexibility and possibility for continuous and open processes at high speed (Samanta, Jassal, and Agrawal 2010; Kan, Yuen, and Tsoi 2011; Gotoh and Yasukawa 2011; Gao et al. 2011; Tian et al. 2011; Ceria et al. 2010; Wang et al. 2008). In APPJ technology, the high electron temperature enhances the plasma chemistry processes, while the plasma gas remains close to room temperature (Xiong et al. 2008). Hence, the APPJ technique is potentially a very useful and effective way to obtain practical surface modifications—including surface etching, chain scission, polymerisation, cross-linking, development of functional groups, surface roughness, etc.—of temperature-sensitive materials such as fabrics (Ceria et al. 2010). However, future developments in carrying out reactions in atmospheric-pressure cold plasma remain a big challenge within the textile industry. The important challenges include (a) generating a stable glow plasma in the presence of liquid or gaseous monomers and (b) keeping the generated radicals active in the presence of contaminants such as oxygen and air (Radetic et al. 2007; Samanta, Jassal, and Agrawal 2010).

9.5 CONCLUSION

This chapter reviewed and discussed the economic, environmental and social accountability of plasma treatment for textile wet processing. Through the discussion in previous chapters, the plasma treatment can be considered as a dry and alternative method for conventional textile wet processing, which involves the use of large amounts of chemicals, water and energy. In view of the saving of resources, plasma treatment is a sustainable process for treating textiles.

REFERENCES

Benerito, R., T. Ward, D. Soignet, and O. Hinojosa. 1981. Modifications of cotton cellulose surface by use of radiofrequency cold plasmas and characterization of surface changes by ESCA. *Textile Research Journal* 51: 224–32.

Bhat, N. V., A. N. Netravali, A. V. Gore, M. P. Sathianarayanan, G. A. Arolkar, and R. R. Deshmukh. 2011. Surface modification of cotton fabrics using plasma technology. *Textile Research Journal* 81: 1014–26.

Buschle-Diller, G., S. H. Zeronian, N. Pan, and M. Y. Yoon. 1994. Enzymatic hydrolysis of cotton, linen, ramie, and viscose rayon fabrics. *Textile Research Journal* 64: 270–79.

Cai, Z. S., Y. P. Qiu, C. Y. Zhang, Y. J. Hwangand, and M. McCord. 2003. Effect of atmospheric plasma treatment on desizing of PVA on cotton. *Textile Research Journal* 73: 670–74.

Ceria, A., G. Rovero, S. Sicardi, and F. Ferrero. 2010. Atmospheric continuous cold plasma treatment: Thermal and hydrodynamical diagnostics of a plasma jet pilot unit. *Chemical Engineering and Processing: Process Intensification* 49: 65–69.

Cheng, C., Z. Liye, and R. J. Zhan. 2006. Surface modification of polymer fibre by the new atmospheric pressure cold plasma jet. *Surface and Coatings Technology* 200: 6659–65.

De Geyter, N., R. Morent, and C. Leys. 2006. Penetration of a dielectric barrier discharge plasma into textile structures at medium pressure. *Plasma Sources Science and Technology* 15: 78–84.

Gancarz, I., G. Pozniak, M. Bryjak, and A. Frankiewicz. 1999. Modification of polysulfone membranes, 2: Plasma grafting and plasma polymerization of acrylic acid. *Acta Polymerica* 50: 317–26.

Gao, Z. Q., J. Sun, S. J. Peng, L. Yao, and Y. P. Qiu. 2011. Surface modification of nylon 6 films treated with an He/O_2 atmospheric pressure plasma jet. *Journal of Applied Polymer Science* 120: 2201–6.

Ghoranneviss, M., B. Moazzenchi, S. Shahidi, A. Anvari, and A. Rashidi. 2006. Decolorization of denim fabrics with cold plasmas in the presence of magnetic fields. *Plasma Processes and Polymers* 3: 316–21.

Gotoh, K., and A. Yasukawa. 2011. Atmospheric pressure plasma modification of polyester fabric for improvement of textile-specific properties. *Textile Research Journal* 81: 368–78.

Hossain, M. M., A. S. Herrmann, and D. Hegemann. 2006. Plasma hydrophilization effect on different textile structures. *Plasma Processes and Polymers* 3: 299–307.

Inagaki, N., S. Tasaka, and Y. Takami. 1990. Durable and hydrophobic surface modification by plasma polymers deposited from acetone/hexafluoroacetone, ethylene/hexafluoroacetone, and ethane/hexafluoroacetone mixture. *Journal of Applied Polymer Science* 41: 965–73.

Kan, C. W., C. W. M. Yuen, and W. Y. Tsoi. 2011. Using atmospheric pressure plasma for enhancing the deposition of printing paste on cotton fabric for digital ink-jet printing. *Cellulose* 18: 827–39.

Kassenbeck, P. 1964. Electrical discharge process and its application to wool. *Institute Textile De France* 18(110): 7–33.

Keller, M., A. Ritter, P. Reimann, V. Thommen, A. Fischer, and D. Hegemann. 2005. Comparative study of plasma-induced and wet-chemical cleaning of synthetic fibres. *Surface and Coatings Technology* 200: 1045–50.

Mihailovic, D., Z. S. Saponjic, M. Radoicic, S. Lazovic, C. J. Baily, P. Jovancic, J. Nedeljkovic, and M. Radetic. 2011. Functionalization of cotton fabrics with corona/air RF plasma and colloidal TiO_2 nanoparticles. *Cellulose* 18: 811–25.

Morent, R., N. De Geyter, J. Verschuren, K. De-Clerck, P. Kiekens, and C. Leys. 2008. Non-thermal plasma treatment of textiles. *Surface and Coatings Technology* 202: 3427–49.

Pane, S., and R. Tedesco. 2001. Acrylic fabrics treated with plasma for outdoor applications. *Journal of Industrial Textiles* 31: 135–45.

Poll, H. U., U. Schladitz, and S. Schreiter. 2001. Penetration of plasma effects into textile structures. *Surface and Coatings Technology* 142–144: 489–93.

Radetic, M., P. Jovancic, N. Puac, and Z. L. Petrovic. 2007. Environmental impact of plasma application to textiles. *Journal of Physics: Conference Series* 71: 012017.

SA8000. 2008. *Social accountability 8000*. New York: Social Accountability International.

Samanta, K. K., M. Jassal, and A. K. Agrawal. 2010. Atmospheric pressure plasma polymerization of 1,3-butadiene for hydrophobic finishing of textile substrates. *Journal of Physics: Conference Series* 208: 012098.

Sarmadi, M., and Y. Kwon. 1993. Improved water repellency and surface dyeing of polyester fabrics by plasma treatment. *Textile Chemist and Colorist* 25(11): 33–40.

Shishoo, R., and T. Herbert. 1999. Plasma conquering the textile industry. *RTD Info* 24: 22–23.

Sun, S. Y., J. Sun, L. Yao, and Y. P. Qiu. 2011. Wettability and sizing property improvement of raw cotton yarns treated with He/O_2 atmospheric pressure plasma jet. *Applied Surface Science* 257: 2377–82.

Tian, L. Q., N. Nie, N. P. Chatterton, C. J. Branford-White, Y. P. Qiu, and L. M. Zhu. 2011. Helium/oxygen atmospheric pressure plasma jet treatment for hydrophilicity improvement of grey cotton knitted fabric. *Applied Surface Science* 257: 7113–18.

Wang, C. X., Y. Liu, H. L. Xu, Y. Ren, and Y. P. Qiu. 2008. Influence of atmospheric pressure plasma treatment time on penetration depth of surface modification into fabric. *Applied Surface Science* 254: 2499–2505.

Xiong, Q., X. P. Lu, Z. H. Jiang, Z. Y. Tang, J. Hu, Z. L. Xiong, and Y. Pan. 2008. An atmospheric pressure nonequilibrium plasma jet device. *IEEE Transactions on Plasma Science* 36: 986–87.

Yperen, M. V. 2006. *Corporate social responsibility in the textile industry. International overview*. Amsterdam: IVAM Research and Consultancy on Sustainability.

Index

B

D

G

J

K

L

M

N

O

P

Q

R

S

T

U

V

W

X

Y

Z

Milton Keynes UK
Ingram Content Group UK Ltd.
UKHW020021071024
449327UK00032B/2882

9 780367 262495